THE HISTORY

OF THE

TYNE ELECTRICAL ENGINEERS, ROYAL ENGINEERS.

From the formation of the Submarine Mining Company of the 1st Newcastle-upon-Tyne and Durham (Volunteers) Royal Engineers in 1884 to 1933.

❏ ❏

The Naval & Military Press Ltd

Published by

The Naval & Military Press Ltd

Unit 10 Ridgewood Industrial Park,
Uckfield, East Sussex,
TN22 5QE England
Tel: +44 (0) 1825 749494
Fax: +44 (0) 1825 765701
www.naval-military-press.com
www.military-genealogy.com
www.militarymaproom.com

Printed and bound in Great Britain by
CPI Antony Rowe, Chippenham and Eastbourne

FOREWORD.

Some two years ago I mentioned to Captain Perowne, our Adjutant, that it would be an advantage to have a little book or pamphlet, giving salient facts about the past history of the Tyne Electrical Engineers, which could be handed to recruits on joining the Unit.

With characteristic energy, Captain Perowne set about collecting suitable data and soon amassed so much information that it was decided to compile a comprehensive history of the Unit instead of proceeding along the lines originally intended. Accordingly, a committee was formed consisting of Major Short, Major Sherlock and Captain Perowne (Lieut. Fraser being subsequently co-opted), and it is to their competent and untiring efforts that this book owes its existence. All those who have served with the Unit, or are otherwise interested in its history, owe, I think, a very real debt to these officers for the great amount of time which they have given to this work and for the very thorough manner in which they have checked and sorted information drawn from many different sources.

Time has already taken its toll of many of the earlier members of the Unit, and, if the information concerning its first years had not now been collected, much of it would have been lost for ever.

The actual printing of the book has been made possible by the help of certain past and present Officers of the Unit, who have acted as guarantors, and to a generous contribution from the Institution of Royal Engineers.

E. H. E. WOODWARD,

Bt. Lieut.-Colonel, R.E.(T).
Commanding
Tyne Electrical Engineers, R.E.

HEADQUARTERS,
TYNEMOUTH.
31st December, 1933.

v.

COLONEL SIR CHARLES MARK PALMER, Bt., V.D., D.L., J.P., M.P.
Founder and Honorary Commandant of the Tyne Division, Royal
Engineers (Volunteers), Submarine Miners.

PREFACE.

When offering this history of the Tyne Electrical Engineers to all those who have served or are serving in the Unit, the members of the Committee recognise that the information it contains is, in places, incomplete, mainly due to the fact that, during the fifteen years which have elapsed since the termination of hostilities, officers and men have become widely dispersed, and, in spite of every effort, no information has been obtainable in some cases and very little in many others. Had it been practicable to examine the various official records and war diaries in the custody of the Committee of Imperial Defence in London, further information would undoubtedly have been obtained, but this, unfortunately, proved quite impossible as no one could be found to devote the time necessary for so prolonged a research.

While the fullest use has been made of all official evidence procurable, the Committee has, to a large extent, been dependent on information obtained from members of the Unit, past and present, and, although the greatest care has been exercised in order to ensure accuracy, the members of the Committee desire, in advance, to express their regret for any mistakes or omissions that may have occurred. They also desire to convey their gratitude to Colonel J. T. Woolrych-Perowne, V.D., T.D., and Lieut.-Colonel E. H. E. Woodward, M.C., T.D., who have read the typescript and proofs, and to all those who have otherwise assisted in the preparation of the work, as, without this kindly help, the history could not have been compiled.

Among various publications consulted by the Committee are :—

" Her Majesty's Army " - - - - - - - W. RICHARDS.
" A History of Submarine Mining in the British Army " Brig.-Genl. BAKER-BROWN.
" A History of Northumberland " - - - - H. H. E. CRASTER.
" The History of the Corps of Royal Engineers "
" The Manual of Submarine Mining, 1904 " - - - - H.M.S.O.
" The R.E. Journal "
" Standing Orders, Rules and Regulations of the T.D.R.E. (V) S.M., 1889 "
" Military Antiquities " - - - - - CHARLES GROSE.
" Electric Lights for Coast Defence " - - - - E.L. SCHOOL.
" The Manual of Military Electric Lighting " - - - - H.M.S.O.
" Hart's Army List "
" Rank Badges and Dates in H.M. Army and Navy " - - O. L. PERRY.

PREFACE—*continued*.

" What is the Territorial Army ? "   -   -   - Colonel G. R. CODRINGTON.
" A Text Book of Small Arms "   -   -   -   -   - H.M.S.O.
" Drill Book for V.R.E.S.M. Pts. I and II, 1889 "
" The London Gazette " 1883—1933
" Air Defence "   -   -   -   -   - Maj.-Genl. E. B. ASHMORE.
" The German Air Raids on Great Britain, 1914—1918 "   -   Capt. J. MORRIS.
" The Work of the R.E. in the European War, 1914—1919 "   -   R.E.I.
" Statistics of the Military Effort of the British Empire during the
     Great War, 1914—1920 "   -   -   -   -   -   - H.M.S.O.
" British Campaign on the Western Front "   -   -   - CONAN DOYLE.

and the Committee desires to acknowledge its indebtedness to the authors or compilers.

No literary merit is claimed for the work. If its perusal affords some pleasure to those who read it, the members of the Committee will feel that they have not laboured in vain.

O. M. SHORT, Major, R.E. (T.). Retd., *Chairman.*

H. SHERLOCK, Major, R.E. (T.)

L. E. C. M. PEROWNE, Capt., R.E.   } *Members of the Committee.*

M. A. FRASER, Lieut., R.E. (T.).

CONTENTS.

CHAPTER I.

1883—1907.

THE TYNE DIVISION, ROYAL ENGINEERS (VOLUNTEERS), SUB-MARINE MINERS.

The Volunteer Force — Submarine Mining Company of the Newcastle and Durham (Volunteers) Royal Engineers — R.E. Staff — Courses at Chatham — Formation of the Coast Battalion R.E. — Constitution of the Tyne Division — Headquarters at Clifford's Fort — Submarine mining — Vessels and boats — Conditions of service — Uniform — Rifle shooting — Annual training — Ambulance work — Competitions and prizes — The Fishermen's section — Nicknames — Electric searchlights introduced — Formation of the Corps of Electrical Engineers — The South African War — Detachments for the Thames — Establishment increased — The Ladies' Challenge Trophy — Band and Pipers — Detachments at Hull, Weymouth and Portsmouth — Further increase in establishment — Ladies' Day at Clifford's Fort — Social occasions — Rifle Club — Sham attacks — The abolition of submarine mining in the Army — First motor searchlight equipment — Operations with balloons — Disposal of submarine mining stores — Conversion of the Division to Electrical Engineers.

CHAPTER II.

1907—1913.

THE TYNE DIVISION, ROYAL ENGINEERS (VOLUNTEERS), ELECTRICAL ENGINEERS, THE DURHAM FORTRESS R.E. (T.F.), THE NORTHUMBERLAND FORTRESS R.E. (T.F.), AND THE TYNE ELECTRICAL ENGINEERS R.E.

Electrical Engineers Volunteers — Lieutenant Burton — Death of Sir Charles Mark Palmer — Altered conditions of service — Mobile searchlights — Formation of the Territorial Force — Coast defence organisation — Durham Fortress R.E. (T.F.) — New conditions of service — Special Service Section — Changes in uniform — Permanent Staff — The Ladies' Trophy won — Northumberland Fortress R.E. (T.F.) — Coronation detachments — Companies for Portsmouth defences — Recruiting — Amalgamation of the Northumberland Fortress and Tyne Electrical Engineers — New establishment — Retirement of Colonel Johnson — Special badges — First annual presentation of prizes — Earl Percy accepts the post of Honorary Colonel — Combined operations continued — A chaplain appointed — Rifle shooting successes — Adverse criticism of the Territorial Force — Training and instructional system — New rifle range at Ponteland — Regimental prizes and trophies — The Unit thirty years old.

CONTENTS—*continued.*

CHAPTER III.

1914—1919. THE GREAT WAR.

THE HEADQUARTERS AT NORTH SHIELDS.

CHAPTER IV.

1914—1919. THE GREAT WAR—*(Continued)*.

THE DEPOT AT GOSPORT.

CONTENTS—*continued*.

CHAPTER V.

1915—1919. THE GREAT WAR—*(Continued)*.

No. 1 (LONDON-AND-TYNE) E. & M. COMPANY.

Formation — Functions — Work in H.Q. and First and Second Army areas — Water supply for first battle of the Somme — Fifth Army workshops — Increase in E. & M. establishment — The Company split up — Duties of the new Companies — Work in preparation for the third battle of Ypres — The battle of Messines — Passchendaele — Attached personnel — Work in forward areas — The second battle of the Somme — " Carey's Force " — The battle of the Lys — The Companies re-unite — Final British advance — Demobilization — The Army of Occupation at Cologne — Disposals Board — Disbandment.

CHAPTER VI.

1915—1919. THE GREAT WAR—*(Continued)*.

THE ANTI-AIRCRAFT SEARCHLIGHT UNITS AT HOME.

A.A. organisation prior to the Great War — The Admiralty takes charge — First Zeppelin raids — Searchlight companies formed for London — The War Office once more assumes responsibility — Mobile searchlight companies formed — The Zeppelin L.15 shot down — Aeroplane defences — Aeroplane searchlight units — Re-organisation of London defences — Development of defences in North and Midlands — Establishments and technical equipment — Companies formed for Tyneside — And at Leeds, Hull and Sheffield — Transfer of Tyne Electrical Engineers from London to Coventry — Further aeroplane searchlight units formed — Further mobile companies — The Cuffley, Billericay, and Potter's Bar Zeppelins — Searchlight company for Birmingham — Aeroplane searchlight companies for York and Leeds — Commercial gas for engines — L.34 brought down off Hartlepool — Extension of the searchlight barrage — Company for Gretna — Aeroplane searchlight companies converted — Company for Nottingham — Distribution of officers and units in April, 1917 — Progress in organisation and tactics — Aeroplane raids — Formation of L.A.D.A. — The A.A. defences re-organised, January, 1918 — Establishments — Appointments to G.H.Q. H.F. — Formation of northern air defences — Dover A.A.D.C. — Superiority of the defence over the attack — Developments in equipment and organisation — Height-finding and track-plotting — Distribution of officers and units, July, 1918 — Further re-organisation — The Armistice — Disbandment — Observations on Home service.

CONTENTS—*continued*.

CHAPTER VII.
1915—1920. THE GREAT WAR—*(Continued)*.

THE ANTI-AIRCRAFT SEARCHLIGHT SECTIONS OVERSEAS.

CHAPTER VIII.
1919—1923.

THE TYNE ELECTRICAL (FORTRESS) R.E.

CONTENTS—*continued*.

CHAPTER IX.
1924—1933.

TYNE ELECTRICAL (FORTRESS) R.E. AND TYNE ELECTRICAL ENGINEERS, R.E.

APPENDICES.

ILLUSTRATIONS.

ILLUSTRATIONS—*continued*.

MAPS AND PLANS.

CHAPTER I.

1883—1907.

THE TYNE DIVISION ROYAL ENGINEERS (VOLUNTEERS) SUBMARINE MINERS.

The Volunteer Force — Submarine Mining Company of the Newcastle and Durham (Volunteers) Royal Engineers—R.E. Staff—Courses at Chatham — Formation of the Coast Battalion R.E. — Constitution of the Tyne Division — Headquarters at Clifford's Fort — Submarine mining — Vessels and boats — Conditions of service — Uniform — Rifle shooting — Annual training — Ambulance work — Competitions and prizes — The Fishermen's section — Nicknames — Electric searchlights introduced — Formation of the Corps of Electrical Engineers — The South African War — Detachment for the Thames — Establishment increased — The Ladies' Challenge Trophy — Band and Pipers — Detachments at Hull, Weymouth and Portsmouth — Further increase in establishment — Ladies' Day at Clifford's Fort — Social occasions — Rifle Club — Sham attacks — The abolition of submarine mining in the Army — First motor searchlight equipment — Operations with balloons — Disposal of submarine mining stores — Conversion of the Division to Electrical Engineers.

ALTHOUGH the existence of volunteer soldiery is recorded even in very early history, the Volunteer Force from which the Territorial Army of to-day traces its uninterrupted descent dates only from the year 1859. In that year, when the relations between England and France were far from secure, Lord Derby's Ministry authorised the enrolment of volunteer corps and, in a few months, a volunteer army of 150,000 men had sprung into being. The employment of this army, which undertook not only to provide its own arms and equipment but also to defray the cost of its own maintenance in times of peace, was to be confined to service at home in the event of " actual or apprehended invasion."

This volunteer force included detachments of various branches of the army and one of the earliest of the volunteer engineer units to be raised was the Newcastle-upon-Tyne Company which was formed in 1860, under the command of Captain P. G. B. Westmacott, from personnel of the Armstrong factories at Elswick.

About this time the defence of our military and commercial ports by submarine mines was engaging the anxious attention of the War Office and, in 1863, this mode of defence was definitely introduced into the service, but it was 1871 before the first submarine mining company of the regular army was established, the unit selected for conversion being the 4th Company, Royal Engineers.

A

Considerable importance was attached to this branch of defensive military engineering and the establishment of Royal Engineer personnel for this particular service rapidly increased. By April 1877 the number of regular submarine mining companies had risen to five and it was not long before the possibility of this duty being undertaken in part by volunteer engineer units began to be considered.

The first movement in this connection appears to have originated in 1878 with an enquiry at the Tyne by Sir Lintorn Simmons, Inspector-General of Fortifications and Director of Works at the War Office, as to the formation of a submarine mining and signalling corps from the Newcastle and Durham Engineer Volunteers. These two corps had become affiliated in 1868 and a complete consolidation was accomplished in 1880, the unit, with an establishment of some 1,300, becoming known as the 1st Newcastle-on-Tyne and Durham (Volunteers) Royal Engineers, under the command of Lieut.-Colonel C. M. Palmer (later Colonel Sir Charles Mark Palmer, Bt., V.D.) with headquarters at Jarrow-on-Tyne.

Colonel Palmer promptly set to work to prove the feasibility of the suggestion and, in the autumn of the year 1883, Sir Andrew Clarke, who had in the meantime succeeded Sir Lintorn Simmons at the War Office, forwarded definite proposals for the formation of " a volunteer submarine mining company for the defence of the Tyne." At first the Government did not seem to appreciate this response to the suggestion of their late Inspector-General of Fortifications, but after a delay of a few months the requisite permission was promised, subject to the condition that " Colonel Palmer would find the necessary craft, etc., at his own expense, and that the men should undergo a course of training on the River Tyne " : and then " if found qualified for this service, his application (to form a Volunteer Submarine Mining Company) would be granted."

Colonel Palmer accepted these somewhat stringent conditions as being quite in keeping with the spirit in which the Volunteer Force had been raised and provided not only the necessary craft, but also a considerable part of the cost of the experiments which were forthwith undertaken at North Shields. Sixty men were selected from the Newcastle-on-Tyne and Durham corps and organised into a submarine mining company under Captain O. Allison [1] and Lieutenant T. J. Allen. The first practice of this company took place in February 1884 in conjunction with a detachment of Royal Engineers under Lieutenants F. G. Bowles and L. Quill, R.E. [2], who were sent to the Tyne expressly for the purpose.

Such was the energy and enthusiasm with which the volunteers undertook their novel tasks that they were very shortly reported as " highly qualified for carrying out this important duty." The War Office thereupon redeemed its

---

[1] O. Allison, Lt. 13/2/78, Capt. 4/11/81, 1st N.D.V.R.E.

[2] F. G. Bowles, Lt. 2/2/76, Capt. 2/2/87, Maj. 14/4/95, Lt.-Col. 1/10/02, Col. (B.G.) 1/10/05. Retired 12/1/13. Served in the Soudan Campaign, 1885.
L. Quill, Lt. 30/7/79. Died at Malta 16/2/88.

promises, the detachment received official recognition, mining stores were provided at public expense for its use and instructors were sent from Chatham.

The Royal Engineer submarine mining staff so appointed consisted of some 26 of all ranks commanded by Lieutenant J. Pring, R.E., who also carried out the duties of adjutant to the unit in addition to other technical duties on the coast from Holy Island to the River Tees. The establishment of regular personnel included one company-serjeant-major-coxswain (later promoted warrant-officer-mechanist-coxswain) who was responsible solely for the boat-work and was a local man appointed on account of his knowledge of the River Tyne. Others of the R.E. Staff were one quarter-master-serjeant-storekeeper, one quarter-master-serjeant-instructor (submarine mining), one mechanist engine driver, one mechanist electrician, one company-serjeant-major and one serjeant, the remaining personnel being sappers with a proportion of junior ranks.

In October 1884 similar companies to that at the Tyne were formed at Greenock from the 1st Lanarkshire Engineer Volunteers and at Liverpool from the 1st Lancashire Engineer Volunteers; these were shortly followed by units raised for the defence of most of the more important commercial harbours.

Arrangements were now made for officers and other ranks of the volunteer submarine mining companies to attend courses of instruction at the Submarine Mining School at St. Mary's Barracks, Chatham. It was not, of course, possible for all the members of a unit to leave their civil occupations for this purpose, but the first course, held in May 1885, contained a proportion of men from the Tyne. The courses, which continued to be held at intervals, lasted two months and were divided into three parts. Part I included the shore work, Part II the water work, and Part III the electrical work. Parts I and II were compulsory for all officers and carried the letters " p.s." in the Army list ; Part III was voluntary and carried the letter " p." All three parts were voluntary for N.C.Os and men. Officers who were unable to attend these courses were trained locally and went to Chatham for the examination only which occupied a week.

In 1885 also, the Coast Battalion, Royal Engineers, was formed to provide an entry to commissioned rank for R.E. N.C.Os and to encourage the enlistment of recruits possessing a higher standard of intelligence. The Establishment of the Battalion was 14 officers, 22 serjeants and 104 men divided into sections, one at each commercial port defended by submarine mines, that at the Tyne being named No. 1 Section Coast Battalion R.E. These sections absorbed the Regular R.E. submarine mining staffs which had previously been posted to the various ports. All the officers were recruited from the warrant and non-commissioned ranks of the Royal Engineers, but the N.C.Os and men retained their regimental position and seniority and were interchangeable with the rest of the R.E. Submarine Miners. The functions of the Battalion were the custody and maintenance of the submarine mine defence and the instruction of the volunteer submarine mining companies at the ports. These companies then had no permanent staff or adjutant posted to them, but the section of the Coast Battalion serving on the station henceforth acted as staff, and the officer of the Coast Battalion as acting-adjutant and staff officer to the officer commanding the volunteer unit in addition to his special duties in charge of the submarine mine defence. Non-commissioned officer instructors of the Coast Battalion ranked on parade as senior to all volunteer N.C.Os, and the acting adjutant, if of army rank not below that of captain, ranked as senior captain.

The regulations for the training and pay of the Submarine Mining Volunteers went through a period of natural growth and it soon became apparent that, owing to the longer training necessary and the higher capitation grants earned, it was not entirely convenient for these units to be administered as part of the existing corps from which they had been formed. Consequently when it was decided, early in 1887, to increase the establishment of the volunteer submarine mining personnel at the Tyne, this was done by the inclusion in the " Peace Establishments " for that year of a second, and independent, company under the title of " Tyne Submarine Mining Company." This company, with an establishment of three officers and sixty other ranks, was to have been separately recruited and, while working in conjunction with the original company formed from the Newcastle-on-Tyne and Durham Engineer Volunteers, was otherwise to have been an independent unit. Before this Unit was formed, however, it had been decided to give all the existing volunteer submarine mining companies their own separate organisation. By the time this came about there were nine groups of Volunteer Submarine Miners in existence, and, of these, all except those at the Tyne were constituted ' Divisions ' and entirely severed from their parent Corps at the end of 1887. For some reason the companies at the Tyne, one of which had been the first to be formed, were not constituted a separate Division until March of the following year, 1888. The precedence of the Divisions was not affected by this and was as follows :—

    (1) Tyne.
    (2) Severn.
    (3) Clyde.
    (4) Humber[1]
    (5) Tees.
    (6) Forth.
    (7) Tay.
    (8) Mersey.
    (9) Falmouth.[1]

The Unit now received the title of the Tyne Division Royal Engineers' (Volunteers), Submarine Miners, and was given an establishment of three companies consisting of 189 enrolled members, with the following proportion of officers :—

    Major ...   .   1.
    Captains   ...   ...   2.
    Lieutenants ...   ...   6.

[1] Shortly afterwards replaced by Militia Submarine Mining Divisions.

LIEUTENANT-COLONEL W. JOHNSON, C.B., V.D.
Commanding Officer, 1888–1911.

To face page 4]

The command of the newly constituted Unit fell to Major Wm. Johnson who, together with Captains T. J. Allen and C. Baguley and Lieutenant J. G. R. Hartness, transferred from the 1st Newcastle and Durham Engineer Volunteers.

The other officers, who were directly commissioned into the Unit on its formation, were Second-Lieutenants J. R. Scott, F. G. Scott, A. Cay, and R. H. Wesencraft. The acting-surgeon was Dr. R. K. Tait, and the Rev. H. Bott, subsequently Vicar of St. Aidan's, Newcastle-upon-Tyne, was appointed acting-chaplain with precedence among the officers as captain. At the same time, Colonel Palmer, the founder of the whole movement, was appointed Honorary Commandant, and later Honorary Colonel. The appointment of Second-Lieutenant A. W. Heaviside in December completed the establishment of officers.

The Headquarters were now removed from Jarrow to Clifford's Fort, North Shields. On a spit of sand projecting from the north bank of the River Tyne and close to the present Low Lighthouse, the Marquis of Newcastle had erected in 1642 one of the two blockhouses that were to guard the Narrows, the surface works of which are recorded as having consisted of earth-filled basket gabions, although the foundations appear to have been more substantial, as was shown in October 1811 when they were laid bare by an exceptionally heavy sea.

In 1672, on the outbreak of the third Dutch war, a new fort, named after Lord Clifford of Cabal fame, was completed on the north bank of the River Tyne at the seaward end of the narrows and the Low Lighthouse was included within its perimeter. The property on which this fort was built belonged to Ralph Milburne of East Chirton and it is said that King Charles never paid him for the land. It was left to Ralph Milburne's widow to put in a claim to the Treasury in 1708 and she was then paid £150.

The buildings evidently suffered various alterations, as drawings of the Fort soon after its construction show a keep as a three-storey building with a central turret.

It was at first armed with thirty culverins[1] and ten demi-culverins on the east and south sides, and was commanded by the Governor of Tynemouth Castle. One account states that there was a salary of £284 7s. 11d. attached to the combined post, but Grose, in his " Military Antiquities," states that this did not exceed four shillings a day. Be that as it may Earls, Baronets, Lords and Generals were pleased to hold the appointment which gradually, however, became no more than a sinecure until, on its falling vacant in 1839, the Government decided not to fill it again.

---

[1] The name Culverin is derived from the French ' coulevrine ' (Couleuvre—Common Snake). Names of reptiles were frequently applied to early cannon. The culverin was a large type of ordnance which was very long in proportion to its breadth. The length of the ordinary culverin ranged from 10 to 13 feet ; the diameter of its bore from 5 to 5½ inches ; and the weight of shot from 17 to 20 lbs. The Demi-culverin (' Coulevrine-moyen ') had a bore of 4½ inches and fired a shot of about 10 lbs.— (New Oxford Dictionary).

The history of Clifford's Fort does not ring with the stirring accounts of many combats. Perhaps the most interesting episode in the annals of the place was the celebrated "siege" of 1804. In that year the North Shields and Tynemouth Volunteers entered upon permanent duty for one month, and the guards at Clifford's Fort, Tynemouth Castle and the Spanish Battery were entrusted to them. At Clifford's Fort the Company had not been in occupation for more than four hours when Major Doyle of the Light Brigade from Sunderland, accompanied by one company of the 61st Regiment (later the 2nd Battalion, The Gloucestershire Regiment), one company of the Northumberland Militia and one company of the Lanark Militia, crossed the Tyne from South Shields where they were in camp. The officers, it appears, had become somewhat vainglorious over their cups and the Major, who had boasted that he could easily surprise any of the forts garrisoned by the Volunteers, was dared to make the attempt. Unfortunately for himself he chose Clifford's Fort as his intended victim.

He landed with his troops in the early morning, and from the Low Lighthouse Sands proceeded as noiselessly as possible, he himself leading on his charger, up the narrow passage close to the fort. But, before they could reach the gates, the Volunteers had made preparations to receive them, their landing having been observed in spite of their caution.

The number of Volunteers inside the Fort was insufficient for the guarding of the embrasures and walls, so an 'express' was sent off for the remainder of the Corps, who happened to be on parade in Dockwray Square. These had already noticed what was going on and they hastened down the bank to the assistance of their comrades forcing their way at the point of the bayonet along the narrow passage mentioned above. There was a sharp scuffle and later the besiegers attempted to take the Fort by storm, but another attack in the rear saved the situation and Captain Robert Shield of the Volunteers, who had been captured by the Northumberland Militia, was rescued. Eventually, the assailants of the Fort were repulsed and retreated sullenly to their flat-bottomed boat 'Buonapart' in which they had crossed the Tyne.

But their cup of bitterness was not yet full, for, on arriving at the Bents, they found that a party of Volunteers headed by Captain Shield, who was thirsting for revenge for being captured, had slipped across the river in sculler-boats, demolished the attackers' camp and carried off their colours. This was worse than defeat as it involved disgrace and it was many a long day before the incident was forgotten. The affair was a feather in the cap of the Volunteers who proved themselves worthy of being entrusted with the Fort. The Fort was not built to resist a land attack and it is interesting to note that one-third of the attacking party were soldiers of the Line.

But a more dangerous assailant from which Clifford's Fort has had to defend itself is that powerful and ever-present enemy, the sea. In 1812 sixty feet of the walls were washed away, and again, in 1852, one hundred feet were

CLIFFORD'S FORT.

Main Gateway.

The Tablet above in archway is now erected within
the Headquarters of the Tyne Electrical Engineers
at Tynemouth.

To face page 6]

carried away in a fierce north-easterly gale, but the erection of the massive breakwaters at the mouth of the river, which were designed to make the Tyne a harbour of refuge, subsequently afforded complete protection.[1]

In 1859, during the agitation for the improvement of the national defences, Clifford's Fort was put in a state of defence at the same time as additional armament was being installed at Tynemouth Castle.

In these and other ways the march of progress brought about many changes in the place and nearly all the old features disappeared. In 1888 additional buildings were erected in the already somewhat restricted enclosure and Clifford's Fort became the home of the submarine mining 'Establishment' for the River Tyne and the headquarters of the Tyne Division Royal Engineers (Volunteers) Submarine Miners.

For more detailed information respecting the technical work involved in the defence of harbours by means of minefields, the reader must be referred to the various official publications[2] but the following brief outline is inserted at this stage in order to give some idea of the duties carried out by a volunteer unit of this particular type.

" A submarine mine," runs the definition in the manual, " consists of a charge of explosive *moored* at, or beneath, the surface of the water, intended by its explosion to . . . damage . . . any hostile craft . . . when attempting to pass within its range." All the submarine mines in the British service were electrical, being fired from a powerful firing battery placed in a protected position on shore.

The whole of the mines in a defence had to be connected with the shore by means of electric cables and were harmless except when deliberately fired by order of the officer in charge of the Defence. As the range of effective action of a submarine mine is limited, it was very important to ascertain accurately when a vessel was sufficiently near any particular mine to be put out of action by it. There were two methods by which the correct moment for firing a mine were ascertained, viz. : either (a) by circuit closer (C.C.) or (b) by observation.

The former method consisted in placing within the mine itself a mechanism which operated an indicator on shore when the mine case was struck by a vessel. The latter method consisted of observing accurately the progress of an enemy vessel through the minefield by means of an instrument known as a position finder (P.F.) installed in a protected cell ashore. In either case, when a vessel was noted to be in contact with or sufficiently near to a given mine, that mine would be fired by closing an electric circuit in the firing station. Mines were accordingly classified as either 'Electro-contact' (E.C.) or 'Observation' mines depending on the method of firing adopted.

[1] The North and South Pier foundation stones were laid in 1854 and the Piers completed in 1895. The North Pier was breached in 1897 and the reconstruction of the outer portion was completed in 1909.
[2] The " Manual of Submarine Mining."—The War Office, 1904.
The " History of Submarine Mining in the British Army."—Lt.-Col. W. Baker Brown, Chatham, 1910.

It will be evident that the duties of the officer or non-commissioned officer in charge of the observing station were of the highest importance and the manual states that " to follow accurately the water-line of a vessel except when she is steaming slowly in a good light and smooth sea is by no means a simple matter . . . . and only an experienced observer will be able to keep the telescope properly sighted."

The personnel of a submarine mining unit had, therefore, to be capable of undertaking a variety of duties of a technical nature including loading the mines with their charges of guncotton, preparing and inserting the firing apparatus, laying-out the mines accurately in their appointed positions in the water, and connecting up the electric cables and testing these both in the submerged junction-boxes (J.B.) and in the test-room on shore. All this demanded a rare combination of sailor, surveyor, engineer and electrician, and, in addition, owing to the obvious necessity of maintaining continuous communication between the shore and the various boats and vessels engaged in laying-out or testing the mine-field, it was important that as large a percentage as possible of the men should understand visual signalling with flag and lamp. In the Regular submarine mining service every man had to pass as proficient in this duty, but the difficulties of insisting on this in a volunteer unit were realized, and the manual lays down that " those N.C.Os and men who are selected for such duties as laying-out squads and J.B. boats hands should, if of sufficient education, be qualified as ordinary signallers."

All this work required a considerable shore establishment, and the buildings in Clifford's Fort included offices, stores, boat-sheds, loaded mine stores, magazines, loading sheds, fitting rooms, fitted apparatus stores, workshops, cable tanks, test-rooms, etc., etc. Here the mines were loaded and connected up and from the east end of the Fish Quay they were loaded onto the mine-laying vessels, the practice minefield being just east and south of the Groyne Light.

The vessels provided for these purposes consisted of the " No. 13 ' Miner,' " a service submarine mining steam vessel about 70 feet in length ; the steam launch " Sir Charles Palmer " which had been used in the original trials in 1884 ; two junction-box boats No. 1 and No. 2 ; one 30 ft. cutter ; one 22 ft. gig ; and a 16 ft. and a 14 ft. dinghy. The coxswain of the two steam vessels were Q.M.S. Dixon of the " Miner," and Q.M.S. Jefferies of the " Sir Charles Palmer." When extensive operations were contemplated, the " Ampere " and the No. 6 " Miner " from Middlesborough would sometimes be made available or additional vessels hired. The steam tug " Ulysses," for example, was frequently chartered for service as a " Miner," while sailing cobles hired from the local fishermen were found extremely useful as mobile J.B. boats.

The Division, which was governed by the Volunteer Act of 1865, consisted of two classes, viz. :—enrolled members and honorary members. The latter class contributed to the funds of the Corps, subscribing five shillings each

CLIFFORD'S FORT.
Soldiers' Quarters, Serjeants' Mess, etc.

CLIFFORD'S FORT.
No. 1 Observing Station, Office, Store and
Technical Block.

To face page 8|

annually, or a life donation of one guinea, but were not enrolled for service. A retired member, however, who had been returned as efficient eight times in the annual returns of the Corps and who signified his willingness to re-enrol should occasion demand was admitted to honorary membership without contribution.

The term of service for which men were enrolled was three years, and particular care was exercised that only men of good character were accepted. It was laid down that submarine mining companies should be composed of persons who had learned a mechanical trade but there was no detailed establishment by trades as exists to-day. Beyond this, it was sufficient, to quote the "Regulations and Orders" of the Division, that its members should be above the age of 17 and below that of 49 on enlistment, 5' 6" or more in height (in their stockings), 33 inches round the chest, strong and capable of lifting heavy weights, and as a rule accustomed to boat work." On attaining the age of 50 years all members except officers were compulsorily retired.

The nomination of candidates for commissions rested nominally with the Lord Lieutenant of the County, but this prerogative was seldom exercised and the Commanding Officer made his recommendations direct to the Secretary of State for War. Officers had to be over 17 and under 60 years of age.

For the dual service on land and sea, a single pattern of uniform was not adequate. On parade, the Division turned out in the scarlet uniform with blue facings, white cords and shoulder-cords common to all volunteer engineers but distinguished by the letters S.M. and the word "TYNE" on the shoulder straps,[1] and the wearing of a silver grenade on the right arm by the N.C.Os and sappers returned as efficient. In "marching order," the helmet was worn with leggings and white buff equipment, though for "walking out" the helmet was discarded in favour of the more jaunty forage-cap. The men were also provided with a special working dress for their duties afloat. This working dress consisted of a blue reefer jacket, blue woollen guernsey, navy pattern trousers, leather knee-boots and navy pattern cap with ribbon bearing the words 'Submarine Miners.' All buttons were of white metal and bore the title of the Unit.

The uniform of the officers followed very closely the patterns worn in the Regular Army, except that all buttons, lace and appointments were of silver. The normal dress for officers for parades, etc., was an undress uniform of blue with a blue forage-cap with silver band and buttons.

Honorary members were permitted to wear the uniforms of sappers of the Corps, but with a distinguishing mark approved by the Commanding Officer. They were not allowed to interfere in any way with the military duties or finance of the Corps but might, on occasion, attend plain clothes drill.

---

[1] The wearing of the word "TYNE" on the shoulder strap was authorized by A.O.281 on June, 1888.

Every effort was made to foster esprit-de-corps and all ranks were adjured to pay " a strict and conscientious attention to their drills and to all the rules of the Corps, to which they should be proud to belong, and the honour and fair fame of which should be one of their highest objects." In support of this injunction, serious attention was paid in the Division to the turn-out of all ranks at all times. In the " Regulations and Orders " there are constant references to this important matter. " When going to or coming from a drill, a volunteer should always conduct himself in a manner befitting the position of a soldier. If with arms, the rifle should be properly carried, not as if it was felt to be a burden or something you were afraid to handle ;" and again, " The Commanding Officer strongly disapproves of smoking in the street. A pipe or cigar in the mouth when in uniform entirely spoils the appearance. Chewing tobacco is not allowed."

It is remarkable that, when the Division was formed, beards were still worn in the Service and they continued for many years to be permissible. " In some cases a good beard can be worn with advantage " reads Standing Order No. 28 of the time, but hastens to add the caution, " If a beard is worn, let it be of moderate length and kept neatly trimmed." There is an almost maternal ring about the following :—" When gloves are worn, the proper place for them is on the hands. (N.B.—Clean hands under clean gloves). The cuff is a good place for the handkerchief."

At the time of its formation the Unit was armed with the Martini-Henry .450 rifle and long sword-bayonet. Although the Corps was not at that time obliged to carry out an annual musketry course, or " class-firing " as it was then termed, the Commanding Officer enjoined all members not to forget that all should be able to use their rifles well when called upon to do so. This advice was well taken to heart with the result that the Unit shortly began to achieve success on the rifle range. The open range then used was at Harton Moor, South Shields, long since dismantled. That due care was not always exercised with their arms by these Volunteers is, however, evident from the fact that the Commanding Officer was empowered to impose a fine of five shillings for " discharging the rifle accidentally " while, for " loading or wilfully discharging the rifle or drawing the sword without orders," or even for " pointing the same, loaded or unloaded, at any person," a heavier fine of ten shillings was levied.

Recruits on enrolment had to carry out nine squad drills and thirty sub-marine mining drills before being authorised to wear uniform.

The technical training of officers and trained men consisted of seventy-eight drills of one hour each and eight days camp each year until reported expert when the compulsory drills were limited to forty-eight hours only.

The first annual training of the new unit was held in 1888 at Willington Quay Shipyard, a somewhat strange place to pitch a military camp. The camps lasted 15 days, of which eight only were compulsory, subject to exemption in special cases. The Corps received a camp allowance of 10/- a day for officers,

THE " MINER."

A LINE OF GROUND MINES.

To face page 10]

5/- a day for sappers, and intermediate allowances for the warrant and non-commissioned ranks, in addition to travelling allowances for the journeys to and from camp, and it was the custom to pay these amounts to individuals less messing and camp expenses. Beyond this, members of the Volunteer Divisions received nothing when attending Army Schools for periods of instruction, but the Corps received a capitation grant of five pounds in respect of each effective officer or man on its strength, to defray general administrative expenses. A man failing to qualify as efficient was liable to make good this amount to the funds of the Corps.

There was also a system of whole-day and half-day parades for technical instruction. A whole-day parade consisted of not less than six hours continuous work and was the equivalent of six drills, a half-day consisted of not less than four hours continuous drill and was equal to four drills. Pay was given at the rates of ten shillings for officers and five shillings for other ranks for a whole-day, and half those amounts for a half-day, up to a maximum of ten pounds per individual officer and five pounds per individual man, but not exceeding an average of one pound ten shillings per head for the strength of the Corps. This pay could only be earned after a minimum number of twelve voluntary drills had been done by each individual.

The Adjutant inspected each company in the course of the year and on these occasions every member had to be present. There was also an annual general inspection of the whole Division, but unless there were at least two-thirds of the enrolled members present it was not counted as an inspection and had to be held again. A fine of two-shillings-and-sixpence was inflicted on any member failing to attend, without leave or proper excuse, this or any other parade ordered by the Commanding Officer. All current orders were notified through the columns of the 'Shields Daily News' and the 'Shields Gazette.'

Early in the year 1889 Dr. Tait was confirmed in the appointment of surgeon and Dr. W. H. Brown then joined the Division and acted as second surgeon. The medical officers conducted courses in stretcher drill and in field dressing and men who had passed a satisfactory examination received certificates to that effect from the Volunteer Ambulance Department. Two men in each Company who were so qualified were selected by the Commanding Officer on the recommendation of the medical officers to act as stretcher bearers and to wear the appropriate badge. Sir Charles Palmer took a particular interest in this branch of the Unit's work as it had been largely through his efforts that the first Volunteer Ambulance Corps had been raised as part of the Newcastle Corps as early as 1875.

The annual training in camp in 1889 was held in the Spanish Battery at Tynemouth[1] and during the following eleven years either there or in Clifford's Fort itself, or at Rockcliffe, Whitley Bay, or at Preston, North Shields.

---

[1] Built 1545 by Sir Richard Lee, to protect the base of the English Fleet in its invasion of Scotland. The Battery was named after its first garrison which was provided by Spanish Mercenaries. In 1643, on the verge of the Civil War, the low stone walls were raised by the addition of brick superstructure and finished with casemates for guns, 300 men being employed on the work.

The technical work appealed strongly to the men who were drawn largely from the ship-building and engineering works of Tyneside and, in spite of the statement of their own Honorary Commandant that it seemed "almost incredible that such highly scientific work could be carried out by volunteers in their spare time," the Division soon came to be recognised as one of the leading volunteer units. The reports on the inspections were all of a highly complimentary character though it would seem that the degree of efficiency achieved by the Division in its military duties on shore did not, in every case, reach the standard of that afloat. This deficiency sometimes made itself apparent in a somewhat humorous manner when the Division was on parade.

On one occasion the Division was in quarter column and the order was given "On the left form line." The 'G' was sounded and the company markers were sent out. The length of No. 3 Company was under-estimated with the result that by the time it reached the new alignment there was not sufficient room for it. The officer commanding No. 2 Company promptly called out in his best parade voice "Move along there No. 3, you're squashing us to ——— !" Another officer had such a staccato manner of giving an order that his company once 'changed arms' to the yapping of a dog on the side-walk. Yet another officer is reported to have extricated himself from the position of having forgotten the order for reforming line when his men were in fours by giving the command "Form-Twos," while a fourth was responsible for getting over the difficulty of negotiating a stile by issuing the somewhat unorthodox order "Scatter you ——— and fall-in on the other side !"

On another occasion, when the Division was drawn up awaiting the arrival of the inspecting officer, the C.O. himself suffered from a fit of absent-mindedness. On seeing the General coming through the gate onto the parade ground, he forgot all about the "General Salute" and hurried forward to greet him and enquired after his health, for all the world as if they were in a drawing-room !

But these were merely incidents which caused some amusement at the time and bear no true relation to the Unit's efficiency, for in submarine mining work they were recognised as second to none. It is recorded that the Tyne Volunteer Division once executed a certain manoeuvre connected with mine-laying in the shortest time the work had ever been done in the service.

Prizes were given annually for excellence in shooting, rowing, sculling, loading, connecting-up and laying-out mines, signalling, electrical work, swimming, bugling, and for a written examination in the year's work which also decided the promotion of N.C.Os. One of the most important events in this connection was the competition between companies of the Division for the cup presented by Sir Charles Mark Palmer in 1890[1].

---

[1] Competed for on only five occasions :—
1890 }
1892 } Won by 'A' Company. Captain J. R. Scott.
1893 }
1894   Won by 'C' Company. Team of 14 under C.S.M. Armstrong.
1895   Won by 'B' Company. Team of 14 under C.S.M. Jessop.

OFFICERS, 1892.

Standing, Left to Right—Second-Lieut. F. E. Macfadyen, Lieut. R. P. Winter, Captain F. G. Scott, Lieut. C. R. Toomer, Lieut. J. Martin, Co. Bn. R.E. (Adjt.), Captain R. H. Wesencraft, Lieut. A. W. Heaviside, Second-Lieut. R. Stephenson.
Seated, Left to Right—Captain J. R. Scott, Major W. Johnson.

TYNE DIVISION, ROYAL ENGINEERS (VOLUNTEERS), SUBMARINE MINERS.
Annual Training on the Rockliffe Ground.
1892.

In this particular year, the duties of O.C. Coast Battalion Section at North Shields and of adjutant to the Tyne Division were taken over by Lieutenant J. Martin, Coast Battalion R.E., who had joined the Royal Engineers as a sapper in the year 1873. This officer was exceptionally well qualified for his new appointment as he had recently held the post of Serjeant-Major Instructor at the School of Submarine Mining at Gillingham, Kent, and, under his expert guidance and tuition, the technical efficiency of officers and other ranks attained an even higher standard than before.

In February 1891 Dr. Tait resigned his post as surgeon and the duty was then entirely taken over by Dr. Brown. At the same time Hon. Lieut. D. Rioch was appointed quartermaster to the Division to fill the post authorized in 1889.

About this time a section of the Division was raised at Cullercoats, composed of fishermen, who were found to be well suited to the class of work they were required to undertake. As soldiers, it is reported, they were not always very efficient and it was not thought advisable to issue them with scarlet uniform or with arms. But for the service afloat they were naturally excellent and it was doubtless largely due to their efficiency in this branch of the Unit's activities that the Tyne Division was able to beat the Clyde Division in competition for a silver challenge trophy in 1893 on the second and last occasion when the event took place.

There is on record the case of one worthy fisherman enlisted into this section who had but one eye! This fact appears to have escaped the attention of the authorities until, on the occasion of an inspection by the Inspector of Submarine Defences from the War Office, the order was given " Eyes-left " instead of the more usual "Eyes-right." The weird contortions of the unfortunate Sapper in his endeavour to comply with this command led to the immediate discovery of his deformity. During an examination in knots and lashings the R.E. officer from Gillingham once questioned a certain fisherman as to the purpose for which a catspaw should be used and received the astonishing reply " To kill mice with ! "

The tale is also recounted of one occasion when a senior officer of the service was inspecting the work of the Unit afloat and found cause to correct a fisherman as to the particular manner in which he was carrying out some technical duty connected with the laying-out of a mine. The fellow listened patiently enough to the remarks of the inspecting officer and, when he had finished, said " You stand aside Mister Soldier, and let us get on with the work."

The company-serjeant-major of the company to which the " Fishermen's Section " belonged was, for many years, C.S.M. Storey. The Section continued to form part of the Tyne Division for a considerable period but eventually disappeared about the end of the century.

The local inhabitants took considerable interest in the work of the Division to which they generally referred as the " Deep Sea Pitmen," the " Fish-Quay Lancers," the " Mussel Shifters," or, more respectfully, as the " Submarines "; though, among the members of the Division, the abbreviation " Sub-Miners " was in more general use.

In 1893 the Rev. S. Pearson, Minister of the Congregational Church, Tynemouth, joined the Unit in the capacity of second acting-chaplain.

For some years experiments had been in progress at Portsmouth, Plymouth and at other regular stations with the use of electric searchlights as an adjunct to the artillery and submarine defences. Now, in 1895, the first searchlight to be used at the Tyne was installed at Clifford's Fort for illuminating the minefield. The power for this light was provided by a Robey horizontal steam engine, with a Fowler boiler, driving a dynamo. In the same year the establishment of the Coast Battalion R.E. was increased to provide personnel for duty with these electric lights at the various ports and was re-organised in eleven sections, the section at Clifford's Fort being increased to 35 other ranks.

In 1897 the use of searchlights for coast defence was considerably extended and extra plant provided. The light from Clifford's Fort was now moved to Tynemouth, the engine and dynamo being installed in an engine room in the castle moat and the searchlight in an emplacement at the Spanish Battery, where it was better sited for the purpose for which it was required. A second light was also installed in the castle moat shortly afterwards, both this light and the No. 1 light at the Spanish Battery being of 30° dispersion.

Similar increases in searchlight installations were made at all the defended ports and the personnel of the Coast Battalion R.E. was insufficient for manning them. In these circumstances the possibility of training auxiliary forces to take a share in these duties had to be considered and, at ports where a volunteer corps already existed, their establishments were adjusted where required and the necessary training begun.

The Tyne Division Submarine Miners contained a large proportion of mechanics of just the type required for the electric light work, and, in addition, the general intelligence and education of all ranks were of a high average with the result that good progress was made in this new branch of the Unit's work.

The year 1897 was also noteworthy for the formation of a new volunteer corps with which the Tyne Division was destined to be closely associated in later years. This was the Corps of Electrical Engineers, which was formed, under the Honorary-Colonelcy of Lord Kelvin, from men of science and leading members of the electrical profession in London, to assist especially in the working of the electric lights installed in connection with the Thames Defences. Two of the original members of this Corps, the two brothers Hopkinson,[1] had previously served for eighteen months each with the Tyne Division Volunteers in order to become acquainted with the working of submarine defences.

This Corps of Electrical Engineers furnished a searchlight unit composed of 8 officers and 55 other ranks for service in South Africa during the Boer War. The Newcastle-upon-Tyne Engineer Volunteers also sent a service section overseas, but, although volunteers were not lacking in the Tyne Submarine Miners for duty in the field, no opportunity arose for their employment and none of them was actually called upon to undertake active service.

The annual training in 1900 was held locally at North Shields and was notable as the first occasion on which the Division was permitted to operate the coast defence electric lights. This was undertaken in conjunction with the personnel of the Coast Battalion Section, and it is reported that the work was, from the very first, carried out with great efficiency.

The continued efficiency of the Division had by now established for the Unit the reputation of being the leading volunteer submarine mining corps in the Kingdom and the appreciation of this by the authorities is to be inferred from the fact that the Division was now frequently called upon to furnish detachments of men for work at ports in many parts of the Country.

The first occasion on which this took place was in the summer of the year 1900, and the success which attended this new departure may be gathered from the following letter from the War Office :—

"Sir,—I am directed by the Secretary of State for War[2] to inform "you that the following reports have been received upon the work of "the detachment of the Tyne Division Submarine Miners, Royal "Engineers Volunteers, which took part in the late mobilization of the "Thames defences. This Corps sent a detachment of two officers and "forty-two non-commissioned officers and men (thirty electricians and "twelve engine-drivers). They were divided up into parties for the "N. and S. Boom and Sheerness side, and although quite strange to "the place, settled down with the assistance of but few local Royal "Engineers to all-night nursing of lights from the first night of their "arrival in an admirable manner. The engine-drivers for oil engines "had not much experience, as it is understood that there are at present "no oil engines at the Headquarters of the Corps, but picked up their

---

[1] J. Hopkinson was the first Commanding Officer of the Corps of Electrical Engineers, being appointed to the rank of major on transferring to that Unit on its formation on 28th April 1897.

B. Hopkinson was born on 18th January 1874, and educated at St. Paul's and Trinity College, Cambridge. He was called to the bar in 1897, and became in 1898 a consulting engineer in partnership with Messrs. Charles Hopkinson and Ernest Talbot. In this connection he was responsible for the design of the electric tramways at Leeds and Newcastle-upon-Tyne.

In 1903 he was appointed Professor of Mechanism and Applied Mechanics at Cambridge. In 1908, he transferred his military activities to the Cambridge University O.T.C., holding at that time the rank of major on the unattached list of the Territorial Force. In this capacity he commanded the Fortress Company R.E. attached to the C.U.O.T.C. until after the outbreak of the Great War.

In 1914 he was appointed to a Professorial Fellowship at King's College, Cambridge. During the War his services were employed from 1915 to 1916 as a staff captain attached to the Directorate of Military Aeronautics and subsequently as Assistant Director of Aircraft Equipment with the Air Board. He was created a Commander of the Order of St. Michael and St. George in 1917. He died in August 1918.

[2] The 5th Marquess of Lansdowne.

" work well. The two officers, Lieut. Bruce and Second-Lieutenant
" Warburton, were in charge of groups of lights numbers 3, 4 and 5,
" and the zeal and efficiency of both officers and men cannot be too
" highly spoken of. The Commander-in-Chief[1] has already expressed his
" appreciation of the services rendered by this Corps taking part in the
" operations, but he desires that this satisfactory report may be com-
" municated to the Officer Commanding the Corps, informing him at
" the same time of the pleasure with which it has been perused by his
" Lordship.—Yours, etc., J. H. LAYE."

In August 1900 the establishment of the Tyne Division Submarine Miners
was further increased to a total of 324 and the organisation altered to five
companies which were quickly recruited up to strength. The technical instruction and experience which could be acquired by men joining the submarine
mining volunteers in days when electricity was yet comparatively in its infancy
always proved a strong incentive to intending recruits and stimulated recruiting
to the advantage of the Tyne Division as compared with other non-technical
units in the district.

The officers commanding companies at the conclusion of this re-organisation
were Captains F. G. Scott, C. R. Toomer, R. Stephenson, E. Towers and A.
Blackburn.

In the autumn of this year (1900) there took place on the rifle range at
Whitley Bay the first competition for the Ladies' Challenge Trophy. The trophy,
valued at £60, was the outcome of a special effort made by a large number
of local ladies under the presidency of Mrs. Phillip Watts and was presented
for " annual competition between the R.E. Units of the Counties of Northumberland and Durham." On this, the first occasion, the trophy was won by
' B ' Company of the 1st Newcastle Engineer Volunteers. This competition
has since been continued as an annual event which always arouses considerable
enthusiasm among the Units concerned.

On the 2nd February 1901 a memorial service was held for Her late Majesty
Queen Victoria at Christ Church, North Shields, at which the Tyne Division
was represented by a detachment and the band. Of this latter no mention
has so far been made. It was originally formed in 1890 and was still at this
period under the direction of its first bandmaster, N. A. Patterson, who had
previously served with the Tynemouth Royal Garrison Artillery Volunteers.
The band consisted at first of twenty-four musicians but was later supplemented by a band of silver bugles.

In 1902 a proposal was put forward for the raising of a pipe band for the
Division, the establishment suggested being three pipers and three drummers.
The scheme was not at first a success and all except Serjt. Piper J. Wilson were
discharged. In 1904 his son was enlisted and the Unit continued for many years

---

[1] Lord Wolseley.

PIPER.

Tyne Division, Royal Engineers
(Volunteers), Submarine Miners.
(*circa* 1900).

To face page 16]

with two pipers. Their instruments were however Highland and not North-umbrian pipes. A special pattern of uniform was approved by the War Office for these pipers, the tartan adopted being that of the Clan Fergusson. This selection is said to have originated from a close friendship between Major Johnson and a prominent member of that Clan.[1]

The annual inspection in 1901 took place at Clifford's Fort on June 28th after a week's preliminary training at that place. The infantry drill was inspected by Colonel G. E. Shute, C.R.E. Newcastle, while Major H. N. Dumbleton, Inspector of Submarine Defences, carried out the inspection of some 150 of the men at their technical work.

On 1st July a party of one officer and twenty-two men were sent to Hull to assist in the operation of the electric lights installed for the defence of the Humber and, on the 20th of the same month, a detachment of about a hundred men went to Weymouth to man the submarine defences of Portland Harbour. A week later, a third body, 130 strong, entrained for duty in the Portsmouth defences though, owing to the naval review in Spithead in connection with the Coronation of King Edward VII, the operations on this occasion took place in Southampton Water.

From 1901, owing to the difficulty experienced by officers in getting away to Chatham for their examinations and on account of the cost to the public involved in sending them there, it became the practice for an instructor to be sent up from the School to conduct these examinations at the Headquarters of the Unit. At the conclusion of such an examination held towards the end of this year, the examining officer, Captain G. P. A. Acworth, R.E., left it on record that " he had seldom had a class of officers who were better acquainted with their duties or who showed greater keenness in their work."

In 1902 the number and functions of the searchlights at Tynemouth were altered. Up to this date the lights had mainly been used for illuminating the minefield, but closer co-operation with the Artillery was being established and in this year the number of lights was increased to four. Of these, Nos. 1 and 2 were at the Spanish Battery and served to cover the mouth of the harbour, both being fitted with reflectors of 16° dispersion. The old No. 2 Light in the Castle moat, now No. 3, was also fitted at this time with a 16° dispersing lens and was used superimposed on No. 1 at the Spanish Battery to increase the intensity of illumination between the pier-heads. After a series of experiments, a fourth light, No. 4, was installed under Tynemouth Castle on the cliff face at the shore end of the North Pier. This was a concentrated beam searching over a determined arc northwards from the pier. All these were now operated from Hornsby-Ackroyd oil engines established in pairs at the moat engine room and in the Spanish Battery.

---

[1] In this connection it is to be noted that two of the officers of the Tay Division Volunteer Submarine Miners on its formation were Major W. H. Fergusson and Captain R. A. Fergusson and it is perhaps to this fact that the pipers of the Tyne Division owe the pattern of their tartan.

The annual training this year was held at Portsmouth, and the annual inspection took place in the cricket field at Preston Avenue, North Shields, on Saturday the 26th July in a torrential downpour of rain. The inspecting officer was Colonel A. O. Green, Commanding the Royal Engineer District, York, assisted by Colonel Shute.

Miniature rifle shooting now became an important, though privately managed, part of the Unit's activities and, as a result of the facilities provided in this direction and on the open ranges at Billy Mill Quarry, North Shields, and at Harton Moor, the Division continued to achieve a measure of success in local rifle meetings.

In order to test the efficiency of the electric light and mine defences of the river, sham attacks were arranged from time to time. One of these was to have been made on the night of the 22nd August 1902 by the Medway Flotilla of the Royal Navy. All the guns and searchlights were manned and the mine-field was in readiness but, probably on account of the misty weather conditions prevailing at sea, no actual raid was made on the harbour though the garrison remained on the alert until the small hours of the morning.

On the 1st April 1903 the establishment of the Division was once more increased, this time to seven companies designated ' A ' to ' G ' of which one was a staff company, bringing the total strength up to 30 officers[1] and 457 other ranks. This increase resulted in the promotion of Major and Hon. Lieut.-Colonel Johnson to the substantive rank of Lieut.-Colonel, and the appointment of Major F. G. Scott as Second-in-Command. The seven companies were commanded at this date as follows :—

" A " Company ...	...	...	Captain C. R. Toomer.
" B " Company ...	...	...	Major F. G. Scott.
" C " Company ...	...	...	Captain C. Johnson.
" D " Company ...	...	...	Captain E. Towers.
" E " Company ...	...	...	Captain A. Blackburn.
" F " Company ...	...	...	Captain G. A. Bruce.
" G " Company ...	...	...	Captain E. Robinson.

In May, a second surgeon was again added to the Establishment, Surgeon Major F. W. Gibbon, from the Durham Volunteer Engineers, joining the Division to assist Surgeon Major Brown in his duties.

The personnel at this time was recruited from fitters, blacksmiths, mechanical and electrical engineers, and engine drivers from all over Tyneside, though it is recorded that Newcastle itself hardly did its share in this respect, contributing only 80 or so of the total men in the ranks. The officers, however, were largely drawn from Newcastle. For the greater convenience of those members living in the city, a house was rented in Wentworth Place, Newcastle, where theoretical instruction was given. Practical work was carried out at North Shields during the summer on three nights each week, and boat work on Saturday afternoons.

---

[1] Excluding medical officers and chaplain. The establishment allowed four officers in each company, with the Lieut.-Colonel and Quartermaster in Headquarters.

On one of these latter occasions, the wind and sea having risen rather suddenly, the practice was cancelled and the vessels ordered to return to their moorings. It was then observed that one of the junction box boats had broken adrift from the box on which its crew had been working and was being carried rapidly out into the North Sea. The frantic signalling of the occupants, using a blue reefer jacket tied to a boathook as a flag, fortunately called the " Miner " to the rescue and prevented the incidence developing into a disaster as the J.B. boats possessed no motive power and were ill-adapted to ride out a storm in the open sea.

Camps were held this year (1903) at Weymouth and at Stokes Bay, Portsmouth, from the 20th June to the 3rd July. The detachment for Portsmouth consisted of 20 officers and 300 men under Colonel Johnson, and the Weymouth contingent of 8 officers and 110 men under Captain Blackburn. At Portsmouth the submarine mining operations were carried out in Spithead, the mines being loaded and connected up within Fort Blockhouse which was then a military station. The technical work of both detachments was inspected by Major Dumbleton, and, on the return of the Division to the north, the annual inspection on parade took place on Saturday the 13th July at Preston Avenue. The inspecting officer was again Colonel Green from York who was assisted by Lieut.-Colonel G. F. Leverson, C.R.E. Newcastle. The total on parade was 436 officers and men, only 21 short of the new establishment. At the conclusion of the inspection Colonel Green stated that the Division constituted a most valuable force and expressed the belief that it would render ' inestimable service ' in time of emergency.

On the 25th July an interesting function took place at Clifford's Fort when a ' Ladies' Day ' was held. The visitors were first shown over the establishment at the Fort and later embarked in the " Miner " to witness the firing of some of the mines in the harbour. The subsequent competition among the local boatmen to recover the multitude of fish which, stunned by the force of the explosions, floated to the surface provided the guests with a further, and unexpected, excitement. ' Ladies' Day ' was held once more in the following year and, although this function was subsequently discontinued, other social events were arranged from time to time. Among these were regular annual company dinners or smoking concerts for all ranks and the holding of evening " Socials " in the Officers Mess at which the officers of local volunteer corps such as the Tynemouth or Durham R.G.A. and the Newcastle or Jarrow Engineers were entertained. The officers also held annually a dinner to which each officer was entitled to bring a guest. The ' Turks Head ' Hotel was the venue for an ' Official Smoker ' which was given on one occasion at a cost of nearly one hundred pounds, and, on another occasion, a whist drive and dance was held in the ' Golden Lion ' Assembly Rooms at South Shields. In camp it was generally found possible to arrange a garden party and, when at Portsmouth or other military or naval stations, cricket matches were sometimes organised against the local garrisons.

The marksmen of the Unit, armed now with the magazine loading Lee-Metford .303 Rifle, achieved a notable success on the 24th September 1903 when they defeated the 5th Volunteer Battalion of the Durham Light Infantry and the South Shields Civilian Rifle Club in the competition for the John Hunt Shield held on Harton Moor Range. This was only the second occasion on which this trophy had been competed for and it was now won from the holders, the Durham Light Infantry, by a margin of 32 points.[1] In order to foster enthusiasm in rifle shooting and to render it possible for the Unit to compete in events such as these, the Tyne Volunteer R.E. Rifle Club had been started and was by this time in a flourishing condition.

Another sham attack took place on the defences of the Tyne on the 24th October 1903 in which the submarine mining vessels from the Tees together with those from North Shields constituted a supposed torpedo-boat flotilla. The night was very dark and somewhat wet and boisterous. The guns were manned by the detachment of the 24th Company R.G.A. stationed at Tynemouth Castle in conjunction with personnel from the Tynemouth R.G.A. (Volunteers) and extra artillery was introduced into the defences for the occasion. It was about nine o'clock when the raiding flotilla attempted to run past the defences but they had not proceeded far up the harbour when they were detected by the searchlights and 'annihilated' by the combined action of gun and mine. Brigadier-General Browne, V.C., the General Officer Commanding the North-Eastern District, subsequently caused a special order to be published congratulating all ranks who had taken part in these operations of which he himself had been a witness.

In November Major Scott, Lieutenant Arnison and Second-Lieutenant Schaeffer with a party of N.C.Os and men underwent a week's training at Portsmouth where they earned congratulations on the manner in which they discharged their duties.

It was at this time that His Majesty the Sultan of Turkey decorated Lieutenant Bertram with the order of the Medjidie (4th Class) for special services rendered when taking out to that country the Imperial Yacht.

The Division held a church parade at Holy Trinity Church, North Shields, on Sunday the 24th April 1904. The holding of a church parade outside the period of training in camp subsequently became an annual event, although the proposal for a second one to be held annually at South Shields had to be abandoned.

This year, 1904, training was carried out in several divisions. On the 21st May, 7 officers and 120 other ranks, under the command of Captain C. Johnson, proceeded to Weymouth and, on Saturday, June 11th, a further contingent of 12 officers and 303 N.C.Os and men left North Shields for a week's training

[1] The team, as representing some of the more successful rifle shots of the period, may be quoted and was as follows :—Major Scott, Qr.-Mr.-Serjt. Tanner, R.E., S. Serjt. Smith, R.E., Coy.-Serjt.-Major Oates, Coy.-Serjt.-Major T. Robson, Coy.-Serjt.-Major Shelton, Coy.-Serjt.-Major Dawson, Serjt. Womphrey, Serjt. Young, Serjt. Stewart, L/Corpl. T. Oates. The Shield was also won by the Division in 1904

in the south of England. On arrival in London, a detachment of 3 officers and 95 men, including the band, were sent on to Sheerness under the command of Captain Bruce, while the main body, under Major F. G. Scott, proceeded to Portsmouth. Here a further sub-division was effected, a company of 4 officers and 88 men being dispatched to the Isle of Wight, where they manned the lights at the Needles, Warden and Cliffe End Batteries. The manning personnel was accommodated at the stations, the officers being lodged in the Totland Bay Hotel. A week was occupied in the usual submarine mining and electric light training after which the Division was re-united under Colonel Johnson and returned to the north and camped during the ensuing Newcastle race week at Preston Avenue, North Shields. At this camp, Q.M.S.T. Oates was appointed acting regimental-serjeant-major, the substantive rank not being included in the establishment of the Unit at that time. Further work was carried out with the submarine defences of the Tyne and some musketry was also done at Harton range. On Sunday 19th June, a drumhead parade service was held which was attended by the public in large numbers. The following Saturday, the 25th, the annual inspection took place, the inspecting officers being Colonel J. E. Blackburn from York and Lieut.-Colonel Leverson. The total strength of the Division at this time represented approximately one quarter of the whole of the Volunteer Submarine Miners in the Kingdom.

It was this year (1904) that His Majesty King Edward VII was graciously pleased in his Birthday Honours to appoint Lieut.-Colonel Johnson a Companion of the Order of the Bath (Civil Division) in recognition of his valuable services to the nation in connection with submarine and electric light defences.

On the 25th July another sham attack was made on the Tyne and the defences were fully manned. The attacking fleet, consisting this time of the two submarine mining vessels of the Division, appeared off the mouth of the river at about a quarter to ten in the evening. The first intimation that the civilian population had that manoeuvres were in progress was the booming of the big guns at Tynemouth Castle. The attackers were immediately illuminated by the searchlights and, although three attempts were made to penetrate the defences, each was equally unsuccessful. The operations were conducted under the personal supervision of General Leslie Rundle from York.

Rather more extensive operations were carried out on three successive nights the 7th, 8th and 9th of September, when the Felixstowe Flotilla of the Royal Navy was pitted against the Tyne Defences. Sufficient men of the Tyne Division were mobilized to man the searchlights and minefield, and the manoeuvres commenced at 9.0 o'clock each night continuing until shortly after 8.0 o'clock the following morning. The defence minefield was laid in preparation and detonators, in circuit with the mines, were fired at the moment when, under active service conditions, the line would have exploded. The attacking vessels, which on this occasion were H.M. Torpedo-boats "Wizard" and "Lightning" and the Gunboat "Halcyon," displayed considerable ingenuity in their efforts to outwit the defence and sometimes the disguises they adopted were difficult to detect, especially as the normal traffic of the river proceeded as usual. On the first night two attempts were made to enter the harbour, the first at about 10.30 p.m. and the second at about 3.0 o'clock in the morning, but both attacks were repulsed by gun-fire before reaching the minefield.

On the following day, however, H.M.S. " Lightning " entered the harbour at 8.0 p.m. and proceeded unchallenged through the Narrows as far as the New Quay, North Shields, where she was made fast to the R.N.V.R. training ship " Satellite." A small party was then landed and, stealing along the base of the cliff and the Black Midden Rocks, succeeded in reaching the Spanish Battery searchlight emplacements, its presence and movements being undetected owing to the numerous spectators gathered on the bank tops. The sentry guarding the emplacements was almost immediately overpowered, the searchlight operators being made prisoners and the lamps put out of action.

Steps were immediately taken to repel the invaders and all the available men turned out fully armed and clambering down the steep sea banks met the naval party as they were leaving the emplacements. In the darkness the invaders mistook the soldiers for another detachment from their ship with the result that they were forced to beat a hasty retreat. For some time the minefield lights remained out of action while the umpires considered the position but eventually it was decided that, as the warship had landed the party before the official hour for the commencement of hostilities, the searchlights could be again brought into use.

The following night an attempt was made to surprise Clifford's Fort but arrangements had been made to prevent the recurrence of the affair of Friday night, and the landing party was immediately detected and retired to their boats. The manoeuvres concluded in a very normal manner with an attack by the two Torpedo-boats at about 10.30 p.m. in which, however, they were both illuminated by the searchlights and forced to put to sea after a gallant duel with the Q.F. guns of Clifford's Fort.

These combined operations were very popular with all ranks and afforded excellent practice for all concerned, while it is safe to say that valuable lessons in coast defence were learnt from such incidents as have just been described. In September the Channel Fleet paid a visit to the Tyne and the opportunity was taken for an interchange of hospitality between the two services.

In November 1904 an interesting event took place when the officers of the Division entertained their Honorary Colonel, Sir Charles Mark Palmer, to dinner at Clifford's Fort to commemorate the twentieth anniversary of the foundation of the volunteer submarine mining movement.

The year 1904 was notable for the decision reached by the Committee of Imperial Defence to transfer all the mine-defences to the Royal Navy while leaving the Royal Engineers with the duty of working the electric lights at the different ports. A special committee was accordingly set up to consider the best method of giving effect to this decision, but it was several years before the transfer was completed. The first manifestation of the re-organisation which this would involve was the introduction of a company system to the Coast Battalion. The 16th Company R.E., which had originally been raised in December 1825 as a survey company for service in Ireland, was one of those selected for conversion and, on the 1st July 1905, absorbed the Coast Battalion sections at the Tyne, Tees, Humber, and Mersey[1] and its headquarters were established at Clifford's Fort.

Training continued during 1905 and 1906 much on the same lines as before though rather greater prominence was given to electric light duties. Detachments were sent each year to Portsmouth and to the Isle of Wight, and in 1905 to Sheerness once more and to Weymouth. On the 27th June, that year, Lieut.-Colonel Johnson with 7 officers and about 230 men travelled to Portsmouth while 2 officers with some 60 other ranks were detached to Weymouth. The Portsmouth detachment camped at Fort Monckton, Gosport, and was followed a fortnight later by a second contingent of about 100 officers and men.

In July 1906 His Majesty King Edward VII visited Newcastle for the purpose of opening the Royal Victoria Infirmary and on this occasion the Unit paraded to assist in lining the streets. Almost all the troops on parade were in khaki and the scarlet uniforms of the Submarine Miners attracted much attention and favourable comment.

It was in this year that the Commanding Officer, Lt.-Col. Johnson, began to consider the possible use of mobile searchlights with the result that the first searchlight equipment to be carried on a petrol-driven vehicle was constructed for the Division by Messrs. J. W. Brooke & Co., motor car and motor boat builders of Lowestoft. The machine was built under the supervision and largely to the designs of Captain G. A. Bruce, an officer of the Unit, who was at that time an electrical engineer in Lowestoft. The Unit was fortunate in having, due largely to the high capitation grant paid to submarine mining units, considerable funds at its disposal and the vehicle and complete equipment were thus made regimental property.

The chassis was 18' 0" in length, with a platform upon which was mounted a 4-cylinder 45 h.p. Brooke petrol motor fitted with both high and low tension magneto ignition. Ample cooling surface was arranged for circulating water. This engine was direct-coupled to a multi-polar dynamo for generating electricity for the 90 cms. searchlight, which was estimated to give some 40,000 candle-power. The propelling power of the chassis was by means of a separate 4-cylinder 15-20 h.p. Brooke petrol motor capable of a speed on the road of 18 miles an hour. It was fitted with three speeds and reverse and the steering was by wheel. The wooden artillery road wheels were shod with solid rubber tyres. It must be remembered that petrol-driven road motors were at this

---

[1] The Mersey Section was later transferred to the 55th Company at Pembroke Dock.

period still in their infancy. The searchlight unit sent out to the Boer War by the Corps of Electrical Engineers had been equipped with traction engines as sources of motive power and the production of this new vehicle accordingly represented a great advance and attracted much favourable notice. The time occupied between stopping the road engine and showing the light was under one minute. Three men only were required to work the whole of the machinery, and there was practically no vibration, an important point where it is required to maintain a steady light. The whole of the machinery was protected by a canopy and seats were arranged for four persons in addition to the driver. The equipment could travel a hundred miles and the light be shewn for ten consecutive hours without fresh fuel.

This equipment was first used in coast defence operations at Portsmouth in June 1906 and was subsequently taken by Second-Lieut. K. S. M. Scott, with a detachment of about thirty other ranks, for a further fortnight's training on Salisbury Plain. The work here was of a particularly interesting nature and constituted the Unit's first association with the air, for the operations were carried out in conjunction with a balloon company of the Royal Engineers.

Major Martin, after the long period of seventeen years as acting adjutant, took his leave of the Division and the Corps of Royal Engineers at a dinner given in his honour by the officers of the Unit on the 2nd May 1907 at which he was presented with a large oval silver tray and tea and coffee service.[1]

It was also in this month that the Division was permitted to make use of the quarters in Tynemouth Castle as an Officers' Mess. The accommodation so provided was considerably more satisfactory than that existing at Clifford's Fort and continued to be used regularly by the officers of the Unit for some thirteen months, after which they were reluctantly compelled to evacuate the Castle in order to make room for Regular officers who were to be stationed there.

In the meantime the arrangements for the closing down of the submarine mining service were progressing and most of the mining stores at Clifford's Fort, including the whole of the electrical plant, had been broken up or sold by public auction. The bulk of this costly equipment fetched only the price of scrap metal though the submarine mining vessels and boats found rather better markets. The " Miner " herself was bought by the Tyne Pilotage Authority who had her in use for some time after which she was exchanged with the Burntisland Shipbuilding Co. for a motor boat.[2] The two junction-box boats passed through several hands before being bought by the Tyne Improvement Commission in 1911 and are still used in the river as wreck-marking vessels, a purpose for which their construction renders them particularly suitable.

---

[1] Major Martin's services were, however, not yet ended for, in the early days of the Great War, he was re-engaged at the age of 61, and subsequently reached the rank of Lieut.-Colonel in the Royal Marine Submarine Miners who had their Headquarters at Clifford's Fort. (See Chapter III.)

[2] The bell of the " Miner " was generously presented to the officers of the Unit by this firm in 1932.

THE FIRST MOTOR SEARCHLIGHT EQUIPMENT, 1906.

To face page 24]

With the sale of all the equipment and stores, the existence of the Tyne Division as a submarine mining unit may be said to have terminated. In this way the British Government, by a stroke of the pen, abolished the whole system of submarine mine defences in the United Kingdom, a system which had been gradually built up during the preceding forty-two years. The exact reasons for this sudden radical change of policy were never fully made public and the whole matter became the subject of widespread controversy at the time.

The absence of submarine mining units was subsequently severely felt in the early months of the Great War and, as will be recorded later, a special force had hastily to be raised in 1915 to perform the very duties which were now regarded as unnecessary. The submarine mining units, however, had no alternative but to submit to the decrees of the War Office and thus, on the 1st June 1907 the Unit, after an existence of twenty-three years, discontinued submarine mining and was re-named the Tyne Division Royal Engineers (Volunteers) Electrical Engineers.[1]

---

[1] London Gazette 11/6/07. A.O. June 1907.

CHAPTER II.

1907—1913.

THE TYNE DIVISION ROYAL ENGINEERS (VOLUNTEERS) ELECTRICAL ENGINEERS, THE DURHAM FORTRESS R.E. (T.F.), THE NORTHUMBERLAND FORTRESS R.E. (T.F.), AND THE TYNE ELECTRICAL ENGINEERS R.E.

Electrical Engineers Volunteers — Lieutenant Burton — Death of Sir Charles Mark Palmer — Altered conditions of service — Mobile searchlights — Formation of the Territorial Force — Coast defence organisation — Durham Fortress R.E. (T.F.) — New conditions of service — Special Service Section — Changes in uniform — Permanent Staff — The Ladies' Trophy won — Northumberland Fortress R.E. (T.F.) — Coronation detachments — Companies for Portsmouth defences — Recruiting — Amalgamation of the Northumberland Fortress and Tyne Electrical Engineers — New establishment — Retirement of Colonel Johnson — Special badges — First annual presentation of prizes — Earl Percy accepts the post of Honorary Colonel — Combined operations continued — A Chaplain appointed — Rifle shooting successes — Adverse criticism of the Territorial Force — Training and instructional system — New rifle range at Ponteland — Regimental prizes and trophies — The Unit thirty years old.

THE Electrical Engineer Volunteers, to which the Tyne Division now belonged, embraced all the former volunteer submarine mining units with the exception of the Tay Division. This latter unit was disbanded as the use of electric lights in the Tay defences had been discontinued. At the same time, the original Corps of Electrical Engineers was converted into the London Division of the Electrical Engineer Volunteers. There were, therefore, now seven Divisions in order of precedence as follows :—

(1) London.
(2) Tyne.
(3) Severn.
(4) Clyde.
(5) Tees.
(6) Forth.
(7) Mersey.

The Tyne Division, having thus been transferred from the particular service for which it had been originally raised, fell from the position of senior volunteer submarine mining unit to that of second among the electrical engineer units. With a reduced establishment of five companies, it now set to work with renewed energy to attain in the duties of electric lighting its former standard of technical efficiency, and also in the construction and maintenance of electrical communications which were now added to its responsibilities.

In this it was destined to be ably assisted by Lieutenant H. E. Burton, Coast Battalion R.E., who succeeded to the post of acting-adjutant at this stage. This officer was particularly well qualified to carry out these duties as not only was he recognised as an authority on electric lighting work, having specialised in that branch on active service in South Africa, but he had also been at Clifford's Fort during the past few years assisting Major Martin with the coast battalion section and gaining useful experience of the particular conditions prevailing in the volunteer force.

In the year 1905 Lieutenant Burton had been asked by the Royal National Lifeboat Institution to supervise the trials of their first motor lifeboat, which was to be installed at Tynemouth. As none of the local fishermen in those days would trust a motor-boat, this officer, with the permission of the C.R.E., had formed a crew of sappers from the 16th Company, Coast Battalion R.E., and with them ran the boat until the fishermen were convinced that she was superior to their rowing life-boat. The sapper crew actually manned the boat for eight months, after which the fishermen agreed to form a crew on the condition that Lieutenant Burton remained in charge of the boat. In this way he came to be appointed honorary superintendent of the Tynemouth boat and remained a lifeboatman for a period of over twenty years.

On the 4th June 1907 Colonel Sir Charles Mark Palmer, founder and Honorary Colonel of the Division, died at his home in Yorkshire at the age of 85 and the funeral, which was attended by several of the officers, took place on the 9th of that month. During his lifetime he had come to be known as " the genial Baronet of Jarrow " which constituency he had represented in Parliament as liberal member since 1885, while his energetic association with the Auxiliary Forces had gained him the almost universal sobriquet of " the Father of the Volunteers." Colonel Sir Charles Mark Palmer was the first baronet, having been so created in 1886 ; he was a Deputy Lieutenant of the County of Durham and a Justice of the Peace besides being a Commander of the Italian Order of St. Maurice and St. Lazarus and the recipient of many other honours including the Volunteer Decoration.

Annual training in 1907 proceeded on very similar lines to that of former years, although the conditions of service underwent certain modifications on the alteration in the duties of the Unit. The capitation grant was reduced from £5 to £4. Recruits were henceforth only required to attend drill on eighteen occasions and trained men on but six. Attendance at the annual inspection and at camp for eight days continued to be compulsory although twelve additional drills could be substituted for two out of the eight.

Camps were held at Portsmouth and in the Isle of Wight as well as locally at North Shields. Practice with mobile searchlights was carried out at Portsmouth, not only with the motor equipment previously referred to, but with other similar sets which had now been provided by the War Office and with the horse-drawn field equipment which had been developed as a result of the experience of the

South African War. The drill book of the period emphasised the importance of removing the horses from the power wagon before starting up the oil engine, but the Tyne Division never experienced any difficulty in this direction as no horses were ever provided !

In December Hon. Captain Rioch retired, leaving the Division once more without a quartermaster.

The history of the Unit now enters a period which is marked by a degree of uncertainty and confusion partly occasioned by the altered nature of its duties but mainly due to the sweeping changes in the general conditions under which the Volunteer Forces of the Country were governed. Already in 1907, extensive alterations in the constitution and organisation of the Auxiliary Forces were under consideration, which were to involve the abolition of the Volunteer Force and the creation in its place of a Territorial Force. These changes were put into effect through the provisions of the " Territorial and Reserve Forces Act " of 1907, by means of which Mr. Haldane, Secretary of State for War, re-cast the whole of the Auxiliary Forces to form a home defence army made up in exactly the same way as the Regular Army, co-ordinating the systems of administration and training and the terms of enlistment. The passing of this Act, which has been accepted as one of the greatest achievements of army re-organisation in the history of this Country, left the fundamental object of the Force unchanged. The Territorial Force was for service at home in the event of actual or apprehended invasion, but, for the first time, provision was made for units to volunteer for colonial garrison duty in the event of war. An important modification was, however, introduced. While the training of the Territorial Force was to continue to be carried out through the usual military commands, the administrative work was in future to be the responsibility of territorial associations which were formed, under the presidency of the Lord Lieutenant, in each County. One result of this change on the Tyne Division was to deprive it of a large measure of the independence, financial and otherwise, which it had hitherto enjoyed, and thus to curtail to some extent the power of experiment which had been a characteristic of the Unit since its formation.

Another feature of the new organisation, so far as it affected coast defence units, was that the personnel required for the defence of any given locality was to be raised from the local population. It will later be seen that this principle, though admirable in theory, was doomed to failure and, indeed, in some quarters its impracticability was foreseen from its very inception. The immediate effect on the Tyne Division Electrical Engineers was very serious. The five companies of which this Division consisted were far in excess of the numbers actually required for the operation of the few searchlights installed at the mouth of the Tyne. In the past the establishment had been allowed to grow in order to provide a source from which efficient personnel could be obtained in the event of war for the defence of other ports besides the Tyne, but the new principle precluded any such employment in the future.

MOBILE SEARCHLIGHTS, HASLAR CAMP, 1907.

To face page 28]

The scheme for Royal Engineers in defended ports was that their organisation should follow as closely as possible that of the fortress companies of the Regular Army which had come into being on the abolition of submarine mining in 1905. Of these, a certain number were allotted for electric light duties at the larger fortresses and others for 'Works' services in the garrison, while at the smaller fortresses companies were formed in which the duties were combined.

The establishments for Fortress Units Royal Engineers (T.F.) for the defence of the Tyne included four companies of which three were works companies and one an electric light company. It is evident that these four companies were intended to absorb the whole of the personnel of the Tyne Division Electrical Engineers but there were two important circumstances which militated against the achievement of this object. The first was that no personnel was officially drawn from the County of Durham for the defence of the Tyne which is just as much a river of that county as of Northumberland; but the second and more important from the point of view of this Unit was that the Tyne Division consisted almost entirely of tradesmen highly skilled in electrical and mechanical work who were neither suited for nor anxious to undertake the constructional work of the fortress works companies. This sentiment is traceable throughout the subsequent history of the Unit, but its immediate effect was that, at the time of which we are writing, it resulted in the three works companies being allotted to the County of Durham and the single electric light company only being raised in Northumberland.

Accordingly, early in 1908, a company of 3 officers and 88 other ranks was formed from the personnel of the Tyne Division Electrical Engineers and was styled the Electric Light Company, Durham Fortress R.E. (T.F.). At the same time, the three other companies for the defence of the Tyne were raised independently in Durham under the name of Works Companies, Durham Fortress R.E. (T.F.), with an establishment of 11 officers and 291 other ranks.

The establishment of officers in the newly formed company was one captain and two lieutenants and the following were immediately posted to fill these vacancies :—Captain C. Johnson, Lieutenant C. Arnison, and Second-Lieutenant M. W. Buck. The full quota of other ranks was made up forthwith with the result that this company figured in the first list, published on May 1st 1908 of units of the Territorial Force recognised as efficient.

There now remained some twenty officers and three hundred other ranks of the Tyne Division Electrical Engineers who could not immediately be absorbed into the Territorial Force and it was urgently represented that the services of this valuable body of technically trained personnel should not be lost. It was therefore agreed that they might provisionally be retained pending the raising of local units to replace them in the defences of the Humber and Portsmouth, but they were to have no lasting status or establishment and no recruiting was to be carried on. A similar state of affairs was occurring in other divisions of Electrical Engineers with the result that numbers of the officers appeared in the

Army List under the heading " Officers of the Electrical Engineers (Volunteers) not yet gazetted to the Territorial Force." The general uncertainty as to the future, however, caused many of the officers to resign their commissions.[1] The Unit was now represented only by the single electric light company which dropped to third place in the precedence list of the eighteen divisions of Fortress Engineers. The first and second on this list were the Lanarkshire (later the Renfrew) and the Lancashire Fortress R.E. (T.F.) respectively, the order of precedence of these Units having reverted to that of the County Corps of Volunteer Engineers as originally formed.

The conditions of service in the Territorial Force were very similar to those which had obtained in the Volunteers but the officers were commissioned afresh and a new attestation had to be made out on behalf of each member of the Volunteers on his transfer to the new Force.

The annual training for the Territorial Force consisted as before of a number of preliminary drills and annual camp. A trained man was required to perform only six preliminary drills all of which had to be company drill. Recruits carried out twenty attendances in their first year of service and of these eight were company drill and the remaining twelve technical instruction. Annual camp lasted as before for 15 days but all of these were now regarded as compulsory in order to qualify for the bounty of £1 which was paid to each man returned as proficient each year. Every man also fired an annual musketry course for which a further 10/- bounty could be earned. A proportion of officers and other ranks formed a special service section which was liable to be called out in the event of national emergency on twenty-four hours notice. The personnel of the Section was sufficient to operate the coast defence electric lights pending the embodiment of the Auxiliary Forces by Act of Parliament. Each officer and man received a gratuity of 10/- per annum. The capitation grants of the Volunteer Force were no longer paid to individual units but were replaced by grants based on the strengths of units and paid to the County Territorial Force Association. Out of these grants the Association provided for all the administrative needs of the units in the County. One of the important duties of the Association was the provision and maintenance of headquarters and drill halls. The Tyne Division was peculiar in having its headquarters situated in a war department establishment at Clifford's Fort and this resulted in considerable difficulty—amounting at times almost to impossibility—in getting necessary work of maintenance or redecoration carried out in those parts of the Fort occupied by the Unit.

In this year, 1908, annual camp took place once more at Portsmouth and in the Isle of Wight and was attended by the whole of the personnel of the late Electrical Engineers in addition to the newly formed electric light company of the Durham Fortress Engineers.

---

[1] Among these was Second-Lieutenant H. E. Webb-Bowen who had previously served in the ranks of the Corps of Electrical Engineers from November 1899 until he was commissioned into the Tyne Division Submarine Miners in June 1903. This officer now transferred to the London Division Electrical Engineers and served in that Unit until August 1917 when he was seconded to the Royal Engineers and appointed Director of Works to the British Salonika Force with the rank of colonel. For his services in this post he was created a Commander of the Order of St. Michael and St. George and was awarded the Distinguished Service Order.

While the work undertaken by the Unit when in camp in the Portsmouth area was necessarily very similar from year to year, the searchlight stations manned varied to a certain extent, partly as a result of the particular camp site selected, and partly owing to defence schemes arranged occasionally by the Portsmouth Command to coincide with the training period. As a result of this the officers and men of the Unit acquired a practical knowledge of the functions of nearly all the lights in the Portsmouth defences, a knowledge which proved very valuable during the Great War. As a general rule only the shore stations were operated, but occasionally detachments were sent to one or more of the sea forts. On two or three occasions also vessels of H.M. Navy attacked the harbour and much useful experience was gained in picking-up and following the attacking vessels.

Owing to the restricted conditions existing at the Headquarters at Clifford's Fort, the period in camp represented the only opportunity for company or battalion drill and this was practiced each morning. The progress made in such a short time was remarkable and was mainly due to the high standard of intelligence among the men, and inspecting officers invariably commented on the steadiness of the ranks on parade and the precision with which the various movements were executed.

The uniform of the Unit now underwent certain changes. The first of these was the removal of the old submarine mining working dress of nautical pattern which now had no further purpose and was discarded. This was replaced by a service dress of khaki serge, puttees and ankle boots and the new Royal Engineer badge and buttons were taken into use. This Unit was one of the last to adopt khaki which had been in use by many volunteer units since the South African War.

In July an application was submitted, in accordance with a resolution passed at a meeting of the officers, for permission to change the ornaments on the full-dress uniforms from silver to gold, and this was later approved. In this way the most conspicuous of the differences between the Regular and Volunteer uniforms was abolished. Pending a decision as to the future of the Unit, the officers, with the exception of those posted to the Durham Fortress Engineers, did not acquire the new uniforms at once but continued for several years to wear the old patterns.

Prior to December the responsibility for the instruction of the Unit had remained with the Coast Battalion Company R.E., the administrative duties being superintended by Regimental-Serjeant-Major Freene R.E., assisted by a civilian (ex-C.S.M.R.E.) in the clerical work. At this date, however, a new system of permanent staff instructors was introduced by which specially selected N.C.Os of the Regular Army were permanently attached to units of the Territorial Force for duty as instructors. All permanent staff instructors

(P.S.I.) had to revert to the rank of serjeant on transfer to the P.S. and they were automatically promoted C.S.M.I. (Company-Serjeant-Major-Instructor) on completion of twenty-one years' service. They could continue in the service up to the age of fifty years.

On the 16th December Serjeant A. Reed R.E., up to this date a staff-serjeant engineer storekeeper at Plymouth, was appointed permanent staff instructor to the Electric Light Company. This Company was grouped with the Works Companies, Lieutenant Burton, Coast Battalion R.E. acting as adjutant and R.S.M. Freene R.E. as regimental-serjeant-major-instructor. The establishment of P.S.I. to the Electric Light Company was completed by the appointment, early in the following year, of Serjeant G. S. Pearks R.E. who joined for duty from Malta.

The Unit continued to be armed with the Lee-Metford "long" rifle and short sword bayonet, though most of the Territorial Force units were re-armed at this date with modified patterns of this rifle incorporating better sights, or with the newer Marks of Lee-Enfield rifle. The year 1909 was notable as the first occasion on which the Unit succeeded in winning the Ladies' Challenge Trophy, after competing unsuccessfully for nine years and, to mark the event, ten medals were presented to the team by the officers. The first representative of the Tyne Electrical Engineers at Bisley was C.S.M. I. T. Womphrey who attended the National Rifle Association meeting there in this year.

Early in 1909 the title of the Electric Light Company was altered to the Northumberland Fortress R.E. (T.F.). There were now nineteen groups of Fortress Companies R.E. in the Territorial Force, the Northumberland Company retaining its precedence and the Durham Fortress Engineers dropping to fourth.

The annual training in 1909 was carried out in the Portsmouth district once more, the camp on this occasion being situated at Southsea Castle.

On the 4th July Captain E. Robinson was appointed adjutant to the Unit under War Office authority. Up to this time, as has been explained above, the officer commanding the detachment of the Coast Battalion R.E. had acted as adjutant. This new system of an adjutant within the Unit was not how-ever destined to last long as, on 3rd September in the same year, the appointment of Captain Robinson was cancelled on financial grounds and the duties were once more taken over by Lieutenant Burton.

In this year the Rev. H. Bott and the Rev. S. Pearson resigned their posts as chaplains to the Unit and Bandmaster Patterson retired and was succeeded by Bandmaster (Serjeant) F. W. Richley.

On the 10th March 1910 Serjeant Pearks was relieved as P.S.I. by Serjeant A. J. Sergeant R.E. from Plymouth.

CORONATION DETACHMENT, 1911.

In 1910 and again in 1911 the Northumberland Fortress Engineers carried out their annual training at Tynemouth operating the defence electric lights in co-operation with the Tynemouth R.G.A. (T.F.). At this time a fifth searchlight was in use at Tynemouth. This was a mobile equipment which was installed, whenever operations were contemplated, near the lighthouse on the South Pier, producing a concentrated beam which was used as a search beam to the east and southward. This light which was largely of an experimental nature was, on occasions, in danger of being washed away and its use was subsequently discontinued, the provision of a permanent installation to replace it not being approved. The remainder of the Unit went south again each year, camping outside Fort Monckton, Haslar, and manning the electric lights in the Portsmouth defences.

The Unit was represented by a detachment of 2 officers and 20 other ranks at the coronation of King George V and Queen Mary in London on the 22nd of June 1911. The Tyne Division Electrical Engineers furnished 16 men under Captain Robinson and the Northumberland Fortress R.E. 4 under Lieutenant Arnison. Both detachments were in camp at the time and proceeded to Kensington Gardens on the 21st June and joined a composite Territorial R.E. battalion under a colonel of the Lancashire Fortress R.E. (T.F.). This battalion was divided into six companies of about 16 officers and 100 other ranks each. The quartermaster of this battalion was Major W. Pearce who was afterwards so closely associated with the Tyne Electrical Engineers during the War. Rehearsal drills commenced at 6.0 p.m. on the 21st and were resumed at 5 a.m. the next morning. The battalion lined a portion of the route followed by the coronation procession on the 22nd, being stationed at the top of Constitution Hill, and was on duty just outside the Wellington Arch in Piccadilly on the occasion of the " Royal Progress " through London on the following day.

It has been observed above that, on the formation of the Territorial Force, it was laid down that the personnel required for the defence of each locality was to be raised in that district. After a trial of three years it was found that this system was not altogether satisfactory as the necessary numbers were not always forthcoming. The Portsmouth and Isle of Wight areas, for example, were largely agricultural and did not contain sufficient men of the required type for the technical work in connection with the manning of the coast defences. At first dockyard employees at Portsmouth were allowed to enlist and this made up the deficiency to some extent, but later on the Admiralty withdrew their sanction for this on the very reasonable grounds that their men, in the event of war, would be required in the dockyard. It was accordingly decided to make use of the surplus numbers available on the Tyne to supplement the locally raised units for the Portsmouth and Isle of Wight defences and two electric light companies were organised for this purpose in the summer of 1911. The establishment of these companies amounted to eight officers (1 major, 2 captains, 3 lieutenants, 2 second lieutenants) and 150 other ranks. At the same time, in view of probable

c

lack of personnel in other ports, a third company was organised from the surplus Tyne personnel with an establishment of three officers (1 major, 1 captain, 1 second lieutenant) and 100 other ranks and was styled the 'Special' company, having no definitely allotted war station.

These three companies were included in a provisional establishment for the Tyne Division Electrical Engineers R.E. (T.F.) and the necessary re-organisation was commenced with effect from the 28th July 1911.

It has been noted above that, on account of the prevailing uncertainty as to the future of the surplus members of the old Tyne Division Volunteers, a number of the officers had tendered their resignations and had retired from the service. The new organisation required more than were actually serving at this time but the requisite number of candidates was available and the establishment of officers was completed before the end of August. Authority was received for several new officers to be commissioned second-lieutenants supernumerary to the establishment and this was done. These supernumerary officers subsequently proved of considerable value in the great expansion which the Unit underwent in the early days of the Great War in 1914.

Owing to the fact that no recruiting had been allowed during the preceding three years there was now also a deficiency of 187 in the required total of other ranks. Recruiting was immediately re-opened and 118 men were enlisted during the ensuing three months. A further 63 men who presented themselves for enlistment had to be refused on various grounds. Great care was exercised in the selection of men, none being accepted without an appropriate trade qualification and every recruit being required to pass a written examination to ensure that he was of a sufficient educational standard.

As all the four companies referred to were now corporate units of the Territorial Force, it was naturally desirable that they should be re-united under one title and this was effected through the medium of the *London Gazette* of 15th November 1911, in which " His Majesty was graciously pleased to approve of the Northumberland Fortress and the Tyne Division Electrical Engineers R.E. being in future designated the Tyne Electrical Engineers R.E." The other divisions of electrical engineers, with the exception of the London Division, then disappeared and the army list subsequently showed only the two remaining Units under the heading :—

" ELECTRICAL ENGINEERS."

    (i)   London.
    (ii) Tyne.

At the same time as these changes were made, the establishment of the company for the Tyne was increased to 3 officers (1 captain, 1 lieutenant and 1 second-lieutenant) and 102 other ranks. The total establishment of the Tyne

LIEUTENANT COLONEL F. G. SCOTT, V.D.
Commanding Officer, 1911–1915.
Colonel, Royal Marine Submarine Miners, 1915–1918.

Electrical Engineers R.E. was now 15 officers and 353 other ranks distributed as follows :—

	Lt.-Col.	Maj.	Capt.	Sub.	O.R.	Total.
Headquarters	1	—	—	—	1(a)	2
Tyne Company No. 1	—	—	1	2	102	105
Portsmouth Companies Nos. 2 and 3 ...	—	1	2	5	150	158
' Special' Company (No. 4)	—	1	1	1	100	103
Total	1	2	4	8	353	368

(a) Qr. Mr. Serjt.

On October 31st 1911 Lieutenant-Colonel Johnson resigned his commission after a period of twenty-three years in command of the Unit. The outstanding characteristic of his tenure of command was the readiness with which he sought for and welcomed additional responsibilities outside the immediate and local duties of his unit. The " History of Submarine Mining in the British Army" (Baker-Brown) has left on record that " the Tyne Volunteers, under their energetic commanding officer, Lieutenant Colonel W. Johnson, were always ready and willing to provide detachments of men for work at any port in Great Britain," and it is clearly due to this characteristic of their commanding officer that the bulk of the Unit was not swept away on the formation of the Territorial Force. Had this taken place, it is improbable that the Tyne Electrical Engineers could ever have attained to the extremely important position in the defensive organisation of the country which they were subsequently to occupy.

Lieutenant Colonel Johnson was succeeded in the command of the Tyne Electrical Engineers by Lieut.-Colonel F. G. Scott V.D., who was confirmed in that rank on the 16th December 1911.

The following was now the distribution of the more senior Officers :—

Commanding Officer	Lieut.-Colonel F. G. Scott V.D.
No. 1 Company	Captain C. Johnson.
No. 2 and 3 Companies...	Major E. Robinson.
	Captain G. A. Bruce.
	Captain A. K. Tasker.
No. 4 Company	Major C. R. Toomer.
	Captain C. Arnison.
Acting Adjutant... ...	Captain H. E. Burton Co. Bn. R.E.
Medical Officers	Lieut.-Colonel F. W. Gibbon V.D., R.A.M.C. (T.F.)
	Major W. H. Brown T.D., R.A.M.C. (T.F.)

The Regular Army permanent staff instructors serving with the Unit at this time were Sjt. A. Reed (No. 1 Company) and Sjt. A. J. Sergeant (Nos. 2, 3 and 4 Companies).

The coronation meeting of the Northumberland Rifle Association was held on the Morpeth ranges on Friday and Saturday the 8th and 9th of September 1911 and several members of the Tyne Electrical Engineers took part, Captain E. Robinson, Lieutenant N. H. Firmin, C.M.S.T. Oates and L/Corporal McN. Wallace figuring in the prize lists. The Tyne Electrical Engineers' team also distinguished itself by taking second place and a prize of four pounds in the competition for the Freemen of Newcastle Challenge Shield.

The annual regimental prize shoot of the Unit took place on the 7th November at Whitley ranges when some 48 members of the Unit entered for the competitions, the championship of the corps being won by L/Corporal McN. Wallace. For this he was awarded a gold medal subscribed for by the officers and also a silver cup presented by Messrs. Edwin Cook and Sons, of Bow, London.

This year the Tyne Electrical Engineers once more succeeded in winning the Ladies' Challenge Trophy[1] the event being contested on the Whitley ranges on Saturday the 11th November.

In December 1911 special War Office authority was obtained for the officers of the Unit to have the word " TYNE " embroidered on the scroll beneath the R.E. grenades worn both on the service dress and mess dress. In the uniforms of the Regular officers of that period this scroll bore the motto of the Corps of Royal Engineers, " UBIQUE," but this honour had not yet been earned by members of the auxiliary forces and in all other Territorial units the space was left blank.[2]

Among other changes in the uniform of the Unit which followed the formation of the Territorial Force in 1908 the ' S.M.' shoulder titles had been withdrawn in favour of the more appropriate $\frac{\text{T}}{\text{R.E.}}$ but authority was now obtained to retain the word " TYNE " previously authorised to be worn on the shoulder straps and a special shoulder title bearing the legend $\frac{\text{T}}{\text{R.E.}}$ was introduced.
TYNE

The first event of interest in the year 1912 was the first annual presentation of prizes which took place in the Albion Assembly Rooms, North Shields, on Saturday the 27th January. The prizes were given away by Colonel H. V. Kent, the Commander North-East Coast Defences. The Coronation (1911) Medal was presented to Major Robinson, Captain Arnison, Quarter-Master-Serjeant

---

[1] The team was as follows :—Captain E. Robinson, Lieut. N. H. Firmin, C.S.M. Womphrey, C.S.M. Oates, C.S.M. Dawson, Serjeants Carr, Campbell, Oates and Richards, and L/Cpl. Wallace. The conditions of the shoot were :—One sighter and seven shots at 200 and 500 yards. Twelve men in each team ; ten best scores to count : T.E.E. 553, N.D.R.E. 536.

[2] Subsequent to the Great War, officers of the Royal Artillery and Royal Engineers of the Territorial Force were authorised to adopt the motto " UBIQUE " and to wear this on the scrolls of their collar badges, but the change was not made in the badges of the Tyne Electrical Engineers.

CAPTAIN THE RT. HON. THE EARL PERCY
(later His Grace the Duke of Northumberland,
K.G., C.B.E., M.V.O., T.D.).
Honorary Colonel, 1912–1930.

To face page 36]

Pickering and C.S.M. Gordon, who had formed part of the coronation detachments. The function was attended by Earl Percy and the opportunity was taken to welcome him as Honorary Colonel of the Tyne Electrical Engineers, which appointment he had recently accepted. Earl Percy was at that time captain in the Grenadier Guards Special Reserve and had fought in South Africa. He had also held a staff appointment in Egypt and he had served a term as aide-de-camp to Earl Grey when the latter was Governor-General of Canada. These qualifications and his family connection with the County of Northumberland rendered Earl Percy eminently fitted for the post of Honorary Colonel.

In accepting the post, Earl Percy wrote to the Commanding Officer as follows :—

" I am in receipt of your letter asking me to accept the Honorary Colonelcy of the Tyne Electrical Engineers.

" I am extremely honoured by this request on the part of the officers of the Corps, and shall of course be delighted to accept the Honorary Colonelcy of a Corps which has such a great reputation, and which has so many services to its credit.

" I much appreciate this attention and consider it a great honour to myself."

The Unit had been for nearly five years without an Honorary Colonel following the death of Colonel Sir Charles Mark Palmer, and, although the new appointment was not confirmed in the ' London Gazette ' until the 22nd March 1912, the knowledge of its acceptance by Earl Percy met with universal approbation.

Combined operations continued to be held periodically with a view to testing the local defences, the personnel taking part on these occasions receiving full pay and allowances at army rates. On Saturday the 17th February 1912 the Tyne defences were manned by the Tynemouth R.G.A. (T.F.) and the Tyne Electrical Engineers with the intention of carrying out manoeuvres lasting over the week-end. Earl Percy took this opportunity of acquainting himself with the nature of the Unit's work and spent the night on duty with the detachments. The attacking fleet consisted of H.M. Scout " Sentinel " and H.M. Destroyers " Vixen," " Arab " and " Leopard."

Foul weather marred the proceedings from the first and the fleet was forced to anchor at sea during a dense fog while a truce was declared which lasted until mid-night. At that hour, the fog having lifted somewhat, operations were resumed, but only one attack was possible before mist again enveloped the water. Between 7.0 and 8.0 a.m. on Sunday morning the destroyer "Arab" entered the river by sheltering on the off-side of an incoming trawler and reached the Groyne before the guns engaged her. About an hour later the rest of the fleet entered the harbour and were also engaged. Owing to the continued bad weather the rest of the operations were cancelled later in the morning.

On the 13th March 1912 the Rev. H. L. Lloyd, Vicar of St. Peter's, North Shields, was appointed chaplain to the Unit being gazetted Chaplain to the Forces (Territorial Force) Class IV with precedence as a captain. This officer was the first chaplain to be officially appointed to the Unit with army rank.

Gosport, being the war station of two of the companies, was now recognized as the regular site for the annual camp. In 1912 this was held from the 23rd June to the 7th July and was attended by 16 officers and 300 other ranks. At the annual inspection the inspecting officer found occasion to compliment the commanding officer upon having " a brilliant set of officers and a fine body of men, well able to carry out the actual work which would be allotted to them in war."

At the Northumberland Rifle Association meeting at Morpeth this year the Tyne Electrical Engineers team[1] were successful in winning the Freemen of Newcastle Challenge Shield and C.S.M. Womphrey distinguished himself by winning the County Prize, while several other members figured in the individual prize lists. This was the first occasion on which representatives of the Unit won any of the competitions at an Association meeting. These successes, constituting a remarkable achievement for a unit which received an annual allowance of only twenty rounds of ammunition per man for musketry classification, were largely attributable to the energy and enthusiasm with which Lieut.-Colonel Scott, Major Robinson and the other officers encouraged voluntary practice among the men.

The annual prize giving took place this year on the 18th December in the Tynemouth Palace when the Countess Percy distributed Territorial Efficiency Medals and the various shooting prizes.

At this time considerable adverse criticism was being aimed at the Territorial Force as a whole on the grounds that it had failed to produce the number of men required and that it was based on no definite national stratigic requirement. Some even went so far as to aver that the voluntary system was moribund and that nothing short of compulsory universal service would provide for the military needs of the country. It was, therefore, a cause for satisfaction to the Tyne Electrical Engineers that they could claim to be within 41 of their full establishment, that their personnel consisted of skilled engineers, and that the Unit was allotted a definite share in the defensive organisation of Great Britain.

A high percentage of the officers were trained engineers, many of whom subsequently attained to positions of eminence in their profession, and the

---

[1] Team :—Major Robinson, Lieut. Firmin, Second-Lieut. Campbell, C.S.M. Robson, C.S.M. Dawson, C.S.M. Womphrey, Serjt. Wallace, Serjt. Richards.

WINNERS LADIES' CHALLENGE TROPHY, 1909.

Back Row, Left to Right—L/Corporal McN. Wallace, Serjt. J. Young, C.S.M. S. Dawson,
Serjt. J. Etherington.

Front Row, Left to Right—C.S.M. Shelton, A/R.S.M. T. Oates, Serjt. J. Richards, Serjt. T. Oates,
Lieut.-Col. F. G. Scott, C.S.M. I. T. Womphrey.

WINNERS FREEMEN'S CHALLENGE SHIELD, 1912.

Standing, Left to Right—Serjt. J. Richards, C.S.M. I. T. Womphrey, C.S.M. S. Dawson,
Serjt. McN. Wallace, C.S.M. T. Robson.

Sitting, Left to Right—Lieut. N. H. Firmin, Major E. Robinson, 2nd Lieut. C. M. Campbell.

To face page 38]

high reputation for efficiency which the Unit continued to enjoy was largely due to the considerable amount of their time which these officers devoted to their military duties. This display of enthusiasm had a most salutary effect on the attendance of the men, although quite 60 per cent. of them resided at places outside the immediate locality of the Headquarters.

At the end of 1912 the number of qualifying drills was altered, trained men henceforth having to attend a total of 18 drills each year of which 6 were company drills and the remaining 12 devoted to technical work. This constituted a considerable advance on the previous arrangement under which it had been possible for a man to be returned as efficient without having attended any technical instruction at all.

By 1913 the training of the Tyne Electrical Engineers followed a regular routine. Individual drills commenced early in the year and included infantry drill, coast defence plant operation for electricians and engine drivers and promotion classes for the officers. All this work took place in the Fort with the exception of the coast defence practices which were carried out on the lights at Tynemouth Castle and the Spanish Battery. Drill nights were now Mondays, Tuesdays and Thursdays and the hours 7.30 p.m. to 9.30 p.m. The practice of publishing regimental orders in the local press was now discontinued and the orders were printed and circulated to each individual by post.

Instructional courses were also held at the Electric Light School at Gosport at which personnel from the Tyne Electrical Engineers attended with highly satisfactory results. When attending these courses men of the Territorial Force were paid and received allowances on the scale laid down for Regular soldiers and were in every way treated in the same manner as their Regular comrades.

Week-end trainings were carried out from time to time, occasionally in co-operation with vessels of the Royal Navy. Each year the Unit executed a route march, generally on a Saturday in June, and these parades invariably had the effect of stimulating recruiting. The annual church parade took place either in May or June, sometimes at St. Peter's Church, North Shields, and occasionally at St. George's, Cullercoats.

On the remaining Sundays in May and June musketry was carried out at the new ranges at Ponteland where every man fired the annual qualifying course entitling him to the musketry bounty.

In this way all but a few members of the Unit had completed their obligatory drills and musketry by June. The Unit then went into camp for the period of Newcastle race week and the week after or before according to circumstances. This system was particularly advantageous, as many of the more important firms from which the personnel of the Tyne Electrical Engineers was drawn closed altogether for the greater part of race week and this greatly facilitated the men getting away for training. In the course of time this date for annual camp came to be regarded as immutable and it was many years before an attempt was made to depart from established custom in this respect. In 1913, Nos. 1, 2 and 4 Companies went into camp at Fort Monckton, Gosport, where they were honoured by a visit of a few days duration by the Honorary Colonel, Earl Percy. No. 3 Company on this occasion remained in the north, camping at Monkhouse Farm, Tynemouth, and training on the local defence lights.

The acquisition by the Northumberland County Territorial Force Association of the range at Ponteland proved a great boon and stimulated rifle shooting not only in the Tyne Electrical Engineers, but also in all the other units in the county. In spite of the increased distance from the Headquarters the attendances at voluntary practices on the ranges continued to increase, as many as fifty members of the Unit finding their own way out there by bicycle and other means on a single day. The regimental prize meeting took place there on the 4th September 1913 in combination with the Northumberland Hussars Yeomanry, the Northumbrian Divisional R.E. and the 5th and 6th Battalions of the Northumberland Fusiliers. The Northumberland Rifle Association held their annual meeting for the first time at Ponteland on Saturday and Sunday the 12th and 13th of the same month. At this meeting, C.S.M. Womphrey won the Chipchase Cup and several other members of the Unit figured in the prize list, winning between them over £50 in prize money.

On Saturday 27th September the annual competition for the Ladies' Challenge Trophy resulted for the third time in a win for the Tyne Electrical Engineers, the other competitors being the Northumbrian Divisional R.E. and the Durham Fortress R.E. who finished in the order named.

The last event of importance in 1913 was the annual presentation of prizes and regimental ball which took place once more in the Tynemouth Palace on December 16th. The prizes and long service medals were presented by Lieutenant-General E. C. Bethune C.V.O., C.B., Director-General of the Territorial Force, in the presence of the Honorary Colonel and a large and representative gathering.

This year the collection of regimental prizes and trophies was augmented by several handsome additions. These included a fine cup presented by Smith's Dock Company Limited of North and South Shields (subsequently allotted for competition in athletic sports), a silver trophy from Mr. (afterwards Sir) James Readhead of Westoe Hall (awarded for the rifle championship of the Unit), the " Rowland Hodge " Cup presented by the Northumberland Shipbuilding Company for technical work, and silver challenge cups generously given by the Honorary Colonel, Mrs. Ernest Robinson, Mrs. Edmund Swift, George G. Richardson, Esquire, late of Willington-on-Tyne and Alexander

ANNUAL CAMP, FORT MONCKTON, 1913.

OFFICERS, MONCKTON CAMP, 1913.

Left to Right—Captain O. M. Short, Lieut. M. W. Buck, Captain A. K. Tasker, Lieut. N. H. Firmin, Captain H. E. Burton, R.E. (Adjt.), Lieut.-Col. F. W. Gibbon (M.O.), Major C. R. Toomer, Rev. H. L. Lloyd, C.F., Lieut.-Col. F. G. Scott (Commanding Officer), 2nd-Lieut. W. G. Ward, Major E. Robinson, Lieut. E. Swift, 2nd Lieut. I. F. Fairbairn-Crawford, 2nd Lieut. C. M. Campbell, Captain C. Arnison, 2nd Lieut. G. L. L. Russell, 2nd Lieut. C. M. Forster.

To face page 40]

Burns, Esquire, of Long Benton. All of these prizes were not immediately competed for but the allocation ultimately decided upon was as follows :—

The ' Percy ' Cup	Quick firing competition.
The ' Robinson ' Cup	Company or Section team shoot.
The ' Swift ' Cup	Officers' handicap (Rifle shooting).
The ' Richardson ' Cup ...	Highest score in the ' County ' Grand Aggregate.
The ' Burns ' Cup	Recruits' rifle competition.

Although rifle shooting now assumed considerable importance in the Tyne Electrical Engineers, their normal work continued to be maintained at a high standard and excellent reports were received on every occasion when the Unit was inspected, either on parade or at technical duties. The popularity of the corps continued to increase and there was never a lack of candidates for enlistment into its ranks. At the end of 1913 however, the Unit was still 13 men short of its full establishment, though it is recorded that over 120 men were turned away during the course of the year having failed to pass the requisite standards of education or technical qualification.

Thirty years had now elapsed since the formation of the experimental submarine mining company at Jarrow and, although many of the characteristics of its early service had disappeared since the abolition of the Volunteers in 1907, the military efficiency of the Unit had increased steadily while the technical work had been maintained at its traditionally high standard.

The term " Sub-Miner " was now no longer heard on Tyneside, but there were few at that period who did not know something of " The Electricals " or, as they were perhaps more popularly known, " The Tynes." A single link with their earlier nautical connection is to be traced in the regimental orders of this period : " All Hands parade at the Fort at 2.30 p.m." !

CHAPTER III

1914—1919. THE GREAT WAR.
THE HEADQUARTERS AT NORTH SHIELDS

Annual training, 1914 — Outbreak of the Great War — Embodiment and mobilisation — Detachments leave for Southern Coast defences — Organisation in the Tyne defences — Recruiting — Duties and Works — Wreck of the " Rohilla " — A quartermaster appointed — The Unit re-armed — Re-introduction of submarine mining — The Royal Marine Submarine Miners — Lieut.-Colonel Toomer Commanding Officer — Anti-aircraft searchlight at Dunston — Northern Command schools — Air raids on the Tyne — A.A. defences on Tyneside — Air raid alarm system — Drafts for overseas — Re-organisation of the Tyne Garrison — Football — Formation of the (Special) Tyne Searchlight Company — Increases in Establishment — Re-organisation of the A.A. defences — E. and M. duties — Defences at Sunderland and Blyth — Defences at Hartley and Marsden — Function of Headquarters — Administration of officers — Miscellaneous duties — Coast defence electric lights — Convoy system — Re-organisation of coast defences — 594th (Tyne) Fortress Company R.E. (T.F.) — 16th Company converted — The " Administrative Centre " at Clifford's Fort — The Armistice — Disembodiment and demobilisation — Miscellaneous war services of officers.

IN 1914, No. 1 Company, under Captain A. K. Tasker, carried out its usual fortnight's annual training at Monkhouse Farm, Tynemouth, from the 20th June to the 4th July, while Nos. 2, 3, and 4 Companies were at Fort Monckton, Gosport, under Lieut. Colonel Scott. An innovation of this year's training which had some bearing on the subsequent activities of the Unit was the formation of a section, 42 strong, from the latter companies for special instruction in telephone work.

Meanwhile, the international situation in Europe had become so grave[1] that, on the 29th July, the British Government declared a state of national emergency involving the putting into operation of the measures for the defence of the ports of naval, military and commercial importance throughout the kingdom. Accordingly, on the evening of that day, orders were received at Clifford's Fort for the special service personnel to be embodied. The Unit was at this time up to its authorised establishment of 15 officers and 353 other ranks and, of these, 10 officers and 172 other ranks had signed the special service agreement by which they undertook liability for immediate service on twenty-four hours'

---

[1] On the 28th June the Arch-Duke Francis Ferdinand, nephew and heir to the Emperor Francis Joseph of Austria, was, together with his consort, assassinated at Sarajevo, the Capital of Bosnia. On the 28th July Austria declared war against Serbia and, in a few days, Germany and Austria were at war with France, Serbia, Belgium and Russia. The entry of Great Britain into the conflict was precipitated by the German violation of Belgian neutrality in her advance towards France.

notice being given by the Secretary of State for War. Although the regulations then in existence provided that " calling-up notices " should be sent out by post, these were supplemented by telegram and telephone with the result that the special service section paraded at full strength by 8.30 p.m. on 30th July. Two officers and 68 other ranks immediately took up their duties in the Tyne Garrison, while the remaining eight officers and 104 other ranks travelled overnight by special train to Portsmouth and were manning defence lights there before dark on the 31st July.

On the 4th August Great Britain formally declared war against Germany, the Territorial Force was embodied by Royal Proclamation and the remaining personnel of the Tyne Electrical Engineers mobilized up to full strength. The manning establishment of the Tyne defences (No. 1 Company) was completed by the evening of the 5th August by the addition of 1 officer and 34 other ranks. The personnel required to complete the Portsmouth and Isle of Wight companies (Nos. 2 and 3) together with the " Special " company (No. 4) left for the south of England on the same day. The subsequent expansion and activities of these detachments of the Tyne Electrical Engineers at Haslar are fully dealt with in Chapter IV.

The command of the Tyne Garrison was at this time held by Brigadier-General F. Baylay[1], Commander North East Coast Defences, with Colonel Wiseman Clarke as C.R.A. and Captain Burton O.C. Electric Lights and Telephones. The headquarters was established in a house in East Street, Tynemouth, facing the " Gibraltar Rock." The officers of No. 1 Company at the outset were Captain A. K. Tasker, Lieutenant C. M. Campbell and Second-Lieutenants K. A. Mountain and F. H. Bowers, the company-serjeant-major being C.S.M. I. T. Womphrey.

Clifford's Fort remained the official headquarters of the Tyne Electrical Engineers and nearly all officers joined here on first commission or on transfer, and recruiting was actively carried on throughout the War, the recruits being drafted in batches to Haslar for training. There was at first no special scale of personnel to assist the Commanding Officer in these adminstrative duties, the establishment for the regimental headquarters consisting only of Lieut. Colonel Scott and the regimental quartermaster-serjeant, Q.M.S. Pickering.[2] The section of the 16th Company, Coast Battalion, though greatly depleted by drafts to the active service companies of the Royal Engineers, remained at Clifford's Fort and much of the adminstrative work of the Tyne Electrical Engineers was at first carried out by personnel of this section in addition to their normal duties.

Accommodation became a matter of some difficulty in Clifford's Fort and the space available, never excessive, was further restricted by the erection of living huts on the parade ground.

[1] Born 6/2/65, Lieut. R.E. 5/7/84, Captain 1/4/93, Major 5/8/01, Col. 4/12/12, C.B.E. Retired 28/6/19.
[2] Q.M.S. Pickering went to Haslar very shortly after the outbreak of the War and was replaced by Q.M.S. I. T. Womphrey. The latter also went to Haslar later on and became R.S.M. there. C.S.M. Womphrey was followed by C.S.M. Carr.

In the first days of the War Clifford's Fort became the scene of the wildest patriotic enthusiasm and great crowds gathered before the gate while a continuous stream of men of all ages passed within to enlist in the Tyne Electrical Engineers. It was thus possible at first to select men of unusually high qualifications and the first drafts of recruits to leave North Shields for the detachment at Haslar set a standard which was never subsequently equalled. More than half, indeed, of the first batch were subsequently commissioned and most of the rest appointed mechanists almost immediately. But in the rush to join the army many men were enlisted who occupied ' key ' positions in local industries and it was not long before most of these were returned to civil life where their services were rightly held to be of greater value to the nation.

From the very outset demands were constantly received by the Tyne Electrical Engineers to carry out or superintend technical work of various kinds not only throughout the garrison but also at other stations in the Northern Command. In addition to their normal coast defence electric lighting and telephone work, No. 1 Company personnel were now undertaking the electric lighting of billets, hutted camps and various military hospitals in the district as well as other electrical and mechanical work of considerable importance, including the erection and maintenance of the electrical plant in the naval wireless station at Tynemouth Castle and the installation of apparatus in the port war-signal station. These works necessitated frequently the dispatch of small parties to considerable distances from the headquarters. Several of the camps (e.g., East Boldon, Hartley and Cleadon) were provided with separate generating units, and N.C.Os and sappers of the company were placed in charge of these. In certain billets and camps where electric supply was not available the Tyne Electrical Engineers condescended to instal gas lighting, but the fittings were carefully selected to resemble electric light fittings as far as possible ! Captain Campbell and Lieutenants K. S. M. Scott and W. Hall were employed at various times on these works under the direction of Captain Burton. Though most of the work carried out was of an electrical or mechanical nature, the company was able to be of material assistance in many other directions. Captain Tasker, by virtue of his professional qualifications as an architect, was particularly valuable in this way and was responsible for the design and for superintending the erection of hutments for miscellaneous services in the various camps throughout the garrison, while the accommodation at nearly all the anti-aircraft gun and searchlight stations which were subsequently required was erected under his supervision. The Durham Fortress Engineers, with headquarters at Jarrow and Gateshead, were at first employed on field defences, erecting huts etc., but soon after the outbreak of the War a portion of that Unit was converted into field companies for service with the New Army divisions and the remainder was employed under the various D.Os R.E. on miscellaneous duties about the garrison. Nevertheless, it is no exaggeration to state that whenever any special or unforeseen work of an engineering nature was required to be done in the Tyne Garrison, the authorities applied first to the Tyne Electrical Engineers and the Unit was ready at all

Searchlight Detachment at Whitby Station after service at the wreck
of the Hospital Ship " Rohilla."

To face page 44]

times to take on whatever was required and in this way upheld the reputation which it had earned in time of peace.

On the night of the 31st October the Tyne Electrical Engineers were able to render signal assistance at the wreck of the hospital ship "Rohilla" off Whitby, Yorkshire. After unsuccessful attempts had been made for two days by various lifeboats to reach the wreck, Captain Burton, whose association with the lifeboat service has already been referred to in an earlier chapter, obtained the permission of Headquarters Northern Command and, with a volunteer crew, took the Tyne-mouth motor lifeboat "Henry Vernon" during the night to the scene of the wreck. It was blowing a gale from the south-east and there were no shore lights owing to war-time restrictions. Shewing no navigation lights for fear of com-plications with naval patrols or enemy submarines and with the glare from the Skinningrove Ironworks at Middlesborough as his sole guiding beacon Captain Burton nevertheless contrived to bring the "Henry Vernon" safely unto Whitby Harbour at about 1.0 a.m. where the boat remained until rescue operations could be attempted. Captain Burton had previously been detailed from Head-quarters, Northern Command, to proceed to Whitby by special train with an old portable petrol-driven searchlight to illuminate the scene of the wreck. As, however, none but himself was qualified to take the lifeboat, and indeed, the crew would not put to sea without him, he had arranged for the searchlight to be taken by a detachment of the Tyne Electrical Engineers under Lieutenant Mountain. This detachment arrived at Whitby almost at the same time as the lifeboat. The searchlight was forthwith erected and brought into action on the cliff-top and during the night illuminated the wreck and was of great assistance in signalling to the survivors. By this means detailed arrangements were made for the rescue of the 50 survivors at dawn. When daylight came this work was effected without a hitch by the Tynemouth lifeboat in spite of the mountainous seas still running. As soon as this most gallant rescue had been accomplished both lifeboat and searchlight returned to Tynemouth where they arrived exactly 24 hours after their departure.

The official recognition of the Admiralty was conveyed to Captain Burton in the following telegram :—

> " Admiralty have received through the Senior Naval Officer present
> " account of your services in proceeding in lifeboat to Whitby in heavy
> " gale and then going alongside Rohilla. The skill and courage shown
> " call for highest praise and their Lordships desire to express on behalf
> " of Naval Medical Service their grateful thanks to you and the whole
> " lifeboat crew for their gallant action.
>
> " Admiralty."

This telegram was shortly followed by a message from H.M. the King of the Belgians, " J'exprime mon admiration du vailant équipage du Henry Vernon—Albert." Captain Burton received the Gold Medal of the Royal National Life-boat Institution as well as the American Gold Cross of Honour for the conspicuous part he played in this rescue. A public subscription was opened as the result of which Captain Burton was presented with a silver tea service and Lieutenant Mountain received a silver cigarette case, while the members of the searchlight detachment were given pipes and tobacco, the presentations being made by the Duke of Northumberland in the Albion Assembly Rooms at North Shields in the presence of a large gathering of people.

The services of the searchlight detachment of the Tyne Electrical Engineers on this occasion were subsequently mentioned in the despatches of Lieutenant-General Sir Hubert Plumer, Commander-in-Chief, Northern Command.

Recruiting continued to be carried forward at high pressure and it soon became apparent that the considerable amount of work connected with the clothing, accommodation and equipment of these men could not efficiently be undertaken without special personnel at the headquarters. The situation was considerably eased, therefore, when in December 1914 the appointment of a quartermaster was once more authorised for the Unit. The first to fill this newly created post was Hon. Lieutenant J. Aitken. At the same time the Unit was re-armed, 50% receiving issues of an experimental rifle known as the " 1914 Pattern " and the remainder " Short Lee-Enfields," though, owing to the prevailing demands for the expeditionary forces, about half of the latter were D.Ps (" drill purposes only").

All this time hostile naval activity in the North Sea and English Channel had been considerable and, following the sinking of many British ships by enemy submarines and the bombardment of Scarborough, Whitby and Hartlepool by an enemy fleet, the Admiralty began seriously to consider the possibility of successful raids being made on the more important mercantile and ship-building ports. In the month of November 1914 H.M.S. " Illustrious " had arrived in the Tyne to act as guard-ship and generally to re-inforce the defences of the river. A removable boom was constructed across the Narrows and a party of Royal Marines landed at Clifford's Fort, where they erected a battery of three guns on the land outside the old R.E. boathouse to cover the minefield which was laid out in the harbour entrance by combined parties of the Royal Navy and Tyne Electrical Engineers. The guns were at first erected on timber frameworks, but these proved to be unstable as the land was " made ground." Special concrete foundations were, accordingly, put in under the direction of Captain Tasker. The boathouse was also repaired and altered to accommodate the marines.[1] The Royal Navy then approached General Baylay with a view to the army taking over the operation of the minefield but, on the report of Lieut.-Colonel Scott and Captain Burton, this was not considered practicable. H.M.S. " Illustrious " was shortly afterwards replaced by H.M.S.

---

[1] The boathouse was subsequently used as quarters by the Tyne Electrical Engineers, and, being outside the perimeter of Clifford's Fort, had a sentry posted over it. On one occasion, the sentry, Sapper Turner, dived into the Tyne in full marching-order (including the full complement of ball ammunition) and rescued a child who had fallen into the river. On regaining the shore it is said that this Sapper once more took up his rifle and proceeded on his beat. Sapper Turner's bravery was subsequently recognised by the Royal Humane Society.

" Jupiter," the marines and battery remaining. The Captain of H.M.S. " Jupiter " also applied to the army for assistance with the minefield but this had again to be refused owing to lack of personnel and equipment. An Admiralty official then arrived at the Tyne to consider the difficulties which had arisen and Lieut.-Colonel Scott proposed that personnel who had seen service in the R.E. Submarine Miners should be collected and specially organised to deal with the submarine mine defences which were now once more shewn to be required. The outcome of this proposal was that, early in 1915, a new service was created for this purpose under the title of Royal Marine Submarine Miners. The headquarters was established at Clifford's Fort and on the 4th February Lieut.-Colonel Scott was transferred to the Royal Marines to take command of the new organisation. At the same time many of the old submarine mining officers were recalled to service, among these being Major Martin who was appointed chief instructor under Lieut.-Colonel Scott at Clifford's Fort, Major Bruce and Captain Ching, formerly members of the Tyne Division. Among the warrant and non-commissioned ranks who had formerly served in the Tyne Division and who were now recalled were Q.M.S. T.Robson and Colour Serjeants W. Harrison, W. McQueen and J. Scougal (all of whom were subsequently commissioned in the Royal Marine Submarine Miners), Colour Serjeants A. Gordon and P. Howett, Serjeants J. Young and McN. Wallace and Corporal W. J. Sparkes. The area covered by this organisation included all the important ports and anchorages from Scapa Flow down the east coast of England, the English Channel (except Portsmouth and Plymouth) and one station at Cap Griz-Nez in North France. The function of the headquarters at Clifford's Fort was recruiting and the instruction of personnel for the various stations on the coast. The actual water work of laying-out mines was carried out by the navy and the shore work of testing, position finding and the actual firing of the mines by the marines.[1] The Submarine Miners were subsequently occupied with the highly secret developments in submarine detection and mining which contributed so largely to the mastery of the U. boat menace.

When Lieut.-Colonel Scott transferred to the Royal Marine Submarine Miners, the command of the Tyne Electrical Engineers devolved upon Lieut.-Colonel C. R. Toomer who was at the depot at Haslar Barracks, Gosport, but all the personnel stationed in the north now came under Major Tasker who became " O.C. Tyne Electrical Engineers, Tyne Garrison."

At this period anti-aircraft defence first came to be considered on Tyneside and in February 1915 the first A.A. searchlight in the garrison was erected at Dunston by men of the 16th Company R.E. under the supervision of Second-Lieutenant W. Hall. This was an oxy-acetylene equipment and was probably the first plant to be specially designed for A.A. work, though we shall see that

---

[1] Some of the R.N. Personnel employed on the minefield in the very early days were notoriously careless with the explosives and cases were detected of men testing detonators which they held in the palms of their hands, or even in loaded mines, within the purlieus of Clifford's Fort ! In the first days of 1917 a line of 500-lbs. mines exploded inadvertently. The report of the inquiry which followed was to the effect that this was caused by lightning and had been accounted for by the fact that the system of ' earthing ' the submarine cables employed by the Royal Navy was not the same as that to which the Army personnel was accustomed.

electric searchlights had been adapted to this purpose in the Portsmouth Garrison very shortly after the outbreak of the war. The Dunston light, which was erected on the roof of the C.W.S Flour Mills, was worked for a few weeks by personnel of the 16th Company and was handed over to the Tyne Electrical Engineers when the Regulars went overseas.

In the same month, Captain Burton was detailed to form a school of instruction in field engineering, signalling and telephony for infantry officers and N.C.Os and accordingly handed over his duties as adjutant of the Tyne Electrical Engineers and O.C. E.L. and T. to Captain J. Collins of the Coast Battalion R.E., Captain Campbell taking over the E. and M. work with Lieutenant K. S. M. Scott as his assistant.

In order that a school such as that which Captain Burton had been instructed to originate should fulfil, at all adequately, its very responsible functions, it required a complete equipment of signalling apparatus, not to mention examples of the most up-to-date hand grenades, trench mortars, etc. It is scarcely necessary to state that the Tyne Electrical Engineers, whose duties at that time were confined to manning the searchlights in the coast defences, could not furnish such equipment; indeed, what they had amounted only to a few antiquated specimens of C II and C III types of telephones. The school started in a small way, Captain Burton collecting such apparatus as he could get from the stores at Clifford's Fort. Short courses of instruction were then given and practical demonstrations in open spaces near Percy Park, Tynemouth. The numbers sent for instruction grew so rapidly that the Grand Hotel was commandeered and was devoted to classes for officers, while additional instructors were provided, one of whom, Lieutenant Hall of the Tyne Electrical Engineers, was of especial assistance in the provision of apparatus, on loan, both from his own private laboratory and from those at Armstrong College, Newcastle. In March the school was named the Northern Command School of Bombing, Signalling and Telephony, and Field Engineering. Instruction in bombing was now commenced. The Mills hand grenade had only just been invented and the first instruction had, so far as practical demonstration was concerned, to be confined to filling a tin can with explosive ingredients with the hope that the required chemical reactions would not take place prematurely. The school was now expanded to include instruction in anti-gas measures, lectures being given by Professor Smithells of Leeds University. In November 1915 the school, having outgrown the accommodation at Tynemouth, was removed to Farnley Park, Otley, Yorkshire, where it continued to grow in importance, remaining under the command of Major Burton until the end of the War. At Otley the staff was joined by another Tyne Electrical Engineers' officer, Captain C. Johnson, who assumed command of the subordinate personnel and took up the duties of quartermaster. Both Captain Burton and Lieutenant Hall made visits of inspection to the battle fronts in France in order to acquire first-hand knowledge. Shortly afterwards, a number of officers previously trained at the school, who had returned home wounded from abroad, became available as instructors and this made it possible for Lieutenant Hall to be released to return to duty at Clifford's Fort. During the period he was employed as an instructor at Otley this officer was responsible for the compilation of a manual of telephony which was taken into use at the school. The establishment then comprised the commandant, two adjutants, a quartermaster, twenty-two officer instructors, 120 infantry for general duties and 120 sappers for fieldworks and maintenance of hutments. The normal numbers under instruction were 250 officers and 300 N.C.Os, though at one time the total strength of the school reached 1150. For his services in this connection Major Burton received the O.B.E. and was three times brought to the notice of the Secretary of State for War. The object of inserting this brief and imperfect sketch at this stage is to show how, from the humblest beginnings, the school developed into one of the great training centres of the country and to trace its origin to that unit of the Territorial Force which is the special subject of this history.

In April Captain Collins had removed with the headquarters of the 16th Company R.E. to the Humber and Captain Campbell was then officially appointed O.C.E.L. & T. Tyne, and these duties he carried out until December 1916. The huge demands for plant for overseas rendered the supply of technical stores a question of considerable difficulty. For example, early in 1915 an issue of a lathe was approved for the E. & M. Workshops at Tynemouth and the necessary indent duly forwarded. The matter was raised several times during the ensuing year and ultimately fell into abeyance but the lathe was duly delivered during the latter half of 1918 !

Meanwhile German air raids over England had commenced, the first on the 19/20th January 1915 over Norfolk and the second over the Tyne. "In the early afternoon of the 14th April the Zeppelin Airship L.9 commanded by Kapitan-Leutnant Mathy was on a naval reconnaissance towards the coast of England. She was to return to her base before dark but when within a hundred miles of Flamborough Head, Mathy, who had a good supply of bombs on board, decided to raid the Tyne, and by means of his wireless he obtained official sanction for the deed. The L.9 appeared off the mouth of the Tyne about 7.0 p.m. and coasted northwards to Blyth before coming inland to swoop down on Tyneside. She was met by rifle fire of the 1st Battalion Northern Cyclists at Cambois. Mathy's first bomb fell in a field at West Sleekburn. This bomb was followed by twenty-two others before the Tyne was reached at about 8.40 p.m. He then unloaded his eight remaining bombs and went out to sea south of South Shields. The only casualties occurred at Wallsend, a woman and child being injured. Two aeroplanes from Cramlington searched the skies in vain."—(' The German Air Raids on Great Britain.'—Captain J. Morris. Page 21).

A surprising fact in connection with this raid was that the local press was allowed to report the full movements of the airship and this was, doubtless, of some help to the enemy as subsequently the most rigid press censorship was enforced.

D

"Up to this time the Naval Authorities had undertaken the defence from the air of the coast line including the Tyne and Humber but the armament available was very meagre, being limited to 33 weapons of small calibre and power which had been mounted at vulnerable points, mainly of naval importance " ('The German Air Raids on Great Britain,' p. 13). After this raid, however, the army once again took a part in the air defence of the country and there was a rush on Tyneside to extemporize anti-aircraft searchlights. The Tyne Electrical Engineers assisted in installing these and provided manning details for them all.

The first electric A.A. searchlight on Tyneside was erected at Carville power station, Wallsend, by Captain Campbell to work with a 3-ins. 30-cwt. anti-aircraft gun installed by the R.G.A. This searchlight was a 60 cms. projector and was under the charge of Lieutenant M. C. James. Both gun and searchlight were in position on June 15th 1915 which happened to be the date of the second air raid on Tyneside.

"In the early afternoon of that date, in weather conditions particularly favourable for oversea work, the L.10 (Kapitan-Lieutenant Hirsch) . . . started from Nordholz for England . . . Hirsch carried on with the Tyne as his objective. He approached the coast at Blyth, well north of his target, obviously to avoid the Tynemouth defences. Otherwise his overland route, although somewhat shorter and more direct, differed little from that followed by Mathy on his visit of 14th April. On making his land-fall at half-past eleven he immediately turned south and steered straight for the Tyne. No bombs were wasted on the open country as they were on the former occasion. The first were thrown at Wallsend. Unwarned, many of the Tyneside industrial establishments had their lights at full blaze ; syren blasts sounded by H.M.S. "Patrol" as a warning were not understood. Damage was done to houses and the North-Eastern Marine Engine Works. After bombing Wallsend and the Hebburn collieries, Hirsch turned his attention to Palmers' works at Jarrow, which presented a perfect target. Seven high explosive and five incendiary bombs fell on the engine construction department, causing severe damage and great loss of life—seventeen men were killed and seventy-two injured. Before leaving, the airship dropped bombs on Willington Quay, East Howdon, Cookson's Antimony Works and Pochine Chemical Works.[1] Hirsch went out to sea via South Shields leaving a scenic railway ablaze near the Harton Colliery staiths. In all 2,500 kgs. of explosive and incendiary bombs were dropped. Two naval machines rose from Whitley Bay but failed to see anything of Hirsch. . . ." ('The German Air Raids on Great Britain,' pp. 38-40).

The Zeppelin had come over Wallsend at about 5,000 feet and was clearly visible from the ground even without the aid of the searchlight and should have been an easy prey. The defence, however, was not fully organised and the opportunity was lost. The gun came into action but too late to bring the Zeppelin down though she immediately altered course.

---

[1] The case of a parachute flare which fell near Carville Power Station is preserved at the Headquarters of the Tyne Electrical Engineers.

50

The German commander records in his diary that he was fired on near Wallsend and claims to have silenced, with four bombs, a coastal battery which, he declares, also opened on him on the way out !¹

Immediately after this disastrous raid serious complaints were made by numerous Tyneside engineering firms that the official warning from Headquarters was not received until a considerable time after the Zeppelin had finished dropping its bombs. Captain Campbell was immediately instructed to investigate the matter and was given authority to organise a system of effective warnings in conjunction with the Newcastle post office engineer, Mr. Elliot. Up to this time the warnings had been sent out verbally by the North Shields telephone exchange and as only one operator was employed on night duty he alone had to make consecutive calls to about a hundred firms with the result that after about half-an-hour he had only issued about twenty or thirty warnings. As a number of large blast furnaces were involved the necessity of immediate warning was important owing to the comparatively long time taken in closing down. It was therefore arranged to equip all these and other important works with a warning bell. A large hand magneto was installed at Tynemouth and shortly afterwards an additional one at Newcastle so that an officer could warn instantaneously and simultaneously by a code of rings a large number of firms and officials. The system was divided into three groups for convenience and to avoid risk of breakdowns. Owing to the zeal shewn by the post office engineers the warning bells were installed and the system working at seventy points within two days. It is understood that this was the first comprehensive and effective warning system to be installed in this country and this system was subsequently followed in other industrial districts with equal success.²

On the 10th June Hon. Lieutenant W. A. Souter took over from Hon. Lieutenant J. Aitken the duties of quartermaster when the latter received a combatant commission in the Durham R.E. (T.F.). On the 3rd September, however, he himself was commissioned into the Tyne Electrical Engineers and was followed in the appointment by Hon. Lieutenant A. Reed who had been a permanent staff instructor to this Unit since 1908 and who had been with No. 1 Company in the rank of regimental-serjeant-major since the outbreak of the war.

In August a detachment of twelve men was dispatched from Clifford's Fort to Southampton to join a draft of London Electrical Engineers and Tyne Electrical Engineers from Haslar who were proceeding to France to undertake the operation of oxy-acetylene searchlights at the front, and in November a larger draft left headquarters to form part of No. 1 (London and Tyne) Electrical and Mechanical Company R.E. (T.F.). The subsequent activities of these detachments are referred to further in Chapters IV and V.

---

¹ This could only have been Frenchman's Point Battery but there is no evidence that this battery was ever attacked by a bomb.

² The warning arrangements on Tyneside were subsequently altered, the signal being given by the Garrison Commander himself to the power station at Carville. All the electric lights were then lowered three times as a warning and finally cut off at the main.

The tasks allotted to the Tyne Garrison were gradually extended to include the defence of all the great munition factories, shipbuilding yards etc. in the vicinity of Newcastle and the rivers Tyne and Wear. In the summer of 1915 the Tyne Garrison was, accordingly, re-organised in three sections and the headquarters moved to the Minories, Newcastle. Brigadier-General Baylay, after a period of sickness, went to Portsmouth to take up the duties of Chief Engineer. The first commander of the re-organised Tyne Garrison was Major-General B. Burton, C.B., C.M.G. On the 15th September 1915 he was succeeded by Major-General R. A. K. Montgomery, C.M.G., C.B., D.S.O.,[1] who remained in command until February 1920. Colonel Wiseman-Clarke continued as C.R.A. Tyne Garrison[2] and Lieutenant-Colonel J. I. Lang-Hyde, R.E.[3] was appointed to the newly created post of C.R.E. Tyne Garrison and remained in that capacity until after the armistice.

The area covered by the Tyne Garrison now extended from north of the Hartlepool defences to the River Tweed and the three sections were centred roughly on Sunderland, Tynemouth and Blyth. Along the coast-line three strong lines of entrenchments, bomb-proof shelters and dug-outs were gradually developed and the construction of certain supporting strong-points and gun positions were set in hand. The designations of the three sections and their commanders were as follows :—

No. 1 Section—Sunderland.—Brigadier-General F. P. English, C.M.G., D.S.O.[4]

No. 2 Section—Tynemouth.—Brigadier-General H. G. Fitton, C.B., D.S.O., A.D.C.[5]

No. 3 Section—Blyth.—Brigadier-General Hon. W. E. Cavendish, M.V.O.[6]

Apart from the fixed coastal batteries at Frenchman's Point, the Spanish Battery and Tynemouth Castle, the forces available for the defences of the coast against enemy landing included 9.2″ guns on railway mountings, two brigades of mobile artillery and no less than sixteen reserve or training battalions of infantry[7] and seven volunteer battalions. The area of coast not covered by the three sections was patrolled by a battalion of Northern Cyclists supported by a yeomanry brigade with headquarters at Morpeth. The total strength of the Tyne Garrison cannot have been far short of 40,000.

---

[1] Subsequently Major-General Sir R. A. K. Montgomery, K.C.M.G., C.B., D.S.O.
[2] Colonel Wiseman-Clarke was succeeded by Colonel Smith in 1918.
[3] Born 17/9/59, Lt. R.E. 26/6/83, Lt.-Col. 1/10/07, C.M.G., O.B.E. Retired 1/10/12.
[4] Relieved January 1918 by Brig.-Gen. J. L. Gibbert, C.B., D.S.O.
[5] Relieved August 1915 by Brig.-Gen. A. J. Kelly, C.B., who was subsequently relieved in November 1917 by Brig.-Gen. H. G. Holmes, C.M.G., D.S.O.
[6] Relieved October 1917 by Brig.-Gen. A. Blair, D.S.O.
[7] These battalions were all draft-finding units for the service battalions of the various regiments overseas. They were frequently very much in excess of their normal establishments, the 3rd Battalion West Yorkshire Regiment, for example, attaining a maximum strength during the War period of 5,003, while many of the others exceeded 3,000. The battalions in No. 2 Section (Tynemouth) were :—
    3rd Battalion West Yorkshire Regiment—Colonel G. Frend, C.B.
    3rd Battalion West Riding Regiment (Duke of Wellington's)—Lt.-Col. R. A. McLeod.
    3rd Battalion South Staffs. Regiment—Lt.-Col. G. Jones-Mitton.
    3rd Battalion North Staffs. Regiment—Lt.-Col. C. H. James (subsequently Lt.-Col. Falls).
    Home Service Battalion West Yorks. Regiment—Col. H. R. Brander, C.B.
    2nd Battalion, Northumberland Volunteer Regiment (when mobilised).

The winter of 1915-16 proved a successful season for the Tyne Electrical Engineers (Clifford's Fort) Association Football Club which, under the captaincy of Sapper Curry (of Newcastle United) and trained by Sapper Thornton, succeeded in winning the Tyne Garrison Army and Navy League competition.

During the winter, additional 60 cms. anti-aircraft searchlights were installed in the Tyne Garrison and in the early weeks of 1916 the detachments coalesced to form a unit styled the " (Special) Tyne Searchlight Company, Tyne Electrical Engineers " under the command of Lieutenant (later Captain) M. C. James and its headquarters were established at 6 Osborne Road, Newcastle-upon-Tyne. For some considerable period (eight or ten weeks) this company failed to receive official recognition from the paymaster with the result that Lieutenant James was obliged to advance the pay out of his own pocket during this period.

On the 1st April 1916, Zeppelins raided Sunderland dropping some twenty-five bombs and doing a great deal of damage, killing 16 people and wounding about 100. Further air-raids crossed the area of the Tyne Garrison on the 2nd and 5th April but there were fortunately no casualities. After these raids certain 3 pdr. A.A. guns were temporarily lent for the Sunderland defences by the R.N.V.R. Mobile Brigade in London and a searchlight from the same source was established at the Bent House, South Shields. Additional mobile guns were also introduced temporarily into the defence from No. 4 Mobile A.A. Battery and lights were lent from No. 9 (Tyne) Mobile Searchlight Company, one of these being erected at Whitley Bay. By the end of April the Tyne A.A. Defences could boast no less than twenty-one searchlight stations, of which nine (those at Bedlington, Birtley, Hazelrigg, Heworth, Hylton Castle, Murton, Seghill, South Gosforth and Stannington[1]) were called "A.A. defence lights " co-operating with defending aircraft from the aerodromes at Cramlington and Whitley Bay, and eleven (at South Shields, Jarrow, Pelaw, Dunston, Winlaton, Newburn, Benwell, Walker, Wallsend, Whitley Bay and Fulwell) were known as " gun lights " and worked in conjunction with anti-aircraft artillery manned by the Tyne-mouth R.G.A. (T.F.). The twenty-first light was a " Special " light erected at Cleadon landing ground. Many of these fixed searchlights were at first supplied with oil-engines of the hot bulb type of which the vapourisers required to be heated up twenty minutes at least before the engines were run. As soon as an attack was anticipated, Anti-Aircraft Control, Tyne Garrison, issued a special warning order which was the signal for the vapouriser blowlamps to be lighted.[2]

This was before the days of sound locating instruments and instructions were given for the direction of approach of the Zeppelins to be roughly ascertained to within five or ten degrees by listening through megaphones which were specially supplied for the purpose, the searching being carried out primarily in the vertical plane.

---

[1] Four of these lights were furnished by No. 33 (Tyne) A.A. Company under the command of Captain J. F.S. Hunter. (See Chapter VI.)

[2] The Walker light was operated from a motor-generator supplied from the town mains.

In very cold weather it was customary to place lighted hurricane lamps inside the projectors to keep the reflectors warm and to guard against cracking when suddenly heated up by striking the arc.

The detachments varied somewhat in the early days. At first one N.C.O. and seven men formed the crew of a gun light and their duties were as follows :— No. 1 engine driver, No. 2 lamp attendant and sighting number, No. 3 extra lamp attendant, No. 4 telephonist at light, No. 5 assistant engine driver, No. 6 telephonist at gun, No. 7 N.C.O. in charge, No. 8 spare man. An A.A.D. light at the same period had a complement of nine comprised as follows :—Three in the engine room, 1 general duties, 1 sentry, 1 observer, 1 telephonist at light, 1 lamp attendant, 1 N.C.O. (a corporal). These scales of personnel were very lavish and were subsequently much reduced. The Tyne Garrison lay within the so-called " ever-ready " zone and a minimum detachment of five (including a sentry) had always to be fully dressed and ready for action on each station. For the remainder, leave was granted from 12 noon on one day to 12 noon the next, but only during periods when there was no full moon and the likelihood of raids consequently less. Absence was granted daily from mid-day until dusk to half of the men on each station but all men on leave had, of course, to return at once on hearing the air-raid alarm. Training by day consisted of two runs each of half-an-hour's duration, one in the morning and the other just before dusk. Other subjects which were practiced daily in this Company were judging-distance, signalling, rifle drill and physical training.

A somewhat unusual feature of the organisation at this period was that the responsibility for the maintenance and repair of the equipment used by the (Special) Tyne Searchlight Company was left in the hands of the O.C.E.L. and T. Tyne who alone had adequate workshop accommodation to deal with it. The A.A. Defence Commander at that time was Lieutenant-Colonel His Highness the Prince de Mahe, R.F.A.[1] and this company came under him for operations, under Major Tasker at Clifford's Fort for discipline and under Captain Campbell for all matters in connection with the plant, a state of affairs which led, not infrequently, to complications !

On 8th August an air-raid was made on the Tyne in which four Zeppelins took part. One of these crossed and re-crossed the River Tyne nine times without dropping a bomb. There was considerable fog which assisted the defences by concealing the ground from the enemy's view but the guns and lights could not do much. Several bombs were dropped in the sea and one fell at Whitley Bay doing little damage. The Whitley gun was the only one to open fire and was believed to have hit one of the airships.

In order to avoid unnecessarily long periods of inactivity during air raid alarms a system was instituted throughout the country by which various degrees of readiness could be maintained in different areas, those more remote

---

[1] Succeeded in February 1917 by Lt.-Col. C. J. H. Swann, R.F.A.

from the centre of danger suffering less dislocation to industry etc. than those more immediately affected. In this connection a large map was constructed for the Anti-Aircraft Defence Command at Newcastle under the direction of Captain K. S. M. Scott, in which the various areas of the Command could be lit up with red, green or white electric lights according to the state of readiness in force at any time. This map was installed in the headquarters at the Minories and was in constant use during all air raid alarms (which were very frequent at this period) and proved of great value to the A.A. Defence Commander.

The establishment of No. 1 Company, Tyne Electrical Engineers, which had been based on the duties in connection with the maintenance and operation of the coast defence searchlights and telephones at Tynemouth at the outbreak of the War, was found totally inadequate to provide personnel for all the purposes which have been mentioned and its present strength was far in excess of its establishment. In June 1916, however, an order was issued by which the establishments of all Territorial Force units were doubled. There followed a re-arrangement of the companies of the Tyne Electrical Engineers.[1] The personnel employed on coast defence electric lights and telephone duties in the Tyne Garrison henceforth constituted No. 3 (E.L.) Company, Tyne Electrical Engineers, under the command of Major E. Swift who was posted for the purpose from Haslar. The headquarters personnel under Major Tasker remained under the title of No. 1 (Depot) Company, Tyne Electrical Engineers, at Clifford's Fort. At the same time the (Special) Tyne Searchlight Company was renamed No. 4 (A.A.) Company, Tyne Electrical Engineers. The distribution of personnel authorised for the Tyne was now as follows :—

Coast defence searchlights	...	5 officers	60 other ranks.
Anti-aircraft searchlights	...	3 officers	121 other ranks.
Telephones...	...	—	16 other ranks.
Administrative details	...	2 officers	24 other ranks.
		10 officers	221 other ranks.

In actual fact, the numbers, particularly in No. 1 Company, were still very considerably in excess of their establishment owing to the large number of men undergoing preliminary training prior to forming drafts for Haslar.

By July the anti-aircraft defences of Tyneside had so far increased that on the 11th of that month No. 4 A.A. Company was split into two parts which thenceforward became Nos. 34 and 35 (Tyne) A.A. Companies R.E., taking their place in the new series of anti-aircraft companies at that time in process of formation. Captain M. C. James was then posted elsewhere[2] and the command

---

[1] The general organisation and distribution of the Unit subsequent to this rearrangement are detailed in Chapter IV.
[2] To the command of No. 10 (Tyne) Mobile Searchlight Company. See Chapter VI.

of the two newly formed units fell respectively to Major Campbell and Captain K. S. M. Scott. Anti-aircraft work thus became almost entirely divorced from the rest of the work at Tynemouth and the subsequent developments in this branch and the later history of Nos. 34 and 35 A.A. Companies are related in Chapter VI. The responsibility for accounting for anti-aircraft stores on charge to these two companies remained in the hands of the quartermaster at Headquarters and Captain Reed assumed the extra title of " Officer-in-Charge of A.A.S/L. Stores " with effect from the 25th August 1916.

The responsibility for the care and maintenance of the coast defence plant which had hitherto been part of the duties of Major Campbell was transferred in December to No. 3 Company at North Shields, under Major Swift, while the E. & M. work in connection with hutted camps and billets was now taken over by a detachment consisting of Staff-Serjeant Smith, Serjeant Mathews and some twenty other ranks commanded by Lieutenant L. Bird, acting directly under the C.R.E. The E. & M. office and stores were established at the Handyside Arcade in Newcastle and the detachment was furnished with a number of motor cycles for transport purposes. Their work at this stage consisted largely of the installation and maintenance of electrical equipment, though Lieutenant Bird, subsequently assisted by Second-Lieutenant J. R. Lang-Hyde, was also responsible for the water-supply of the very large camp at Alnwick. In all this small detachment dealt with 100 to 150 billets or hutments throughout the Tyne Garrison and were responsible also for the electric prisoner-of-war alarms at the concentration camps at Redesdale and Shotley Bridge.

Before the War the defence of Sunderland was not contemplated but this was developed in August 1916, two coast defence searchlights being installed by the Tyne Electrical Engineers to co-operate with a battery of 4·7″ naval guns at Roker. These lights were put in and manned until the termination of hostilities by a detachment from No. 1 Company under Lieutenant C. B. L. Fernandes and were operated from a motor-generator supplied from the town mains.

In this month also, the construction of the Link House Battery[1] at Blyth was commenced to perform the duties of examination battery to Blyth Harbour and to afford greater protection to the submarine depot ship " Titania " which was stationed there. These works were executed by the Durham Fortress R.E. (T.F.). Considerable difficulties were encountered in obtaining sound foundations in the shifting sand of the dunes in which they are sited. The sand also proved a great nuisance here, as in other parts of the coast line of entrenchments, by repeatedly filling up the excavations, but this was partly overcome in due course by the cultivation of coarse grass. When the constructional work was completed in February 1918 two 90 cms projectors and Crossley high-speed engines were installed and, later, manned by a detachment of the Tyne Electrical Engineers under Lieutenant W. W. Wilson.

---

[1] At first known as " Fort Coulson " after Captain M. W. Coulson, of the Durham Fortress Engineers, who was engaged in the construction of the Works.

In February 1917 further extensions to the fixed coast defences were set in hand at Hartley and Marsden to the north and south respectively of the River Tyne. Up to this date it had been understood that in the event of a serious attack on the coast the Royal Navy would have cruisers from Rosyth off the Tyne within 18 hours of warning being given. The Admiralty now notified the War Office that this responsibility could no longer be borne by them but they undertook to provide two twin 12″ gun turrets from H.M.S. " Illustrious " for erection as fixed batteries on shore. Sites were selected at Hartley and Marsden and the construction of " Roberts " and " Kitchener " Batteries was commenced. Three months was the time estimated for the completion of these great works but, owing to shortage of materials and other delays, the work was not completed until after the War.[1] The electrical equipment of these batteries was specified and the installation supervised by Lieutenants Bird and B. H. Leeson. This included a Belliss and Morcom generating set at each battery operating various lighting, heating and ventilating equipment as well as the shell-hoists. The electrical equipment on the actual guns was carried out under the supervision of the Royal Army Ordnance Corps.

In March Captain Collins left the 16th Company R.E. on transferring to the Durham Fortress R.E. (T.F.) and his place as nominal acting-adjutant of the Tyne Electrical Engineers was taken by Captain W. Barr, R.E. (Coast Btn.).

It is important that the relation between the Headquarters at North Shields and the Depot at Haslar Barracks, Gosport, should be clearly understood. It was from Haslar Barracks that all the new companies and sections were formed and the story of the great expansion of the Unit as a whole must, therefore, necessarily be told when dealing with the Gosport Detachment in Chapter IV. The fact must not be lost sight of, however, that it was at Clifford's Fort that most of the officers[2] and recruits first joined and, after the issue of clothing and equipment and undergoing a brief period of preliminary instruction in military duties, they were drafted to Haslar for technical training and posting to units on service. Though the Commanding Officer of the Tyne Electrical Engineers was actually employed at Haslar, all orders etc. of a regimental or general nature were published over his signature from Clifford's Fort and it was through the headquarters at North Shields and through Northern Command Headquarters at York that all questions affecting the appointment and promotion of officers and the multitudinous and complicated returns of all sorts which were required by higher authority relating to the Tyne Electrical Engineers as a whole were handled.

Early in 1917 the administration of the R.E. (T.F.) officers which hitherto had been carried out by the ' T ' Branch at the War Office was taken over by A.G.7 and the commissioning of officers who, up to that date, had been nominated by the Presidents of Territorial Force Associations, shortly afterwards followed

---

[1] These Batteries, said to have been completed at enormous cost, were never manned and were subsequently dismantled in August 1926 when the guns, mountings and all the equipment were sold to the Hughes Bolckow Shipbreaking Co. at Blyth.

[2] See Note " A " at end of chapter.

the procedure for special reserve and temporary commission officers. The officers were divided into distinct branches, of which the Tyne Electrical Engineers was one, and were not interchangeable.[1]

In April 1917 the officers at Headquarters, Clifford's Fort, were as follows :—

No. 1 (Depot) Company ...	Major A. K. Tasker.
	Captain W. Hall.
	Lieutenant L. Bird.
	Coy.-Serjt. Major Carr.
No. 3 (E.L.) Company ...	Major E. Swift.
	Captain W. A. Souter.
	Lieutenant W. W. Wilson.
	,, G. B. L. Fernandes.
	2nd Lieut. E. R. Brigham.
	,, ,, W. H. Elliott.
	,, ,, E. L. Hampton.
	,, ,, H. Algar.
	,, ,, J. N. Robertson.
	,, ,, B. H. Leeson.
	Coy.-Serjt.-Major Burgon.
Quartermaster	Hon. Captain A. Reed.
Regm.-Qr.-Mr.-Serjt. ...	I. T. Womphrey.

The training of personnel passing through the Depot Company was carried on under the direction of Captain Hall assisted by 2nd-Lieutenants Robertson and Leeson. Infantry drill was held on the Fish Quay and many technical models were also made to assist in the instruction. Lectures were arranged from time to time for the benefit of other units in the Garrison, as an example of which may be quoted Captain Hall's lectures on " Night marching by the stars " which proved of great interest to the infantry.

The whole of the communications in the defence were under the control of the Officer i/c Signals (Major Price) at the Minories, but a section of No. 3 Company, Tyne Electrical Engineers, was organised as a fortress signal section under Lieutenant Robertson. This section included personnel allotted to each section of the Garrison as well as to the headquarters signal office and comprised dispatch riders, telephonists, telegraphists, carrier-pigeon personnel, clerks etc. The officer commanding this section was answerable to his company commander for administration only and to the O. i/c Signals for operations and work. A very complete system of field air-line was constructed which connected almost every company dug-out in the coast-line entrenchments with headquarters.

---

[1] See Note " B " at end of chapter.

The Unit also continued to be called upon to undertake various special electrical and mechanical works in the Garrison among which may be mentioned the installation of the compressed air plant and remote control of a diaphone at Tynemouth Castle, for use in connection with the convoy system of protecting shipping; this was said to be audible fifteen miles out to sea.

On one occasion the assistance of the Tyne Electrical Engineers was sought to investigate a curious state of affairs which had arisen at the Esplanade Hotel, Whitley Bay, which was at that time in use as a billet by a certain infantry unit. The commanding officer of this unit was particularly keen on fire drill and practices took place frequently, but it was not the custom for the water to be turned on at the hydrants on these occasions. During fire drill one day, however, an over-zealous private opened the stop-cock and the N.C.O. i/c was astonished to observe that no water flowed. The matter was reported and, in due course, the unfortunate private was punished for disobedience to orders and the Tyne Electrical Engineers called in to remedy the failure of the water supply. It was not until a considerable amount of excavation had been carried out by the sappers that it was discovered that the pipes supplying the fire hydrants in and around the hotel had never been connected to the main!

After the various expansions which have been mentioned the coast defence duties at Blyth (under Lieutenant Wilson), Tynemouth (under Lieutenant Leeson), and Sunderland (under Lieutenants Fernandes and Brigham) continued to be carried out without unusual incident up to the end of the War and the work resolved itself into the monotony of long hours and ceaseless vigilance. The Hornsby-Ackroyd oil engines at Tynemouth Castle and in the Spanish Battery had to be run continuously on account of the long time required to start them up. During the hours of "official night," therefore, the searchlights were kept constantly burning but were only shewn out upon an alarm being given from the A.O.P. (Advanced Observation Post) in the lighthouse on the North Pier. With a view to economy in fuel, an attempt was made in 1917 to run the Hornsby-Ackroyd oil engines on petrol and they were fitted with special carburettors for the purpose. The experiment proved a failure as the engines were essentially slow-speed machines and could not be adapted to operate on such a fast burning fuel as petrol. At Blyth and Sunderland the lights were not kept burning continuously as the generating plant was of a type that could be started up very rapidly.

Throughout the Garrison the alarm was frequently practiced, it being found possible for the lights to be shewn out in the short space of ten seconds, while at no time were any of the defence lights out of action during the hours of darkness.

At Tynemouth a private house, No. 4 Prior's Terrace, was taken over as a billet for the manning details for the Castle and Spanish Battery lights. The detachment at Blyth occupied hutments erected behind the battery and the Sunderland party were in billets at Roker.

Occasionally the examination battery at Tynemouth came into action in order to 'bring-to' some ship which had failed to make the correct signals in reply to those of the examination vessel.[1] The procedure was to fire two rounds in close succession across the bows of the offending ship. If she did not stop immediately (as she almost invariably did!) a third shot was to be fired to sink her which was the signal to all the guns of the fortress to engage her as an enemy target. On one occasion (on the 4th January 1917) this process ended disastrously for the defence. The vessel on this occasion was one of H.M. drifters which for some reason did not comply with the instructions of the examination officer. A shot was fired which unfortunately struck the lighthouse on the South Pier, carrying away the lantern and killing the lighthouse keeper.

Enemy submarine action was continuous in the North Sea and penetrated occasionally within reach of the shore defences. Thus, on the 31st December 1916, the examination vessel "Protector," lying off the mouth of the Tyne, was mysteriously blown up, the loss being seriously aggravated by the presence on board of eleven pilots who all lost their lives in this disaster. On the 22nd March 1917 the "Rio Colorado," laden with 5,500 tons of grain from South America, was torpedoed by a German submarine in broad daylight off the mouth of the Tyne. On the 8th December three more ships were sunk, two off the Tyne and one off Whitby, and on the 14th a German submarine was spotted off Roker pier at Sunderland and immediately opened fire at the town causing considerable material damage before it retired unscathed. It was such incidents as these that led to the adoption of the 'convoy' system for the better protection of shipping. A convoy to or from the Tyne consisted of some forty to eighty vessels protected by three or more destroyers and an airship. The convoys were marshalled just outside the piers by a fleet of R.M. motor launches (M.L.) and generally departed just before dark, the incoming convoys arriving just after dusk. The Tyne was an important marshalling port in the convoy system and the arrival of a convoy of, say, eighty vessels was an interesting spectacle.

Arising out of the British naval raid on Zeebrugge in May 1918 elaborate preparation were made in the Garrison to deal with a possible enemy raid of a similar nature on the Tyne. Special defence schemes were drawn up which were to be put into force on receipt of the code order "Zeebrugge." In this connection it was pointed out to the authorities that a considerable area of water on the south side of the North Pier at Tynemouth could not be illuminated by any of the searchlights at the Castle or in the Spanish Battery and presented a danger in the event of a landing being attempted on the pier. An experiment was forthwith carried out by the personnel of No. 3 Company with an oxy-acetylene searchlight but the time required to light this lamp rendered it quite

---

[1] The examination service at the Tyne was conducted throughout the War by Commander W. W. C. Frith, O.B.E., R.N.R. Between 2nd August 1914 and the 30th December 1918 no less than 139,461 vessels were dealt with by the outer examination service here.

unsuitable for the purpose and it was almost immediately replaced by a powerful incandescent electric light which was dug-in on the north side of the Haven and operated from the Spanish Battery engine room and E.L.D. station.

When the demand for man-power increased towards the end of the War, it became apparent that the organisation of the R.E. coast defence units was extravagant in personnel and, in August 1918, the coast defence companies R.E. were re-organised on a basis of eight men per light, No. 3 (E.L.) Company, Tyne Electrical Engineers, changing its title at the same time to the 594th (Tyne) Fortress Company R.E. (T.F.). The establishment for the four lights at Tynemouth was thus reduced to a total of 32 men[1] to man the defences (at the longest period of darkness) from 4.0 p.m. to 8.0 a.m. This was carried out in two reliefs of 14 detailed as follows :—

Two engine rooms	...	...	...	...	2 drivers.
					2 electricians.
Four searchlights	...	...	...	...	4 electricians.
Two directing stations		...	...	...	6 operators.
			Total	...	... 14

The remaining four men were employed as cooks or on other administrative duties. The mechanist in charge of the Tynemouth plant was Staff-Serjeant Watt and the N.C.Os in charge of the two reliefs Serjeants Brown and Abernethy. Serjeant Turner was the N.C.O. in charge at Blyth.

In the same month (August 1918) the 16th Company, Coast Battalion R.E., which had hitherto remained split up between the defended ports of the Northern Command, was concentrated at Tynemouth and assumed the functions of a fortress works company for the Tyne Garrison.

It was also agreed at this time that the headquarters of the Tyne and London Electrical Engineers should become the parents of all coast defence units throughout the kingdom. No. 1 (Depot) Company, Tyne Electrical Engineers, therefore, also disappeared and the Headquarters received the title of "Administrative Centre," which it retained until demobilization was complete. When this change in title took place a special establishment was authorised, but the personnel was never obtained and the various other reserve depots which had existed for raising E.L. personnel were never closed down as intended.[2]

---

[1] To say nothing of the dog " Jack," an airedale, who, coming whence none knew, nightly paraded with the first relief and marched with them as far as Tynemouth Castle. He would then go off on his own, to return, punctually to time, to accompany the second relief on to the Works and march the first off. It is said that he understood every word of command given and was never known to be late on parade, even on one occasion finding his own way to Roker to support, by his presence, the detachment at that place as it marched on to the Works. When the War was over, " Jack " was still to be seen hanging about outside Clifford's Fort or Tynemouth Castle vainly waiting for parades which never took place, and died, it may be supposed of grief, very shortly afterwards.

[2] It was at this date (August 1918) that Lt.-Col. Toomer was succeeded in the command of the Tyne Electrical Engineers by Lt.-Col. E. Robinson. This change resulted in the promotion of Major Tasker to the rank of Lt.-Col.

The armistice was concluded at 11.0 o'clock on the 11th November 1918 but news that an armistice was being drawn up was received by naval wireless at Tynemouth at 7.30 a.m. on that date. This piece of news spread rapidly and was known all over Tynemouth and North Shields long before the official message signifying the signing of the armistice came through to Newcastle.

Vigilance in the coast defences was not immediately relaxed though,[1] as soon as the armistice was concluded, the work of demobilization was set in hand and continued at high pressure until the whole of the Territorial Force personnel had been demobilized. This work had been very carefully planned beforehand by the authorities concerned and was carried through without a hitch in spite of the added complications arising from re-engagements for service with various minor expeditions and in the armies of occupation. Nearly all the officers and other ranks of the many companies and sections of the Tyne Electrical Engineers passed through Clifford's Fort on their way to the various dispersal stations at Ripon, Kinross, Fovant, Purfleet, Prees Heath and at the Crystal Palace. This work also involved the disposal of members of the Tyne Electrical Engineers who were at that time patients in military hospitals throughout the country and the return, through the repatriation camp established at Winchester, of various colonials who had joined or become attached to the Unit.

In January 1919 the lights at Blyth and Sunderland were closed down and Lieutenant Fernandes then took over the duties of E. and M. officer from Lieutenant Bird, on the demobilization of the latter, and continued in this post for approximately eighteen months.

The remaining coast defence personnel quickly dwindled away and, by April, when the headquarters of the Depot returned from Gosport, there was only a small administrative staff remaining at Clifford's Fort.

---

[1] This was fortunate as, after the signing of the armistice, an attack was made upon British ships lying in Scapa Flow by a German submarine manned entirely by German naval officers all of whom perished by being blown up by a mine operated by the Royal Marine Submarine Miners. It was the brother of C.Q.M.S. J. J. Innes, of the Tyne Electrical Engineers, who was on duty at Scapa at the time and actually watched the passage of the " U " boat through the minefield and exploded the mine which destroyed her.

NOTE " A."

No fewer than one hundred and twenty-nine officers joined or re-joined the Tyne Electrical Engineers between 4th August 1914 and 11th November 1918. Several of these had seen previous commissioned service in other units and a large number were commissioned from the ranks of the Tyne Electrical Engineers, the London Electrical Engineers and other branches of the service. Those who transferred to the Tyne Electrical Engineers with previous commissioned service were the following :—

Captain E. H. E. WOODWARD, M.C. enlisted in the 1st Volunteer Battalion of the Gloucestershire Regiment in January 1906 being commissioned second-lieutenant in the same unit in March of that year. He resigned his commission as a lieutenant in March 1910 rejoining the 4th (City of Bristol) Battalion of the Gloucestershire Regiment in September 1914. He was appointed temporary captain in December and served with the 2/4th Battalion the Gloucestershire Regiment in France from May to July 1916. He was wounded on the 19th July 1916 near Laventie, receiving the Military Cross for his conduct on that occasion. The official citation in the ' London Gazette ' reads as follows :— " He displayed great coolness and courage throughout the day." Captain Woodward transferred to the Tyne Electrical Engineers on the 23rd January 1917.

Captain P. E. NEWNAM was commissioned second lieutenant in the Northern Cyclist (1st Line) Battalion in July 1913. After service in the Tyne Garrison he transferred in May 1917 as captain to the 4th Battalion, East Yorkshire Regiment, in France and served with that battalion until he was invalided home with trench fever in August 1917. He reverted to lieutenant in order to join the Tyne Electrical Engineers on the 26th September 1917.

Captain R. B. T. PINKNEY joined the 20th (Service) Battalion, Northumberland Fusiliers (1st Tyneside Scottish), in November 1914 as a second-lieutenant and served with that unit at home until June 1916 when he was promoted captain. He transferred to the Tyne Electrical Engineers on the 22nd June 1916.

Second-Lieutenant D. MYLES served for one month in the 5th Battalion, Northumberland Fusiliers, before transferring in that rank to the Tyne Electrical Engineers on the 25th November 1914.

Lieutenant R. PHILLIPS was commissioned second-lieutenant in the 23rd (Service) Battalion, Northumberland Fusiliers (4th Tyneside Scottish) in December 1914 and saw overseas service with that unit until February 1916. He was then in command of Trench Mortar Battery 102/1 from February until July, when he was invalided home. He transferred to the Tyne Electrical Engineers on the 1st May 1917.

Lieutenant L. S. WINKWORTH enlisted in the ranks of the Tyne Division R.E. (Volunteers) Electrical Engineers in April 1907 and served in the unit until April 1913 attaining the rank of Corporal. In April 1915 he joined the Cambridge University O.T.C. and was gazetted second-lieutenant in the 2/9 Battalion Manchester Regiment at Southport later in the same month. In the rank of lieutenant he served from August 1916 to February 1917 as an instructor at the Southern Army School of Instruction at Brentford and transferred to the Tyne Electrical Engineers on the 2nd February 1917.

Second-Lieutenant L. HORTON served with the 13th (Reserve) Battalion, Royal Warwickshire Regiment, from May 1915 until he transferred to the Tyne Electrical Engineers on the 4th October 1916.

Lieutenant J. A. OGILVY was commissioned in the 5th (City of Glasgow) Battalion, Highland Light Infantry, in May 1915 and served at home with that unit, being gazetted lieutenant in September of that year. He reverted to second-lieutenant on transferring to the Tyne Electrical Engineers on the 29th December 1916.

Second-Lieutenant H. G. BUXTON, after serving in Alexandria in the Princess of Wales' Own Yorkshire Regiment, was transferred in the rank of second-lieutenant to the London Divisional Engineers in September 1915 and to the Tyne Electrical Engineers in September of the following year.

Second-Lieutenant W. RINTOUL served with the Northern Signal Companies R.E. (T.F.) from September 1915 until transferring to the Tyne Electrical Engineers on 24th July 1917.

Second-Lieutenant J. L. BATEY served for one month with the 15th Battalion Northumberland Fusiliers, before transferring to the Tyne Electrical Engineers on the 22nd February 1916.

At the outbreak of War, a number of candidates for commissions were serving in the Tyne Electrical Engineers and many qualified engineers enlisted in the ranks of various branches of the service and had, subsequently, to be called out to take up commissions in technical units, where their qualifications could be put to better use. Among those officers who were commissioned into the Tyne Electrical Engineers in this manner were the following :—

TYNE ELECTRICAL ENGINEERS.—J. Young, R. W. Anderson, H. O. Rogerson, T. Henderson, H. Sherlock, D. A. Williamson, W. H. James, R. H. Rooksby, J. O. Baird, R. G. Ellis, E. J. Parnall, A. S. Burdis, L. Bird, A. Howell, R. E. Jardine, F. S. Corby, H. E. MacDonald, G. B. T. Scullard, F. Thompson, J. Welsh, C. W. Rood, C. A. Armitage, H. Poland, and J. A. Metcalf.

LONDON ELECTRICAL ENGINEERS.—C. G. Huntley and R. O. Porter.

SIGNAL COMPANY 63RD (ROYAL NAVAL DIVISION).—C. A. Stephens, J. N. Robertson and B. H. Leeson.

NORTHUMBRIAN DIVISIONAL R.E.—F. W. Crawford, A. G. Dickson, W. Dixon, E. B. Preston, C. Armstrong.

COUNTY OF LONDON YEOMANRY (and R.F.C. Egyptian Expeditionary Force).—H. D. St. J. Lidiard.

10TH BATTALION, MIDDLESEX REGIMENT (The Public Schools Battalion) (and No. 3 Coy. R.E. as a dispatch rider).—D. E. Ross.

18TH BATTALION, LONDON REGIMENT (The Artists Rifles).—J. B. Murray.

EAST YORKSHIRE REGIMENT.—E. F. Smith.

CANADIAN INFANTRY.—W. Watson.

INFANTRY.—J. A. Power.

ARMY SERVICE CORPS (MECHANICAL TRANSPORT).—W. A. Gladwin and F. W. Bond.

NOTE " B."

During the War, many officers of the Tyne Electrical Engineers were seconded or attached to various branches of the services for special duty. Among those whose work did not bring them into close association with the activities of the Tyne Electrical Engineers and whose duties are, therefore, not referred to elsewhere, were the following :—

Major N. H. FIRMIN, subsequent to the re-organization of the anti-aircraft defences at home at the end of 1917, left No. 9 (Tyne) Mobile Searchlight Company and returned to Haslar, being seconded for duty under the Admiralty on March 3rd 1918. The services of this officer had been requested by Dr. Charles H. Merz (the eminent consulting engineer) who had recently been appointed Director of Experiments and Research. The Department of Experiments and Research was a new one and controlled Admiralty research in all its branches, but was particularly concerned with the development of anti-submarine warfare. The Director controlled also the Admiralty Board of Inventions and Research. Major Firmin was appointed secretary to the Director and in October 1918 was made Assistant Director of Experiments and Research (For Administration). As such, he was responsible, for the administration of the London office and for ensuring that the different experimental stations and committees working directly under the Director, or under his general supervision, were properly organised. Major Firmin continued in this capacity until he was demobilised early in 1919. For his services in this connection Major Firmin's name was specially brought to the notice of the Lords Commissioners of the Admiralty and he was made an Officer of the Order of the British Empire (Civil Division).

Captain W. M. BUCK, being in China at the outbreak of the War, was attached to the 40th Fortress Company R.E. at Hong Kong and served with that unit, subsequently in the rank of major, throughout the war.

E

Lieutenant I. F. FAIRBAIRN-CRAWFORD was one of the very few officers who had had any considerable experience of aircraft and flying prior to the commencement of the War. In a civilian capacity he had visited the principal aeroplane, airship and engine factories in France, Germany, Italy and Holland and had negotiated, on behalf of the Admiralty, with the Italian Government for the supply of two large semi-rigid airships of the Forlannini design which were completed and underwent their trials in July 1914 but which were taken over by the Italians on the outbreak of the War. At the end of 1913 he was appointed general manager of the aircraft department of Messrs. Armstrong Whitworth and Company Ltd., and held this post until the date of mobilisation. On mobilisation he proceeded with the Portsmouth detachment but, after three weeks service in the coast defences, he was seconded, at the urgent request of Messrs. Armstrong, Whitworth and Company, for the purpose of organising a large aircraft factory at Newcastle and supervising the manufacture of service aeroplanes for the Royal Flying Corps. In connection with this service Captain Fairbairn-Crawford proceeded on several occasions to France, being temporarily attached to various squadrons for the purpose of obtaining the latest requirements for new designs of aircraft. At the end of 1916 this officer was transferred to the Admiralty to supervise and control the construction of rigid airships at the Barlow airship establishment near Selby. During the War Captain Fairbairn-Crawford was responsible for the production of over 1,500 service aeroplanes and three large rigid airships including the famous R.33. He was demobilised in October 1919.[1]

Captain J. YOUNG was seconded on the 22nd August 1917 for employment with the Ministry of Munitions. He remained seconded, subsequently under the Air Ministry, until transferred to the reserve of officers on the 13th October 1920.

Lieutenant R. L. WOOD, a mining engineer by profession, was seconded to the Royal Engineers and posted to the 252nd Tunnelling Company on its formation in France on the 22nd October 1915. The company at once commenced mining operations opposite Beaumont Hamel, in the Somme area, and remained on the same front until the end of the War. During this period the company advanced slowly to a position in front of Bapaume ; retreated, in March 1918 to the point from which it started ; and finally advanced nearly to Cambrai. Captain Wood commanded the company during the retreat and during the final advance of October and November 1918. He was mentioned in dispatches and awarded the Military Cross in May 1917 for his work in connection with the removal of enemy mines. He was demobilised and restored to the establishment of the Tyne Electrical Engineers on the 24th April 1919. (N.B.—In May 1920 Captain Wood was appointed to command No. 1 Battalion

---

[1] Captain Fairbairn-Crawford had a distinguished career as an athlete. Besides being past amateur champion of Ireland for the one-mile, half-mile and quarter-mile, he was amateur half-mile champion of England in 1907 and ran for England in the 800 and 1500 metres at the 1908 Olympic Games at the London Stadium. He was, however, beaten in the final of each race after leading the whole way almost to the winning post ; the winner being the American, Shepherd, who at the same time broke the world's half-mile and 800 metre records.

Northumbrian Divisional R.E. (T.A.) with the rank of Lieut.-Colonel. He was appointed C.R.E. 50th (Northumbrian) Divisional R.E. (T.A.) on the 5th May 1922 and promoted Brevet Colonel on the 10th March 1925).

Captain M. C. JAMES was seconded on the 2nd December 1916 for service under the Director of Docks, Brigadier-General Sir R. L. Wedgewood. He proceeded to France and was employed as a staff captain in the docks at Havre where he did " invaluable work in connection with the organisation of the engineering department and the development of equipment, etc., at the French ports." After some time, he was seconded for service under the Ministry of Shipping, retaining the rank of staff captain, and became District Superintendent of Ship Repairs for the whole of the north-east coast of England, under Lord Pirrie, until the end of May 1919 when the work under this ministry ceased.

Lieutenant R. W. ANDERSON volunteered for service abroad when volunteers were required for the signal service R.E. and was sent to the 27th Division B.E.F. France in October 1915 but, as this division was under orders for Serbia, Lieutenant Anderson was returned to undergo a signal course at Otley; after this he joined the 29th Divisional Signal Company, which had just returned from Gallipoli, in the Beaumont Hamel Sector of the Somme line in April 1916. He was appointed artillery signal officer to supervise the communications of the brigades and batteries in the division and to act as laison officer between the artillery and the signal company. He officiated in this capacity during the first battle of the Somme in July 1916 and was mentioned in the dispatches of Sir Douglas Haig on the 19th November for his services in this connection. In October 1916 he was promoted temporary captain and appointed to command the Heavy Artillery Signal Section of the VIII Corps which was then holding the left sector of the Ypres salient. In this post Captain Anderson was occupied with the preparations for the Wytschaete attack of June 7th 1917 and was awarded the Military Cross for the part he played in this affair. The official citation reads as follows :—" For conspicuous gallantry and devotion to duty on many occasions when superintending the repair of cable lines under heavy shell fire. His fearlessness and energy have been of the greatest value to the heavy artillery." In October 1917 he was appointed to command the VIII Corps Signal Company, which post he held in the Chalons-sur-Marne sector of the Ypres salient until June 1918 when he was promoted acting-major and posted to command the 63rd (Royal Naval) Divisional Signal Company in the Beaumont Hamel sector, which had been occupied again after the retreat of March 1918. Here Major Anderson was able to make use of some of the cables which he had buried in the same sector in 1916. He was subsequently responsible for all the signal arrangements of the division during the British advance finishing up near Mons on November 11th. For his services in this connection Major Anderson was again mentioned in dispatches by Sir Douglas Haig. He was demobilised in February 1919.

Captain C. F. Scott was seconded to the R.E. in the fall of 1918 and posted to Malta where he is understood to have served as D.O.R.E. and on the coast defence electric lights until being demobilised at the end of 1919.

Lieutenant W. Sharp was serving, immediately prior to the outbreak of the War, on the staff of the Central Uruguay Railway. On June 1st 1915 he was seconded with the rank of temporary captain to the railway company which was in process of formation at Longmoor. This company, after laying sidings for the munition factory at Blaydon-on-Tyne, embarked for Egypt early in December 1915. On arrival at Port Said the company moved down the canal to Ferdan where Captain Sharp was employed surveying and laying light lines eastwards from the canal prior to the battle of Romani. On the 29th July 1915 he was posted to command the newly formed 53rd Railway Company on the maintenance and traffic working of the Kantara standard gauge lines. After the battle of Romani the railway was pushed ahead across the Sinai Desert at such a speed that the personnel of the 53rd Company was inadequate to cope with the traffic on the ever lengthening line, and, on the 27th February 1917, the railway operating division was formed to which he was appointed with the rank of acting-major. By the time the railway had been laid round the end of Mount Carmel into Haifa the personnel engaged on the line had risen to a total of over 5,000. Major Sharp was gazetted brevet major on the 3rd June 1917, was three times mentioned in dispatches during his service in Palestine and was appointed an Officer of the Order of the British Empire (Military Division). (See also Chapter VIII, Note " D.").

Lieutenant E. H. Gibbon was seconded on the 15th February 1915 as a pilot to the Royal Flying Corps (Military Wing) and served at home with No. 9 Reserve Squadron at Dover, No. 13 Reserve Squadron at Hownslow and No. 19 Reserve Squadron at Croydon before being posted to No. 5 Squadron R.F.C. with the B.E.F. in France. After three unfortunate crashes in the performance of his duty Captain Gibbon returned home, was boarded medically unfit for further service with the R.F.C., and reverted to the Tyne Electrical Engineers on the 22nd August 1916.

Lieutenant G. I. N. Deane was also seconded for service with the Royal Flying Corps (Military Wing) in May 1915 and joined No. 14 Squadron, 5th Wing, R.F.C. in the Delta, Egypt, as a pilot officer in July. In September he was appointed equipment officer, 1st class, and officer i/c workshops at Aboukir. In March 1917 he returned home and was appointed officer i/c workshops of the R.F.C. Aeroplane Repair Park, Northern Aircraft Depot, Sheffield. In September he was posted as senior officer instructor in charge of the aircraft section at the Technical Officers' School of Instruction, Henley-on-Thames. In this post he remained until demobilised in March 1919. Captain Deane's name was included in the list of officers brought to the notice of the Secretary of State for valuable services in connection with the War, 29th August 1919.

Lieutenant T. Henderson was seconded to the R.F.C. (Military Wing) on the 23rd July 1915 and proceeded overseas with No. 10 Squadron from Netheravon, being first stationed at Aire, near St. Omar, and subsequently at Choques, near Bethune. Here Lieutenant Henderson's duties consisted of long-distance reconnaissance extending as far as Valenciennes and occupying usually about three hours flying over the lines. The average speed of the machines (B.E. 2c) was only about 65 m.p.h. A.A. shelling was considerable, 138 patches having to be put on the fabric of Lieutenant Henderson's machine after one such reconnaissance. During the battle of Loos he was employed on artillery co-operation, and was shot down by Immelmann, the famous German ace. He was mentioned in dispatches for his conduct on this occasion. Lieutenant Henderson was also responsible for the original location of the renowned long-range German gun, " Big Bertha." In December 1915 when returning from a reconnaissance his machine received a direct hit by A.A. shell-fire but Lieutenant Henderson succeeded in effecting a safe landing within our own lines. In 1916 this officer returned to England and acted as an instructor until June when he was posted to the Middle-East (Egypt) participating in the battles of Romani, Katia, El Arish and Nedhl. On one occasion lack of fuel forced him to land forty miles inside Turkish territory where he and his observer defended themselves against a troop of Turkish cavalry until relieved by one of our own aircraft three days later. In October 1916 Captain Henderson was one of a small party known as the Hedjaz Expeditionary Force, organised to assist the Arab revolt in the desert of Arabia. On arrival at Rabegh, however, it was discovered that they were " infidel Christians " and were not permitted to set foot in the Holy Mohammedan Hedjaz. Some weeks later the Arabs were again calling for help and the expedition landed and made an aerodrome at Rabegh. There were at that time no maps of this area and Captain Henderson set to work to compile one for the use of the expedition. For this service he was accepted as a Fellow of the Royal Geographical Society. The Turkish Forces were finally invested in Medina and cut off from their main army in Palestine. At the end, Captain Henderson proved to be the only member of the expedition, with the exception of the famous Colonel Lawrence, who served throughout the entire campaign and he was allotted the task of writing the official narrative of the expedition. For his services in this theatre Captain Henderson was awarded the M.C., again mentioned in dispatches and awarded, by King Ibn Hussein of the Hedjaz, the Order of El Nadha " for bravery in the field." Towards the end of 1917 Captain Henderson returned to England and was employed as an instructor and in the survey for all prospective aerodrome sites in S.W. England. He was also responsible for the design of a patent course-setting bomb sight, an improved model of which was subsequently adopted as standard equipment in the R.A.F. (See also Chapter VIII, Note " D.")

Captain D. A. Williamson was seconded to the Royal Flying Corps (Military Wing) on the 27th November 1916. After being stationed at Reading he served as a pilot with reserve and home defence squadrons at Turnhouse, Catterick and Tadcaster before being posted in May 1917 to No. 22 Squadron R.F.C. in

France. This squadron was then operating in the Somme area, being opposed to Baron von Richthofen's famous " Circus." No. 22, which was the first squadron to be re-equipped with Bristol Fighter machines, afterwards moved north behind Arras where it took part in the battle of Messines (7th to 14th June 1917). The squadron then moved further north into the Ypres salient, where it found itself once more opposed to von Richthofen. Captain Williamson was mentioned in dispatches and subsequently reverted to the Tyne Electrical Engineers in September 1917.

Lieutenant C. D. DANBY also served as a pilot with the Royal Flying Corps being seconded to the military wing on the 13th August 1915. After the usual period of training at home this officer was posted to a squadron in France and was awarded the M.C. " for excellent work under bad weather conditions when taking photographs both before and during operations." He was appointed flight commander on the 27th May 1916 and promoted captain on the 1st June. He was killed in action on the 18th July 1918.

Captain F. B. C. SUTTHERY (see also Chapter VIII, Note " D ") and Captain W. HALL were seconded in October 1918 to the Royal Marines at Chatham where a special battalion was in process of formation preparatory to manning the so-called " mystery towers." These floating towers represented the very latest developments in anti-submarine defence but never were actually put into service as the War was over before the training of the manning details was complete.

Lieutenant C. A. C. AITKENS transferred first to the Signal Service R.E. and on the 15th June 1916 to the Special Brigade R.E. In this latter unit he was employed in the operation of liquid-fire flame projectors. He died on the 10th July from wounds received in action during the first battle of the Somme.

Captain H. S. RIPLEY was seconded to the R.E. and appointed on the 10th of January 1918 to a post on the staff of F.W.8 under General Scott-Moncrieff. Captain Ripley was by profession an electrical engineer and was selected for the duty of resident engineer to supervise the construction of a power plant near Dumfries where a special carbonisation process of manufacturing trench fuel from raw peat was being developed. This trench fuel would have been of great importance if the campaign on the western front had extended through the winter of 1918 and extreme pressure was put on the work which commenced in January 1918 and finished just before the armistice. The fuel was actually used for some time after the armistice but in July 1919 the factory was closed down and handed over to the disposals board. Captain Ripley also had control of the financial side of the undertaking and, as the clearing of the accounts occupied a considerable time, it was November 1919 before this officer could be demobilised. For his services in connection with the Dumfries Peat Factory, Captain Ripley was awarded the M.B.E. (Military Division).

Lieutenant H. G. BUXTON was seconded to the R.E. and is understood to have served with a field company in France.

Lieutenant R. L. HUNTER.—After commanding No. 6 Section in France this officer was detached for duty with trench mortars. He was awarded the M.C. on the 26th September 1917 " for conspicuous gallantry and devotion to duty when in charge of a bombing party. Although heavily shelled and gassed he shewed the greatest determination and courage in getting his party, with all the ammunition, through the barrage to our front line. Later, he was successful in demolishing enemy wire which had previously been a great source of trouble. He also displayed great fearlessness and gallantry in observing under heavy shell fire from a part of our front line which had been demolished."

Lieutenant R. P. FINDLAY was seconded in April 1917 for service with the 123rd Field Company R.E. 38th (Welsh) Division, Third Army. Promoted captain 24th March 1918. Was awarded the M.C. for gallant conduct in connection with bridging operations at the battle of the Selle (17th to 25th October 1918). The crossing of this river is said to have involved more arduous work for the Divisional R.E. than that of any other during the War, the bulk of the work being carried out by night in absolute silence as the nearest enemy post was only fifty yards away. Transferred to 1st (Northumbrian) Brigade R.F.A. (T.) on the 8th September 1920.

Lieutenant C. RUSSELL was seconded to the R.F.C. (Military Wing) on the 29th October 1917. Relinquished commission on account of ill-health on the 23rd November 1918.

Lieutenant C. H. R. E. RAVEN was seconded, on the 28th November 1916, for employment with the Ministry of Munitions in the Air Inventions Department. He was seconded for service with the R.A.F. in which he was granted a temporary commission as captain (Grade A) on the 1st April 1918.

Lieutenant E. HARRISON was seconded on the 24th December 1917 to the R.F.C. (Military Wing) as a flying-officer.

Lieutenant F. T. HAMILTON was seconded to the R.E. and appointed temporary staff-captain to the Director of Fortifications and Works at the War Office on the 23rd February 1917. Here he was employed in Department F.W.8, which was responsible for the purchase, inspection and shipment of all R.E. stores. This department dealt with the requirements of all theatres of war as well as of home forces. Lieutenant Hamilton was particularly engaged in the provision of electric lighting and power equipment including searchlight generating sets and all water-supply plant. This officer remained in this post until the end of the War and was awarded the M.B.E. (Military Division) in 1919. (N.B.—This officer served in the ranks of the Electrical Engineers' Searchlight Unit during the South African War, 1900-1901).

Lieutenant F. L. ADENDORFF was seconded to the R.E. on the 16th May 1916 for employment with the light railways in forward areas. He was awarded the M.C. on the 26th July 1918 " for conspicuous gallantry and devotion to duty. Whilst in charge of light railways he shewed great courage in controlling the operation of ambulance and ammunition trains under heavy shell fire and later, when all the forward lines had been cut, he evacuated all stores and personnel to safety." He reverted to the Tyne Electrical Engineers in 1919.

Second-Lieutenant H. S. WATSON resigned his commission in the Tyne Electrical Engineers on the 23rd January 1918 on ceasing to be employed with the Territorial Force. He was gazetted lieutenant and temporary captain in the Royal Air Force on the 1st April 1918, and promoted a major on the 27th May 1919. He was employed on the administration branch R.A.F. at the Air Ministry, as staff officer 2nd class, under the Chief Electrical Engineer. In August 1919 he was brought to the notice of the Secretary of State for valuable services in connection with the War and was transferred to the unemployed list R.A.F. on the 1st February 1920.

Lieutenant H. E. MacDONALD was seconded to the R.E. in January 1918 for training in field works. On the 30th August 1918 he proceeded to France and was posted to the 79th Field Company R.E., attached to the 18th Division, Fourth Army. He served with this company through the final advance and, from January 1919, was employed in Belgium under the Disposals Board until he was demobilised in May of that year. (N.B.—He transferred to the 50th (Northumbrian) Divisional R.E. (T.A.) on the 20th June 1921.)

Second-Lieutenant E. F. SMITH was seconded to the 3rd Field Survey Battalion, B.E.F., for service as sound ranging officer. In this capacity he served in ' H ' and ' C ' sections of the battalion, in the Somme area, from February the 11th 1918 until the armistice.

CHAPTER IV.

1914—1919. THE GREAT WAR.
THE DEPOT AT GOSPORT.

Mobilisation — Army signals — Organisation of Portsmouth Command
— Coast defence duties — The spy mania — Extemporised anti-aircraft
searchlights — The Electric Light School — Workshops — New establishments — Oxy-acetylene searchlights for France — Detachment for
Scottish coast defences — Contingent for E. and M. Company — Development of A.A. defences — A.A. searchlight sections for overseas — Expansion of the Unit — Air raid on Portsmouth — Regimental technical schools
— The A.A. searchlight and sound locator school — Early experiments —
Administrative arrangements — Sport — Reorganisation of coast and A.A.
defences in the U.K. — Women's organisations — Limit of expansion —
Honours — War savings — Demobilization — Return to Clifford's Fort.

ALTHOUGH Clifford's Fort remained throughout the War the official headquarters of the Tyne Electrical Engineers, it was from the Gosport companies that the great expansion of the Unit was destined to take place. During the four and a half years that the Depot was established at Haslar Barracks, the strength of the Unit was multiplied nearly fourteen times and the four companies which mobilised in 1914 gave birth to more than sixty subordinate units scattered from Cromarty in Scotland to the banks of the Piave in Italy. It would be difficult to overestimate the importance of the work done at Haslar Barracks or to pay too high a tribute to the energy, initiative and administrative ability that created, from a handful of officers and men, an organisation so great and far-spread.

Throughout the War Haslar Barracks was the centre from which all the activities of the Unit radiated. For years an almost endless stream of recruits and untrained men flowed in and, after an intensive course of training, passed out again as complete sections and companies for service at home and overseas. Although the stories of most of these off-shoots are dealt with in subsequent chapters the following narrative remains so full of simultaneous incident that it is not easy always to preserve the true chronological sequence. Where this is impossible, it will sometimes be necessary to digress to some extent in order to deal completely with each subject in turn.

It has already been recounted in the preceding chapter how 8 officers and 104 other ranks of the special service sections of Nos. 2 and 3 Companies, under the command of Major C. R. Toomer, entrained at Tynemouth on the 30th July 1914 and travelled overnight to their war stations on the southern coast defences. On arrival at Stokes Bay railway station, just 39 hours after receipt of the mobilisation telegram, the detachment was met by Major J. R. Garwood, R.E.[1], C.I.E.L. (Chief Instructor of Electric Lighting), and several of the N.C.Os were detailed immediately for duty in the Portsmouth Garrison.

The telephone section, which had left the train at Fareham, under Second-Lieutenant W. G. Ward, was posted to Fort Fareham and shortly afterwards transferred to Fort Purbrook where it was, for a time, engaged in establishing telephone communication for the defences which were being constructed by the Wessex (Divisional) R.E. (T.F.) on the eastern side of Portsmouth. In October the telephone section was removed to Milldam Barracks, Portsmouth, where the headquarters of the Officer-in-Charge of Army Signals, Southern Coast Defences, was established. Shortly afterwards a fault developed on one of the submarine cables communicating with the Sea Forts. The scheme then in force provided for the maintenance of these cables by the Post Office, but, at that time, the cable ship was fully occupied on other parts of the coast and could not be made available for use at Portsmouth. In these circumstances, the Officer-in-Charge of Army Signals, Captain A. B. Ogle, R.E.[2], determined to attempt to detect and repair the fault with the means at his disposal. He accordingly called for volunteers from among those members of the Tyne Electrical Engineers' telephone section who had formerly served as submarine miners, and, having acquired a suitable boat from the Dockyard, commenced to under-run the offending cable with a view to locating the source of the trouble. In due course it was discovered that the cable had been broken, probably by a vessel anchoring over it, and, in spite of the choppy sea running at the time, both ends were recovered and a satisfactory joint effected. The manner in which this was carried out, without the aid of the proper appliances or correct materials, reflected the greatest credit not only on the skill of the actual jointers but on the whole boat's crew, and furnished yet another example of the value of the experience which had been acquired in the submarine mining service.

During the ensuing months the entire military telephone system for the southern coast defences (which included Portsmouth, Southampton and the Isle of Wight) was handed over to the Tyne Electrical Engineers, though a few Regular N.C.Os remained in this service. The control of the system was then taken over from Captain Ogle by Lieutenant Ward, who continued in the post of Officer-in-Charge of Army Signals (subsequently in the rank of captain) until the end of the War.

The remainder of the personnel which arrived on the 31st July 1914 proceeded at once to Haslar Barracks, which was to be their home for the next four and a half years. Haslar Barracks, situated to the west of Portsmouth Harbour and adjoining the Royal Naval Hospital at the head of the Haslar Creek, was the peace station of the 4th Company R.E. and housed the personnel

---

[1] Born 24/11/73, First Commission 19/2/93, Lieut.-Colonel 31/8/19. Retired 31/8/19. D.S.O.
[2] Born 12/4/83. 2nd Lieut. 21/12/01, Lieut. 3/8/04, Captain 21/12/12, Major 21/12/16, Lieut.-Colonel 16/5/28. Colonel 16/5/32.

of the electric light school. That night a number of the coast defence electric light stations in the Portsmouth defences were manned, further positions being taken over during the succeeding days.

The Special Company (No. 4), which had hitherto not been allotted to any particular war station, arrived at Portsmouth on the 5th August for electric light duties, to remedy a serious deficiency of personnel in the Hampshire (Fortress) R.E. (T.F.). On the same date the remaining personnel of Nos. 2 and 3 Companies joined for duty at Haslar, bringing the total strength on the southern coast defences to within seventeen of its full complement. The deficiency was caused by medical rejections and by the absence of a certain number of men at sea and was almost immediately made good by recruits enlisted at Clifford's Fort.

The G.O.C. Portsmouth Garrison at this time was Major-General Blewitt[1] and the Chief Engineer Colonel C. Godby[2]. The Command included Southampton and the Isle of Wight and was divided into five districts, in each of which was a C.R.E. The Tyne Electrical Engineers came chiefly under the C.R.E. Portsea, who dealt with the Portsmouth coast defences including the three Sea Forts. The organisation of the Tyne Electrical Engineers at Gosport at this stage was as follows :—

Officer Commanding Detachment...Major C. R. Toomer, V.D. (No. 4 Company).

Second-in-Command... ... ...Major E. Robinson (No. 3 Company).

No. 2 Company ... ...Captain O. M. Short.

No. 3 Company ... ...Captain E. Swift.

No. 4 Company ... ...Captain C. Arnison.

The Tyne Electrical Engineers now manned the electric light stations at Fort Monckton, Haslar, Lump's Fort and Calshot Castle, all on the Hampshire mainland; St. Helen's and Sandown in the Isle of Wight and the three Sea Forts in Spithead. The remaining stations were manned by the Hampshire (Fortress) R.E. (T.F.) and by Regular personnel of the 4th Company R.E.[3] Owing to a delay in the ferry service the detachment detailed for duty at Sandown experienced difficulty in catching the Isle of Wight steamer at Portsmouth. Had it not been for the initiative of Lieutenant F. D. Pyne who commandeered a tram and, after turning out the civilian passengers, compelled the driver to proceed direct to Clarence Pier, the searchlights would not have been in action that night.

---

[1] During the War period the G.Os.C. Portsmouth Garrison were as follows :—Major-General Blewitt, Major-General F. C. Heath-Caldwell, C.B. (late R.E.), and Major-General Douglas Smith.

[2] Born 31/10/63, First Comm. 1/10/82, Colonel 19/7/11, Retired 5/9/19, C.B., C.M.G., D.S.O. Succeeded as C.E. Portsmouth by Colonel T. Ryder Main, C.B., C.M.G. In 1917 the latter officer was succeeded by Colonel F. Baylay who retained the appointment until he retired from the Service in June, 1919.

[3] The modern defences of Portsmouth Harbour, including Spithead and the Isle of Wight, may be said to have originated as the result of a Royal Commission on national defence appointed in 1859. Among other works, Lump's and Sandown batteries and forts Purbrook, Grange and Brockhurst were designed by Captain Crossman, R.E., while the Horse Sand, No Man's Land and Spit Bank Forts in Spithead were the work of Captain E. H. Steward. Gilkicker Battery was originally constructed by Lieutenant-Colonel Fisher, but the whole of these important works were superintended by Colonel W. F. D. Jervois, R.E., then Deputy Director of Works at the War Office.

The electric lights for the inner defences of Portsmouth Harbour, including those in the three Sea Forts, were controlled from Square Tower in Point Battery, and this duty was performed after the first few weeks of the War entirely by officers of the Tyne Electrical Engineers.

It was not the custom in peace time to maintain garrisons in the Sea Forts, and, when the first detachments arrived on the 31st July, they had to make what arrangements were possible for messing and accommodation. Space was so restricted that many of the men slept in hammocks slung over the beds of their fellows. The plant was overhauled and tested and the general organisation arranged before nightfall.

The searchlight plant throughout the garrison was manned every night during the hours of " official " darkness by two reliefs in the summer and three in winter. For some considerable time the lights were run in the ordinary manner, any special orders being transmitted from Square Tower which was in direct telephonic communication with Garrison Headquarters.

The fairway entrance to Portsmouth Harbour was between Spit Bank Fort and Southsea Castle. Ships proceeding to sea passed between a pair of light-ships approximately two miles seaward from Spit Bank Fort. As a protection against submarine attacks, a boom composed of a double tier of wooden piles was constructed from the Isle of Wight to No Man's Land Fort and thence to the port lightship ; a similar boom extended from the starboard lightship to Horse Sand Fort and thence to a point about a quarter of a mile to the east of Southsea pier. In the event of any of H.M. vessels leaving or entering the harbour during the night, the lights were manipulated in accordance with carefully prepared plans, which obviated the risk of illuminating the vessel concerned. These orders were all transmitted from Square Tower, and the operation was, in every case, carried through with entirely satisfactory results.

After about eighteen months, it was decided that the lights should be run behind closed shutters and only come into operation in the event of an alarm or for some special purpose. Eventually, the lights were not run at all except upon alarm, and in order to save fuel certain engines only were kept running. Those not running were kept ready to start up immediately if required. The plant consisted mostly of 120 cms. projectors, with remote control, supplied from generators driven by oil engines of the Hornsby-Akroyd type, though some steam generating plant was still in use, notably at Fort Monckton. The Hornsby-Akroyd engines required some twenty minutes to half-an-hour to start up from cold, and, though very reliable and most suitable for continuous running, were not so satisfactory for the intermittent working which was demanded of them.

In each of the Sea Forts, the R.E. were responsible not only for the operation of the searchlight plant and the interior electric lighting installation, but also for a small borehole pump supplying drinking water from a well in the centre of the fort.

HORSE SAND FORT, SPITHEAD.

The strength of the garrison of each fort naturally depended on the scale of its armament and equipment. On Spit Bank Fort, for example, the garrison consisted of about 100 R.A. and 40 R.E. in winter, but in summer, when the number of reliefs was reduced from three to two, the R.E. complement would be about thirty. The two larger forts, Horse Sand and No Man's Land, had larger garrisons. Life in the forts was monotonous. The War Department steam vessel came alongside twice daily in fine weather for the delivery of rations and stores and to take men to and from shore leave. In rough weather it was not always possible to reach the Sea Forts and stocks of emergency rations had always to be kept in store.

In the early days of the War a number of vessels, unaware that hostilities had broken out, attempted to enter harbour after official nightfall and this caused a considerable amount of activity to the defence until the nationality of the ship was ascertained. Alarms for practice purposes, and of a genuine nature, were given from time to time and false alarms were not infrequent in the early days. Perhaps the most memorable of these was caused by an empty biscuit tin drifting down the main channel, which at first appeared, in the searchlight beams, to be the periscope of a submarine. On another occasion, a dead cow, floating along with its legs uppermost, produced a similar effect, while it is related that the weird combination of a shoal of fish, a packing-case and a boat-hook proceeding together *against* the flow of the tide caused so much speculation that a boat manned by personnel of the Tyne Electrical Engineers was sent away from one of the Sea Forts to investigate while every gun that could be brought to bear was trained on the phenomenon until its harmlessness was assured !

In common with other parts of the country inhabitants of the Portsmouth area were seriously affected by the spy mania, especially during the early period of the War, and the authorities were inundated with reports of suspicious actions which frequently proved to have no foundation in fact. In the majority of cases these reports referred to " signalling "—presumably to the enemy— and although this menace was serious enough to demand attention, the resulting investigations not infrequently had their humorous side. Moving shadows on blinds proved to be caused by innocent gentlemen engaged in performing their " stoutness " exercises, and roving lights in a dark house by the householder searching with a candle to find the position of his fuse-box ! On one occasion, an intermittent light from an upper window in an hotel facing the Solent was thought to merit investigation and a small party under the command of a subaltern was dispatched for that purpose. The officer demanded access to the room, the position of which had been carefully located, but this was refused on the grounds that the occupant was a lady. A threat to enter by force resulted in a short parley with the occupant, at the end of which the door was opened and the officer was confronted by his own Colonel's wife in a dressing-gown, hair curlers, and a villainous temper ! The " signalling " proved to be caused by a broken lath in the venetian blind which moved up and down on windy nights.

On another occasion two officers of the Unit were having tea in a well-known restaurant in Southsea when one of them was requested by the manager to speak to someone at the door. The "someone" proved to be a special constable who warned him to be extremely careful as his companion was suspected of being a German spy!

Rumours of an alarming nature were circulated from time to time and, as an example of these, reports in connection with a battleship whose departure from the Dockyard was postponed two or three times may be mentioned. One was that a German submarine was lying outside waiting for her, but this was almost immediately discounted by a second report to the effect that the submarine had been caught in the nets and, after being left there for several days, had been brought into harbour full of dead Germans. Another was that an infernal machine had been found in the ship which, within an hour, would have entirely destroyed the vessel. This particular rumour was nicely rounded off by a further report that the traitor had been discovered and shot out of hand! The scaremongers were finally discredited by the safe departure of the vessel.

In April 1915 certain modifications to the defences of the Solent were approved and the construction of two new batteries at Egypt Point and Stone Point were put in hand. The searchlight installations and personnel at Sandown and Calshot Castle were then transferred to these new stations under Captain F. D. Pyne and Lieutenant R. Sharp respectively. These stations were handed over to the Hampshire Fortress R.E. (T.F.) on the 8th May 1916.

All these duties in connection with the coast defences were, of course, those for which the Unit had been training and preparing ever since the application of electric light to the purposes of defence against sea attack had first been considered some twenty years previously.

It was not long, however, before a new field of activity was opened up by the menace of raids from the air.

Although the subject had received some consideration prior to 1914, next to nothing had been done by the time the Great War broke out to provide air defences, even for the more vitally important naval and military localities.

It was but natural that, when the need for anti-aircraft searchlights began seriously to be realized, the authorities should turn for assistance to the searchlight units, and it was not long before anti-aircraft lights of an extemporised nature were erected in Portsmouth Garrison and manned by detachments of Tyne Electrical Engineers. Among the earliest lights were those at Lumps Fort, Hilsea Lines, Porchester Castle, Fort Elson, Fort Brockhurst, Fort Grange, Fort Gilkicker, Fort Monckton, Horse Sand Fort and on the Culver Cliffs, I.O.W.

The A.A. searchlight erected at Lumps Fort in the first week of August 1914, by the Tyne detachment stationed there, was the first to be erected in the Portsmouth defences. For the first few nights of operation this light was kept running continuously throughout the period of "official night," and the beam swept the sky, searching after the manner of the coast defence search beams. It was very soon realised, however, that this method of operation was entirely wrong and subsequently the beam was not exposed until an alarm was given. The detachment at Lumps Fort was also probably the first of the Tyne Electrical Engineers to come under fire as, early in August 1914, a shell coming from the direction of the sea passed overhead, striking the gable of a neighbouring house. Fortunately the shell did not explode. It subsequently transpired that this was a 6″ shell fired from Horse Sand Fort, an error which no doubt cost some individual dearly !

In the case of the light at Hilsea supply was taken from the Portsmouth Corporation tramway system and fed to a motor-generator installed near the searchlight station, while at Forts Brockhurst and Grange supply was taken at one time from two (so-called) portable steam-driven generating sets which were loaned from H.M. Dockyard at Portsmouth. In spite of the serious nuisance occasioned by smoke these sets remained in action until more economical petrol-paraffin sets could be provided to replace them.

Experiments had previously been carried out over a lengthy period with the object of lighting the aerodrome at Fort Grange at night by means of searchlights. Night flying was at this time very much in its infancy and, with aeroplanes in the state of development then reached, landing at night was a hazardous procedure. Although the illumination of a large area was accomplished, the use of searchlights proved unsatisfactory from the pilot's point of view.

In the case of the light at Fort Monckton supply was provided from a fixed steam-driven generating set until replaced by a generator direct coupled to a Crossley I.C. engine. These high-speed engines constituted a great advance on the hot-bulb type of oil engines in use in the coast defences, the chief advantage being the short time required to come into action. An example of this was furnished during an inspection of a detachment at Portchester Castle under Lieutenant J. L. Batey, when the equipment, consisting of a 40 H.P. Crossley engine operating two 60 cms. naval pattern projectors, was brought into action with arcs struck in the short space of twenty-seven seconds from the time the order was received. Most of the A.A. lights were of 60 cms. naval pattern and were manipulated by "joy sticks" fixed on mountings about 10 or 12 yards from the projectors to which they were connected by four wires each pair of wires governing the movement of the beam in a horizontal and vertical plane respectively.

The electric light school, together with the R.E. workshops at Stokes Bay, had also gradually been taken over from the 4th Company R.E. by personnel of the Tyne Electrical Engineers. Major Toomer was subsequently appointed acting C.I.E.L. in place of Major Garwood, R.E.

The whole of the major repair work in connection with the maintenance of searchlight plant on the southern coast defences was conducted from Stokes Bay and works parties were employed on the electric wiring of hospitals in the district (including those at Netley and Brockenhurst) and miscellaneous engineering services including the sinking of artesian wells.

Lieutenants Pyne[1], S. E. Monkhouse and J. Young were all employed at different times on these services under Captain W. C. H. Prichard, R.E.[2] When the latter officer went overseas to take charge of A.A. searchlights on the Western Front he was succeeded as Inspector of R.E. Machinery by Captain Monkhouse. The duties of this post, in which he was assisted by Lieutenant H. Sherlock, included the inspection of R.E. installations in the Channel Islands. In all, 83 other ranks were employed at Stokes Bay and on the various detached works mentioned above.

When selecting N.C.Os and men for employment at the Stokes Bay workshops preference was given, as far as practicable, to skilled tradesmen who were already members of the band, or who could play musical instruments. This enabled the band to remain intact at the Depot throughout the War (under the direction of Bandmaster F. W. Richley). In choosing the original personnel required to relieve the 4th Company R.E. at the workshops, the candidates were subjected to trade tests by Captain Prichard who afterwards described some of them as being " artists " at their work. Among those who were regarded as best qualified for the class of work required were several men who, in civil life, were colliery millwrights. Owing to the difficulty in obtaining from the makers spare parts required urgently for the maintenance of plants it often became necessary to make them at Stokes Bay and, amongst the finer work of this description which was carried out by the personnel of the Tyne Electrical Engineers, was the making of patterns and the casting and machining of parts for the vapourisers of internal combustion engines.

On the 4th February 1915, Lieut.-Colonel Scott left the Unit on being transferred to the Royal Marines, Submarine Miners, and was succeeded in the Command of the Tyne Electrical Engineers by Lieut.-Colonel Toomer, Lieut.-Colonel Robinson becoming O.C. of the Detachment at Haslar Barracks.

In the same month a new establishment was given to the Unit for Portsmouth amounting to **31** officers and **431** other ranks, based on the actual duties carried out there. The distribution of this personnel at this time was as follows :—

Coast defence electric lights ...	12 officers	169	other ranks.[3]
Anti-aircraft lights	12 ,,	114	,,
Telephones	1 ,,	30	,,
Administrative details... ...	3 ,,	35	,,
Stokes Bay	3 ,,	83	,,
Total... ...	**31** officers	**431**	other ranks.

---

[1] Subsequently appointed O.B.E. (Military Division) for valuable services throughout the War.
[2] Born 10/11/83. First Commission 21/12/01, Lieut.-Colonel 12/8/28. Retired 1/4/29. D.S.O.
[3] Including 4 at Weymouth.

LIEUTENANT-COLONEL C. R. TOOMER, V.D.
Commanding Officer, 1915–1918.

To face page 80]

Late in July, volunteers were called for for service with oxy-acetylene searchlights at the Front and on August 20th a draft of sixty corporals, second corporals and sappers left Haslar Barracks in charge of Lieutenant N. H. Firmin and proceeded to a concentration camp at Southampton. Here they were joined by a further draft of twelve men from Clifford's Fort and a detachment of thirty-nine London Electrical Engineers.

On Tuesday, August 22nd, orders were received to proceed overseas forthwith. While the detachments were formed up ready to move off a diversion was caused by a man of a labour battalion who had got into difficulties in a lake some five hundred yards away. Lieutenant Firmin hurried with some of the men to the scene and Corporal Ireland, arriving first, entered the water in full marching order and succeeded in rescuing the man. Without changing his wet clothing this N.C.O. then rejoined the draft which marched off immediately to the docks for embarkation.

The detachment reached Rouen the following day and here they parted company with Lieutenant Firmin who was ordered to return to England.

Three weeks were spent in Rouen in various occupations, including the construction of rifle ranges, after which the draft was divided up into detachments of thirteen, each comprising 2 corporals, 2 second-corporals, 8 sappers and 1 driver. The whole entrained together and reached Abbeville where each detachment was detailed to a field company R.E. for duty. On arrival at their destination the oxy-acetylene searchlight plant was issued and a day spent in mastering the operation and management of the somewhat delicate apparatus. This consisted of a 14″ projector of the type used for theatrical work with cylinders of oxygen and of dissolved acetylene. All the components were carried in strong wooden cases, and, for road transport, a light spring wagon drawn by two mules was provided. The actual production of the light was a complicated affair involving the most careful preparation beforehand of plumbago crucibles filled with a rare earth rammed and baked to a hard smooth surface. A great deal of trouble was experienced with this process which was much aggravated by the practical impossibility of preserving the required dryness under field conditions.

These lights were intended for the detection of raiding parties crossing " no man's land " during the hours of darkness, but, with the weak beam produced, it was only possible to pick up and observe a man at a range of three hundred yards. Accordingly it was necessary to instal the plant as far forward as possible and this was done in most cases in sap heads forward of our own barbed wire. The carrying of the bulky equipment along the narrow trenches proved a very arduous and dangerous task. The crew of each light consisted of 1 corporal, 1 second-corporal and 4 sappers. The lights were actually used with some success on many occasions during the ensuing nine or ten months, but the isolated nature of the work introduced considerable risks and the light detachments had always to be covered by specially sited machine guns, while

81

exposing the light at all called forth immediate bursts of fire from the enemy lines. So dangerous indeed was this work regarded, that these detachments of Tyne and London Electrical Engineers earned the universal sobriquet of "the suicide brigade" and the service was ultimately discontinued. While it lasted, however, the detachments were always attached to whichever field companies were in the line, with the result that it was not uncommon for them to remain in the front line for periods of twenty-eight days on end, to be followed by only four days rest before going up once more.

When the use of the oxy-acetylene lights was eventually given up, the personnel remained for a time with the field companies on general field engineering work until, by degrees, they drifted away, some to tunnelling companies, others to the electrical and mechanical companies, while the bulk of the men were ultimately collected and returned to England to join the various anti-aircraft searchlight companies and sections of the Tyne and London Electrical Engineers.

In October 1915 authority was obtained for the appointment of a quartermaster to the Depot at Haslar in addition to the quartermaster at the Headquarters at the Tyne, and the first to fill this post was Hon. Lieutenant A. J. Sergeant who had been with the Unit in the capacity of permanant staff instructor since 1910 and with the Haslar detachment since the outbreak of the War.

On the 9th November a company, which was styled No. 5 Company, Tyne Electrical Engineers, consisting of 4 officers and 100 men, was dispatched at 24 hours' notice from Haslar for service in the Scottish coast defences where their services were urgently required on account of a deficiency in personnel in the local units. This Company was under the command of Major Arnison with Captain J. Young and Lieutenants C. Russell and H. O. Rogerson as company officers. At Edinburgh the Company was split up, Captain Young, with Acting Company-Serjeant-Major J. Chambers and thirty men, proceeding to Ardeer near Irvine, Ayrshire, on the Clyde, to man the newly prepared defences for Nobel's large explosive works situated at that place; the remainder of the Company reinforced personnel of the City of Edinburgh Fortress R.E. (T.F.) in the defences of the Firth of Forth, the importance of which lay in the presence of the great naval dockyard at Rosyth.

Major Arnison remained at North Queensferry in command of the two detachments, paying fortnightly visits of inspection to Ardeer. The Tyne Electrical Engineers assisted to man the lights at North Queensferry and on the Island of Inch-Garvie at the Forth Bridge and also dispatched a further detachment of about twenty men by sea to Dalmeny and Hound Point on the south bank of the Firth of Forth. This latter detachment was under the command of Lieutenant Russell and was attached to the 34th Company R.E. at that time under Lieutenant J. M. McSweeney (Coast Bt. R.E.).

At Ardeer the lights were installed in the open, their operation being unduly complicated by the necessity for extreme measures of caution to avoid any possibility of explosion. These measures were not confined to the operation of the plant, for it is recorded that it was forbidden even to kick a football about the purlieus of the factory! There was, however, in the works, an excellent miniature rifle club of which the whole of the Ardeer detachment were made honorary members, a privilege of which they made full use.

The Dalmeny detachment was accommodated in hutments concealed from view from seawards by the sand dunes reinforced by sandbags. The plant at this station consisted of three coast defence projectors throwing dispersed beams and operated from Scripps-Booth 4-cylinder petrol-paraffin high-speed engines. This equipment could be started up and the arcs struck within thirty seconds of receiving the order to expose. A peculiarity about these engines was that they had been designed for motor boat work and were intended to be operated when lying at an angle towards the stern of the boat, and the lubrication system did not function properly when they were installed on horizontal beds as they were in the engine room at Hound Point. This trouble was first detected by Corporal J. A. Metcalf[1] and finally cured by him by cutting holes in the walls of the sump chamber, thus allowing the oil to keep one level.

The distribution of No. 5 Company at this time was as follows :—

Forth	...C.D. lights ...	...	2 officers	65 other ranks.	
	Admin. details	...		4 other ranks.	
Ardeer	...C.D. lights ...	}	1 officer	{	12 other ranks.
	A.A. lights...				12 other ranks.
	Admin. details	...	1 officer	7 other ranks.	
	Total...	...	4 officers	100 other ranks.	

On the 23rd December Major Arnison and Lieut. Russell with Serjeant Wheatley and some forty other ranks from the Queensferry and Inch-Garvie detachments moved to Kinghorn, Fife, with the object of manning the newly authorised defence lights at that place. On arrival it was found that although the emplacements had been constructed there was no plant, and work had not even been commenced on the engine room. Accordingly, this detachment set to work to assist the contractor with the constructional work which consisted, at that stage, largely of blasting rock! The work here was not completed and the searchlights had not been installed by the time the Tyne Electrical Engineers' personnel was withdrawn in June of 1916. In the meanwhile, on 2nd April 1916, warning had been received that German Zeppelins had set out with the object of bombing Edinburgh and the naval establishments on the Forth. There were at that time no anti-aircraft defences at either of these places and preparations were made for the available searchlights to be brought from their

---

[1] Subsequently commissioned in the T.E.E.

coast defence emplacements for use against the Zeppelins. This was not actually done, as, although the airships L14 and L22 passed over the Forth (within reach of defenceless Kinghorn) they never came within range of the Forth Bridge defences and the only reply to their bombs over Edinburgh was the rifle fire of an infantry unit quartered in the Castle.

In June, subsequent to the doubling of all Territorial Force establishments, the works at Ardeer and on the Forth were handed over entirely to personnel of the local units and No. 5 Company, Tyne Electrical Engineers, returned to Haslar after an absence of barely eight months.

In the same month that the detachments for Scotland were originally formed, that is to say in November 1915, a contingent of Tyne Electrical Engineers personnel for overseas had proceeded from Haslar Barracks to France, forming, together with a detachment from Clifford's Fort, half of an electrical and mechanical company, the other half being furnished by the London Electrical Engineers. The subsequent history of this Company, which was styled No. 1 (London and Tyne) Electrical and Mechanical Company R.E. (T.F.), is told in Chapter V.

In order to provide personnel for these purposes it was found necessary to relieve the detachments on the Sea Forts, their places being taken, from the 9th November, by members of the Hampshire Fortress R.E. (T.F.), who, at this date, had a total of 39 officers and 604 other ranks employed in coast defence and anti-aircraft searchlight duties in the Portsmouth Garrison.

Anti-aircraft searchlights were now beginning to be required for the defence of London and other localities at home in addition to the naval ports, and, to man them, A.A. searchlight companies were raised as part of the Territorial Force. Early in December 1915 the War Office telephoned urgently to the Depot to know whether any details could be spared for this service, and were agreeably surprised to find that a considerable number of men had been trained and were immediately available. These included a large proportion of men who had been recruited and trained in excess of establishment—the total number amounting to some hundreds. The first draft left Haslar for London on the 13th December under the command of Captain C. M. Forster, and formed the nucleus of a new Company under the title of No. 2 (Tyne) Searchlight Company.[1] By Christmas, 1915, further drafts had brought this company up to a total strength of 175 of all ranks.

Further companies were dispatched at intervals to London, Hull and to other districts where repeated Zeppelin raids were causing urgent demands for air defences.

In this way, a large number of searchlight companies and detachments came to be formed for A.A. defence service in different parts of the country, approximately 300 A.A. searchlights being manned by personnel of the Tyne Electrical Engineers.

[1] No. 1 had been formed by the London Electrical Engineers.

OFFICERS AT HASLAR, FEBRUARY, 1916.

Back Row, Left to Right—Lieut. H. Sherlock, 2nd Lieut. H. G. White, Lieut. J. R. Abbott, 2nd Lieut. J. Robinson, 2nd Lieut. C. B. Williamson, 2nd Lieut. H. G. Campbell, 2nd Lieut. E. V. Baldwin, 2nd Lieut. J. B. Murray.

Third Row, Left to Right—2nd Lieut. W. G. Edwards, 2nd Lieut. J. Lawther, 2nd Lieut. R. P. Findlay, 2nd Lieut. C. G. Huntley, 2nd Lieut. E. F. Rendell, 2nd Lieut. J. A. Power, 2nd Lieut. R. P. Winter, Lieut. C. F. Scott, Lieut. A. B. Williamson, Lieut. D. Myles.

Second Row, Left to Right—Lieut. A. J. Sergeant (Q.M.), Major N. H. Firmin, Lieut.-Col. E. Robinson, Lieut.-Col. C. R. Toomer, V.D. (Commanding Officer), Major O. M. Short, Captain R. Sharp, Captain D. A. Williamson.

First Row, Left to Right—2nd Lieut. C. B. S. Micklam, 2nd Lieut. W. Fox, 2nd Lieut. E. Harrison.

OFFICERS AT HASLAR, FEBRUARY, 1916.

Left to Right—Captain R. Sharp, 2nd Lieut. C. B. Williamson, Lieut. H. Sherlock, 2nd Lieut. R. P. Winter, 2nd Lieut. E. V. Baldwin, Lieut. C. F. Scott, 2nd Lieut. E. F. Rendell, Lieut. D. Myles, 2nd Lieut. J. Lawther, 2nd Lieut. R. P. Findlay, Lieut. A. B. Williamson, Captain D. A. Williamson.

To face page 84]

A more detailed account of the organisation and operation of these companies is to be found in Chapter VI. At the end of the year 1915, the 4th (F.) Company R.E. was converted into an advanced park company, and, under Major Garwood, left for service with the British troops at Salonica and the whole of the R.E. accommodation and equipment at Gosport was then taken over by the Tyne Electrical Engineers.

On the War Office re-assuming, on the 10th February 1916, all responsibility for A.A. defence at home, a number of naval ratings of the R.N.V.R. who had been engaged on A.A. searchlight work were transferred to the Tyne Electrical Engineers.

In May, 1916, the first A.A. searchlight section was formed at Haslar for duty abroad and further sections proceeded overseas at intervals until the end of the War. A fuller account of these sections is given in Chapter VII.

Owing to the constant demands for A.A. defence from many parts of the Country, the work allotted to the Unit increased rapidly. It is not easy accurately to trace the successive increases in strength, but on June 1st a general order was issued doubling all T.F. establishments to produce what became known as the " Second-Line " formations of the Territorial Force. The strength of the Tyne Electrical Engineers, including those at the Tyne, at Haslar and the various detachments referred to above, then amounted to a total of 60 officers and some 1,200 other ranks.

The distribution of the Tyne Electrical Engineers subsequent to this re-arrangement was as follows :—

No. 1 (Depot) Company ... ... North Shields.
No. 2 (Tyne) Searchlight Company... London.
No. 3 (E.L.) Company ... ... Tyne Garrison.
No. 4 (A.A.) Company ... ... Newcastle-upon-Tyne.
No. 5 (E.L.) Company ... ... Forth and Clyde Defences.
Nos. 6-11 (Depot) Companies ... Haslar.

There also existed at this date the undermentioned units of Tyne Electrical Engineers which are referred to in greater detail in this or subsequent chapters :—

No. 1 (London and Tyne) Electrical and
    Mechanical Company R.E. (T.F.) ... France.
Oxy-Acetylene Searchlight Detachments ... France.
No. 9 (Tyne) Mobile Searchlight Company... London (H.Q.).
No. 22 (Tyne) Aeroplane Searchlight Section Hornchurch (H.Q.).
No. 33 (Tyne) Anti-Aircraft Company ... Cramlington (H.Q.).

To cope with this great expansion the school at Stokes Bay was kept working at very high pressure, providing a regular flow of trained men who were drafted to various parts of the Country as ordered by the War Office and afterwards by G.H.Q. Home Forces. At the peak, instruction was being given at Stokes Bay during 22 hours of each day (including Sundays).

By the autumn of 1916, approximately 2,000 officers and other ranks had been trained at the school in A.A. and coast defence duties and drafted to various stations at home and overseas.

On the 25th September 1916 the only attack on Portsmouth during the War took place from the air. " In connection with a submarine inroad into the Channel, several Zeppelins were out on this night in the southern part of the North Sea and the Straits of Dover ; of these, only Mathy in L31 came over land. He steered a novel course as far as the Isle of Wight. He then turned north, striking the Hampshire coast about Lee on Solent, and hovered over Portsmouth harbour for some twenty minutes, throwing bright flares. He was first picked up by the A.A. searchlight at Fort Monckton and the guns of the fortress engaged him, though mostly considerably out-ranged. For some reason, never explained, he dropped no bombs. It is just possible that the electric bomb release, which is known to have given trouble in the airships about this time, may have failed to act."[1]

In order to reduce the risk of casualties from bombs falling on crowded quarters, the general scheme for action, in the event of air raid warning, provided for the immediate evacuation of barracks and the scattering of troops over the nearest available open spaces. Accordingly, on this occasion, the men at the Depot were marched on to the adjacent golf course and instructed to lie down until the raid was over.

Mathy passed along the middle of Sussex and finally out to sea by Dungeness dropping bombs at some shipping off Dover. He then went off eastwards, on what proved to be his last journey home, as, on the night of October 1st, he lost his life in the destruction of his ship at Potters Bar.

At the time of this raid several of the anti-aircraft searchlights, including that at Fort Monckton, were still manned by the Tyne Electrical Engineers, but shortly afterwards they were handed over to the Hampshire Fortress R.E. (T.F.), when all the anti-aircraft searchlights in the Portsmouth Garrison were grouped together to form No. 48 (Hants) A.A. Company R.E. (T.F.).

In October 1916, G.H.Q. Home Forces decided to take over the E.L. school at Stokes Bay for the training of personnel from various units throughout the Country who were required for A.A. searchlight duties at home or overseas. This resulted in the Tyne Electrical Engineers regimental school for technical instruction having to be transferred elsewhere, and a depot training centre was accordingly established at Monckton hutments early in 1917 under the command of Captain E. H. E. Woodward, M.C. The staff, by 1918, consisted of 5 officers, namely Lieutenants E. B. Preston, C. B. S. Micklam, Second-Lieutenants J. N. Robertson, W. H. Adams, H. C. Barry, and 40 other ranks, but, in the absence of any recognised establishment, the numbers fluctuated according to the training requirements for the time being.

---

[1] " Air Defence "—Major-General E. B. Ashmore (Page 25).

TYNE ELECTRICAL ENGINEERS' DEPOT, HASLAR.
TECHNICAL INSTRUCTIONAL STAFF, 1918.

Officers, Left to Right—2nd Lieut. H. C. Barry, 2nd Lieut. J. N. Robertson, Lieut. E. B. Preston, Captain E. H. E. Woodward, M.C., Lieut. C. B. S. Micklam, 2nd Lieut. W. H. Adams.

Owing to limitations of time the instruction at this training centre was confined to giving the men a good grounding in the underlying principles of the technical side of anti-aircraft and coast defence searchlight work, the routine and tactical part of the subject being left to be acquired later at war stations.

Every man, on joining, was selected for instruction either as an electrician or an engine driver. They were then formed into classes of 15 and proceeded at once with a preliminary course of a fortnight's duration. An N.C.O. instructor was attached to each class which he took right through the complete course. At the end of the preliminary course, the men were again divided for specialised training either in anti-aircraft or coast defence duties.

The anti-aircraft and coast defence courses which followed generally lasted one month and consisted of lectures and practical work with various types of projectors, lamps and generating plant. In the case of anti-aircraft personnel, a certain amount of instruction was also given in A.A. drill.

Anti-aircraft night practices took place, and, in the absence of actual aircraft for co-operation, Coates fire balloons were utilised for target purposes and were found to fulfil the elementary requirements of beginners' classes very well. These balloons, which frequently travelled to considerable distances, had attached to them a label instructing the finder to return them to the Depot, and offering a reward of one-and-sixpence if he did so.

At the conclusion of these courses examinations were held, and the men were required to take standard tests which were laid down with a view to their qualifying for the different rates of engineer pay at that time in force.

As far as possible, it was endeavoured to avoid sending any men out on stations unless they were able to qualify for at least a " proficient " rating as an engine driver or electrician, and most of the personnel, including all who had been trained in anti-aircraft work, were to some extent qualified for both duties. At times, however, the call for men became so urgent that this was not possible.

The plant available for instruction consisted of eight anti-aircraft sets of various makes and types, including Tilling-Stevens and Dennis Steven petrol-electric lorries which were at that time being used to equip the anti-aircraft sections proceeding overseas. The stationary plant consisted of Coventry-Simplex, Rover, Austin, Keighley, Dorman and Barlow high-speed engines, as well as the Tangye and Hornsby-Akroyd oil engines and steam plant installed in the coast defences. Projectors, both 90 cms. and 120 cms., of such diversity of manufacture as London Electric Company, Imperial Light Company, Siemens Ltd. and Cromptons, comprised the instructional equipment.

The school at Stokes Bay, on being taken over by G.H.Q., was staffed by Territorial personnel selected from various units and seconded to the Royal Engineers for this special employment. Those selected from the Tyne Electrical

Engineers were :—Captain Monkhouse,[1] who was subsequently promoted temporary Lieutenant-Colonel (Commandant)[2]. Captains H. G. White, T. S. Rea and A. Woods ; Lieutenants W. A. Gladwin, F. W. Bond and W. Dixon (Instructors). Captain H. Sherlock was appointed adjutant, whilst the subordinate staff included a large proportion of N.C.Os from the Tyne Electrical Engineers with R. Jordan as regimental-serjeant-major, Staff-Serjeant J. A. Metcalf as mechanist machinery instructor and Serjeant Claremont in charge of the model operation room. The first course assembled here on the 4th November 1916. Shortly after the establishment of this school its title was changed to the " R.E. School of Electric Lighting," and the officers on the school staff were authorised to wear blue forage caps with khaki covers, the N.C.Os and men posted to the staff being distinguished by special shoulder titles bearing the letters " S. of E.L."

Subsequently the school became known as the " Anti-Aircraft Searchlight and Sound Locator School " and was eventually moved to Ryde, I.O.W.

Largely due to the efforts of Lieutenant-Colonel Monkhouse, in conjunction with the staff of a branch experimental station established near the school by the Air Inventions Department of the Ministry of Munitions, the first sound locators were designed, constructed and tried out before being adopted for service use.

The first instrument was designed at Stokes Bay and made in the R.E. workshops early in 1916 by Company Serjeant-Major A. H. Chard, of the Tyne Electrical Engineers. It consisted of one single cone-shaped wooden trumpet tapering from a diameter of about 4 ft. to a small orifice to which a rubber listening-tube was attached. The trumpet was fixed to a swivel mounting on a tripod stand, and the results obtained were sufficiently encouraging to warrant further investigations being made. These led to the construction of a pair of twin trumpets tapering from a diameter of about 2 ft. These were mounted one on each side of a projector with their openings pointing along the direction of the searchlight beam, and were fixed so that they formed one unit with the projector itself, so that when the projector was elevated or depressed, and traversed, the trumpets moved with it.

A pair of stethoscopes was attached to the trumpets and used by the projector controller. This experiment was an improvement on the single trumpet but, of course, was limited in its success as it made no allowance for time lag and refraction from the source of sound. It was, therefore, appreciated that without introducing complicated mountings the sound trumpets could not be satisfactorily fixed to the projector in such as way as to allow them the amount of independent movement that was required, and it was decided to fix the trumpets on separate mountings away from the projector. This had the additional advantage of reducing interference to the listener caused by hissing

[1] Captain Monkhouse was succeeded as I.R.E.M. by Captain (subsequently Major) F. D. Pyne, who was appointed Chief Electrical Engineer, Southern Command, in March 1918.
[2] Promoted Brevet-Major 3/6/18.

arcs, but, on the other hand, it introduced the necessity of increased personnel for manning purposes. It was soon found that two pairs of trumpets gave better results, one pair being capable of movement in a horizontal plane and the other in a vertical plane, and the design of the first sound locator which was constructed on these lines, as the outcome of investigation by the R.E. and Air Invention Department Committee (of which Colonel Monkhouse was a member), was very much the same as that of the pattern ultimately adopted for service use. A bead-sight and ring-sight, fixed on a parallelogram mounting, was developed at the school to allow for the time lag of sound. The idea of fixing the sights on a parallelogram mounting, so as to raise them to a convenient height for men of different stature, eminated from Captain G. E. Scholes (Lancashire (Fortress) R.E. (T.F.)) who was an instructor at the school. He eventually received a monetary award from the Government in respect of his design. The foresight of the sound locator consisted of a number of concentric rings of various diameters, and the bead of the back-sight (which was fixed) was sighted on to the " leading edge " of whichever ring of the fore-sight was appropriate for the speed of the target. This arrangement was subsequently superseded by the finally adopted and simpler method of having only one ring on the fore-sight and the bead-sight mounting constructed in such a way as to enable it to be moved to and from the ring-sight to suit the target speed. The additional personnel for the sound locator consisted of two listening numbers and a sighting number, the latter being provided with a one-way telephone to enable him to give directions to the searchlight controller.[1]

The Unit continued to expand rapidly. By April 1917 the Tyne Electrical Engineers consisted of 24 companies and 10 sections of which Nos. 5 to 11 Companies were stationed at Haslar. These were Depot companies of which the personnel was constantly changing, and, although every officer and man on joining was posted to one or other of them, they became little more than a convenient organisation for administration and pay. New officers were being commissioned into the Unit and recruits and untrained men arriving in considerable numbers so that further accommodation was required. Forts Monckton, Gilkicker and Gomer were made available for this purpose and, eventually, large hutted camps erected at Fort Gomer and on Monckton Common were taken over.

The widespread distribution of personnel which the organisation of the Tyne Electrical Engineers then involved increased the importance of the administrative headquarters at Haslar Barracks to such an extent that it now completely eclipsed that of the coast defence electric light duties for which the Detachment had originally been formed. As the manning of the anti-aircraft defences throughout the Country was a matter of extreme urgency in view of the repeated air raids, arrangements were made whereby men could be posted to the various defences immediately upon their becoming available.

---

[1] Great enthusiasm was shown by the classes attending the School and the tale is told of one young officer, more zealous than the rest to acquire the new art of detecting aircraft by night, who was discovered in a darkened room with a pocket flash-lamp diligently searching for a banana suspended from a fishing rod which he manipulated with his other hand.

This necessitated the suspension of the normal routine and it became the recognised practice for men to be dispatched to their stations considerably in advance of the authority for their posting being received from the R.E. Records Office at Chatham. The formation of such a multitude of small technical units also created an unprecedented demand for military mechanists (both electrical and machinery), and one of the great difficulties of the Depot was the provision, in sufficient numbers, of men qualified for this duty. A rough estimate places the number of these mechanists supplied by the Tyne Electrical Engineers for their own and other units up to the beginning of 1917 as not far short of two hundred. The difficulty of finding these men was aggravated in the first eighteen months of the War by the loss of a number of excellent mechanists who had been appointed to commissions to meet the great demand for technically qualified officers. As in the case of the personnel for No. 2 (Tyne) Searchlight Company in December 1915 however, the demand was foreseen by those responsible at the Depot and there was never a failure to supply men either for these or for the many other specialist duties which were constantly being demanded both for abroad and at home. In many instances, the Tyne Electrical Engineers, alone of all the various depots for R.E. personnel which were at that time established throughout the Country, were able to supply men of suitable qualifications for the many and widely differing requirements of the War Office. In the course of time, this unfailing readiness to assist in whatever direction might be required and the promptness with which personnel was invariably made available by the Depot for any specified purpose came to be recognized by the War Office as characteristic of the Tyne Electrical Engineers and added materially to the reputation which the Unit had established for itself in its earlier history. The work of collecting and dispatching these drafts, sometimes at very short notice, imposed a heavy strain on the officers and clerical staff of the Depot headquarters. In this connection it must be remembered that there never was a proper establishment providing adequate headquarter personnel for the Depot, and although applications and recommendations were continually being submitted for the authorisation of the necessary scales, these had not finally been approved even by the end of the War. It was not until January 1st, 1917, that the Depot officially had an adjutant, Captain R. Sharp[1] being appointed on that date and continuing to act until after the armistice.[2]

The Officer Commanding the Depot, Lieut.-Colonel E. Robinson,[3] to whose ability for organisation and administration the Unit was mainly indebted for its extraordinary expansion, was frequently in London attending conferences and on other duties of an urgent nature and the bulk of the executive work thus fell on the shoulders of Major O. M. Short.[4] Such was the pressure that for months on end the officers of the depot headquarters found it impossible to

---

[1] Appointed O.B.E. (Military Division) 3/6/19.
[2] On the 6th July 1917, C.Q.M.S. C. W. Rood, who up to that date had been head of the clerical staff at Haslar, was commissioned in the T.E.E. and appointed to act as assistant adjutant.
[3] Appointed O.B.E. (Military Division) 3/6/19.
[4] Appointed O.B.E. (Military Division) 6/1/19.

take any recreation and, on occasion, worked all through the night. They were at all times loyally and ably assisted by the quartermaster, Lieutenant A. J. Sergeant[1] and by the warrant and non-commissioned officers and subordinate personnel working with them. Of these R.Q.M.S. Anderson and R.S.M. I. T. Womphrey are deserving of special mention. The latter combined in a rare manner his very humane characteristics with those of a strict disciplinarian. It has been said that he sat each evening in his office where it was known that any man to the last-joined bugler might visit him and lay before him any matter whatsoever, and that he was always ready with sympathetic assistance to a man in genuine trouble while reserving for the malingerer the sharp treatment he deserved. It should, however, be mentioned that this course seldom became necessary, and it is satisfactory to record that, in spite of the large and shifting population of the Depot, instances of crime were unusually rare.

In November 1917 the air defences of Great Britain were re-organised. The A.A. searchlight companies R.E. were considerably reduced in number throughout the Country and the remaining personnel was thereafter attached to the A.A. companies R.G.A. with whom they worked. Under a scheme evolved at Haslar Barracks the N.C.Os and men were attached to the R.A. for rations, accommodation and discipline, but were paid weekly from the Depot at Haslar Barracks where the pay handled in a week (averaged over a period of six months) amounted in 1918 to more than five thousand pounds. The financial work at the Depot was then divided into two separate departments, the one, under Captain Sharp, dealing with the personnel at Haslar, while Captain D. A. Williamson was responsible for the pay of the personnel attached to the R.G.A.

Every opportunity was taken to provide sport for the men at the Depot, and, with the large numbers from which to select teams, the Tyne Electrical Engineers, under the direction of Captain (subsequently Major) F. D. Pyne, were enabled to compete with success in local Leagues and matches in the garrison. The Unit contained several men who, in times of peace, had played for famous civilian clubs and these proved of great assistance in training the teams which represented the Tyne Electrical Engineers at both rugby and association football, cricket, hockey and boxing. Among these may be mentioned W. Low (later captain of the Newcastle United Football Club and centre half back for Scotland), W. Moore (full back for Sunderland), G. Elliott (centre forward for Middlesbrough and England), J. Robertson (Cardiff City) and S. Dixon (Newcastle United).

In the Southern Command association football competition of 1917 they were notably successful, winning their way into the final round where they were defeated by one goal to nil by the Royal Garrison Artillery.

---

[1] Appointed M.B.E. (Military Division) 3/7/19.

In order to encourage friendly rivalry in sport between the London and Tyne Electrical Engineers, Lieut.-Colonel A. E. Le Rossignol and Lieut.-Colonel Toomer jointly presented a silver mug to be competed for by the two Units at football and cricket. This trophy was four times contested, twice at each game, and on each occasion when a decision was reached was won by the Tyne Electrical Engineers.[1]

On the 30th August 1918 Lieut.-Colonel Toomer was recalled to his civil occupation in the ship repairing industry, which was then of great national importance, and was succeeded in the command by Lieut.-Colonel Robinson.

In the same month, with a view to effecting a saving of personnel, many of the fortress units of the T.F. were reduced and a proportion of the officers and other ranks were attached to the depot of the Tyne Electrical Engineers. Later, the London Electrical Engineers and the Tyne Electrical Engineers were constituted the parent units of all A.A. and C.D.E.L. R.E. personnel in the service; but the depots of the various other units had not been entirely closed down before the armistice put an end to hostilities. The Tyne Electrical Engineers, however, became responsible for all personnel required for anti-aircraft and coast defence electric lighting in the Scottish, Northern and Western Commands and in the Portsmouth Garrison.

Among the officers who were attached to the Depot at Gosport as a result of this were Major W. Pearce, of the Hampshire (Fortress) R.E. (T.F.), who was appointed instructor in infantry drill and military duties to the Depot (in which he was assisted by Lieutenant P. E. Newman (T.E.E.)) and Captain A. Reid of the Renfrew (Fortress) R.E. (T.F.), who was put in charge of a large emergency camp near Monckton hutments and who was later employed as demobilisation officer.

When the demand for man-power was at its height, various women's organisations were formed and a large contingent of the Women's Auxiliary Army Corps, under Miss Hay and Miss MacDonald, was attached to the Depot and executed various administrative duties as clerks and mess waitresses, whilst the Women's Legion provided M.T. drivers, etc., thus releasing men for more active duties at home and overseas.

At one period, shortly before the armistice, owing to the exceptionally heavy casualties suffered by the infantry in France, a large draft was dispatched from Fort Gomer intended for transfer to that arm, but were re-appropriated by the Inspector of Searchlights on arrival overseas. By November 1918, the Unit numbered no less than 143 officers and approximately 5,000 other ranks, of whom upwards of 50 officers and 700 other ranks were serving overseas in

---

[1] 13th February 1918 won by T.E.E.   Football   5 – 0.
    17th April   1918   „   „   Football   6 – 5.
    16th July   1918   „   „   Cricket   113 – 61.
    21st August   1918   „   „   Cricket   201 for 4.
                         Match drawn.   L.E.E.   70 for 8.

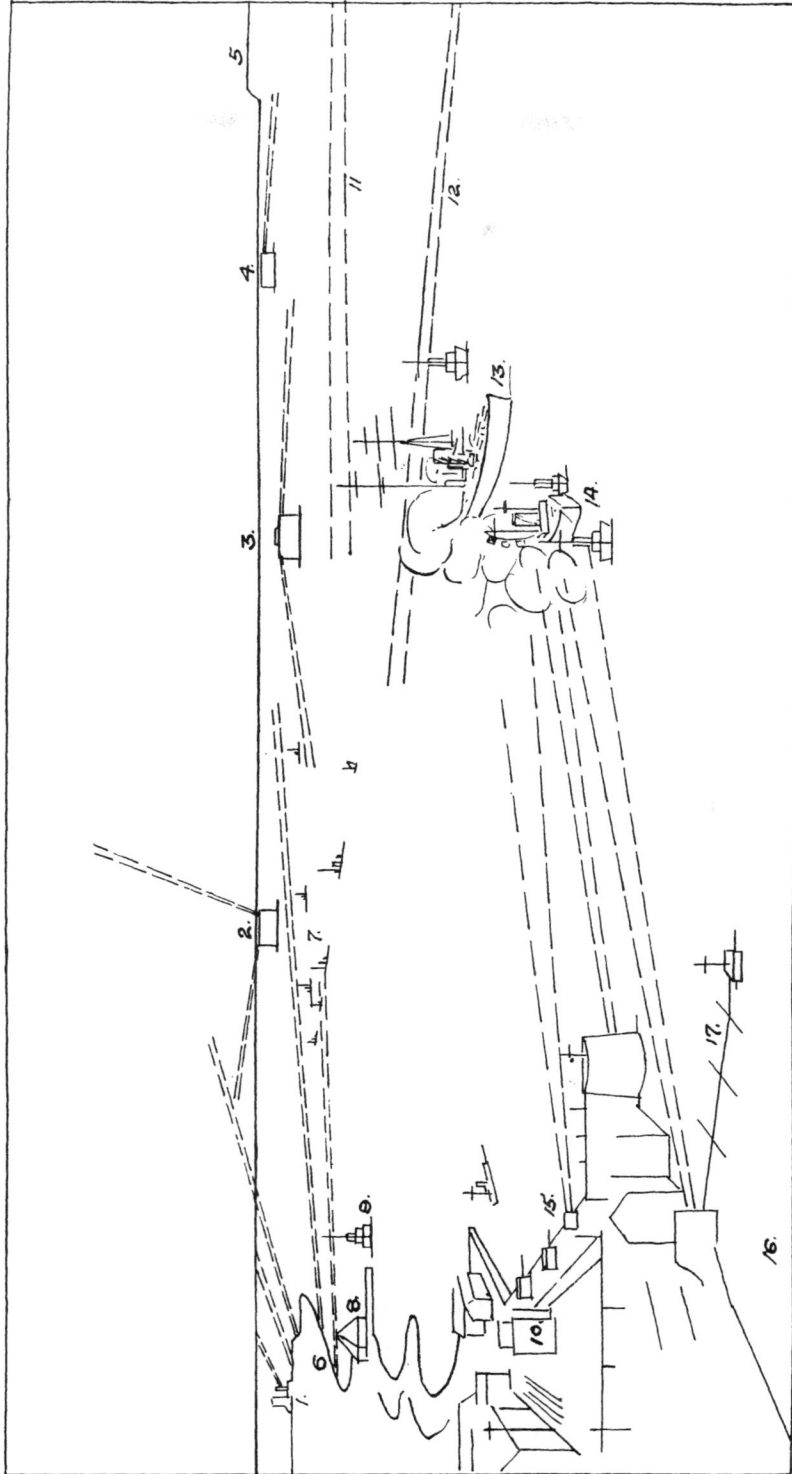

KEY-PLAN

to Painting by W. L. WYLLIE, R.A.

"SEARCHLIGHTS AT PORTSMOUTH"

Painted in 1920. The original was presented to Lt.-Col. C. R. Toomer, V.D., by the officers of the Tyne Electrical Engineers on his relinquishing command of the Unit.

KEY.

1. Southsea Castle.
2. Horse Sand Fort.
3. Spit Bank Fort.
4. No Man's Land Fort.
5. Isle of Wight.

6. Esplanade, Southsea.
7. Night destroyer patrol going out.
8. Clarence Pier, Southsea.
9. Isle of Wight packet leaving Clarence Pier.

10. Square Tower.
11. Searchlight beam from Fort Monckton.
12. Searchlight beam from Haslar.
13. H.M.S. "Repulse" entering harbour.

14. Merchantman entering harbour.
15. Point Battery.
16. Entrance to Portsmouth Harbour.
17. Inner boom.

To face page 92]

anti-aircraft or other specialist R.E. services. The following honours, awards, mentions etc., were received by officers, N.C.Os and men of the Tyne Electrical Engineers during the War[1] :—

O.B.E. (Military Division)	7
O.B.E. (Civil Division)	2
Military Cross	11
M.B.E. (Military Division)	6
Brevet Major	3
Air Force Cross	1
Military Medal	5
Croix de Guerre	2
Military Order of Aviz (Portugal)	1 (3rd Class).
Order of St. Stanislas (Russia)	1 (3rd Class).
Order of El Nahda	1
Meritorious Service Medal	13
	53
Mentions in the field—13 officers and 4 o.rs.	17
Mentions at home—26 officers and 48 o.rs.	74
	91

During the War, the Government urged the public to invest in war savings certificates, and, in order to support this appeal, Lieut.-Colonel Robinson instituted a Tyne Electrical Engineers' war savings association on the lines of similar associations formed by other branches of the service. Collections were made at all pay parades and at the cessation of hostilities the savings of the Tyne E.E. were the fourth highest in the British Army.[2]

A feature of the scheme was the naming of service aircraft after the various units of the army who had invested not less than £2,000 in war savings certificates. As a result of this, two aeroplanes were named after the Tyne Electrical Engineers, each in recognition of an investment of £2,000, before this scheme was discontinued. Thereafter, it became the practice for an 18 pdr. shell and case, suitably engraved, to be presented to units in respect of each further £2,000 subscribed and five of these were received by the Tyne Electrical Engineers.

---

[1] See Note at end of Chapter.
[2] Tyne E.E. War Savings Association. (War Savings and National War Bonds).

Portsmouth	...	...	...	...	£24,453 2 0
Tyne	...	...	...	...	2,891 19 0
		Total subscribed	...	...	£27,345 1 0

N.B.—Fourth in the British Army, the first three being Guards Reserve Bde., Inns of Court and Artists Rifles. First in Southern Command.

In order to contribute towards the welfare of those members of the Unit who were abroad, Lieut.-Colonel Robinson, at an early stage in the War, also inaugurated the " Tyne Electrical Engineers Comforts Fund," to which those who were serving at home were invited to subscribe at pay parades. The fund was well supported and was greatly appreciated by those who benefited from it.

Personnel at the Depot also supported the R.E. Prisoners of War and R.E. Kitchener Memorial Funds.[1]

After the signing of the armistice on the 11th November 1918 Haslar Barracks became a demobilization station for the various units of the Tyne Electrical Engineers. To that place were returned, in convoys, the petrol-electric lorries which had formed part of the equipment of the overseas sections, and from there drafts of men returned to civil life, priority in obtaining discharge depending on the occupation of each man. The Depot was thus kept working at high pressure until April 1919 by which time most of the demobilization of the officers, W.Os., N.C.Os and men at Haslar and attached to the various units, had been practically completed. By the end of that month the Tyne Electrical Engineers had evacuated Haslar and the barracks and hutments had finally been handed over to the O.C. 4th (Fortress) Company R.E. The small remaining staff, consisting chiefly of clerks, returned to the Headquarters of the Unit at Clifford's Fort, North Shields, on the 5th June 1919.

NOTE " C."

For convenience in reference, the honours awarded to members of the Tyne Electrical Engineers during the Great War, 1914-1919, are, as far as possible, tabulated below. The circumstances in which most of them were earned are referred to in more detail in various parts of the text.

ORDER OF THE BRITISH EMPIRE.

MILITARY DIVISION.

Officers—7.

1.	Lt.-Col. E. Robinson	... (Depot headquarters).
2.	Major O. M. Short	... (Depot headquarters).
3.	Major C. M. Forster	... (G.S.O., A.A., G.H.Q., G.B.).
4.	Major F. D. Pyne	... (E. & M. duties Southern Command).
5.	Bt. Major W. Sharp	... (Palestine railway).
6.	Captain R. Sharp	... (Depot headquarters).
7.	Captain W. Fox ...	... (A.A. searchlight equipment officer— France).

---

[1] Tyne E.E. Comforts Fund £677 15 6
R.E. Prisoners of War Fund 1,482 14 5
R.E. Kitchener Memorial Fund 415 10 0
£2,575 19 11

Members—6.

1. Captain K. S. M. Scott ... (A.A. searchlights—Tyne Garrison).
2. Captain J. F. S. Hunter ... (A.A. searchlights—Birmingham).
3. Captain H. S. Ripley ... (Resident Engineer H.M. fuel factory—Dumfries).
4. Lieut. E. Birrell ... ... (A.A. searchlights—Sheffield).
5. Lieut. F. T. Hamilton ... (S/Capt. F.W.8. War Office).
6. Captain (Qr.-Mr.) A. J. Sergeant ... ... (Depot headquarters).

Medal—1.

1. C.S.M. J. Scott ... ... (354th E. & M. Company).

CIVIL DIVISION.

Officers—2.

1. Major N. H. Firmin ... (A.D. Experiment and Research, Admiralty).
2. Lieut. O. W. E. Hedley ... (Donor of the " Ethel Hedley " Hospital).[1]

MILITARY CROSS.—11.

1. Major J. O. Baird ... (No. 1 (L. & T.) E. & M. Company—France).
2. Major E. F. Rendell ... (A.A. searchlights—France).
3. Major R. W. Anderson ... (Signals—France).
4. Captain E. H. E. Woodward (Infantry—France).
5. Captain R. L. Wood ... (Tunnelling company—France).
6. Captain C. D. Danby ... (R.F.C.—France).
7. Captain R. L. Hunter ... (Trench mortars—France).
8. Captain R. P. Winter ... (A.A. searchlights—Italy).
9. Captain R. P. Findlay ... (Field company—France).
10. Lieut. T. Henderson ... (R.F.C.—Arabia).
11. Lieut. F. L. Adendorff ... (Light railways—France).

AIR FORCE CROSS.—1.

1. F/Lieut. T. Henderson ... (R.A.F.—Mesopotamia).

BREVET-MAJOR.—3.

1. Captain C. M. Forster ... (G.S.O., A.A., G,H.Q., G.B.).
2. Captain S. E. Monkhouse... (A.A. Searchlight and Sound Locator School).
3. Captain W. Sharp ... (Palestine railway).

[1] Also awarded the " Medaille du Roi Albert " by the King of the Belgians for having provided and equipped another of his houses as a hospital for wounded Belgian soldiers.

MILITARY MEDAL.—5.

1. Serjeant Gair ... ... (No. 44 A.A.S.S.—France).
2. Corporal Rudge ... ... (No. 26 A.A.S.S.—France).
3. Corporal Swan ... ... (No. 8 A.A.S.S.—France).
4. Spr. H. G. Anderson ... (Infantry—France).
5. Spr. G. Bage ... ... (No. 17 A.A.S.S.—France).

MERITORIOUS SERVICE MEDAL.—13.

1. R.S.M. I. Womphrey ... (Depot headquarters).
2. R.Q.M.S. T. Anderson ... (Depot headquarters).
3. Serjt. C. Driver ... ... (No. 26 A.A.S.S.—France).
4. C.S.M. J. Scott ... ... (354th E. & M. company—France).
5. Serjt. A. Donkin ... (No. 11 A.A.S.S.—France).
6. Serjt. T. Young ... ... (No. 27 A.A.S.S.—France).

Seven others.—Names unknown.

MENTIONED IN DISPATCHES (OVERSEAS).

Officers—13.

1. Major J. O. Baird ... (E. & M. company—France).
2. Major E. F. Rendell ... (A.A. searchlights—France).
3. Major R. W. Anderson ... (Signals—France).
4. Bt. Major W. Sharp ... (Palestine Railway).
5. Captain R. L. Wood ... (Tunnelling company—France).
6. Captain T. Henderson ... (R.F.C.—France and Middle East).
7. Captain D. A. Williamson... (R.F.C.—France).
8. Captain E. V. Baldwin ... (A.A. searchlights—France).
9. Captain W. Fox ... ... (A.A. searchlights—France).
10. Captain R. O. Porter ... (A.A. searchlights—France).
11. Captain W. A. Gladwin ... (Railways—North Russia).
12. Lieut. R. P. Findlay ... (A.A. searchlights—France).
13. Lieut. R. E. Jardine ... (A.A. searchlights—France).

Other Ranks—4.

Names unknown.

FOREIGN DECORATIONS :—

CROIX DE GUERRE (FRANCE).—2.

1. Captain W. Fox ... ... (A.A. Searchlight Equipment Officer—France).
2. Lieut. D. Myles ... ... (A.A. searchlights—France).

ORDER OF ST. STANISLAS (RUSSIA), 3rd Class.—1.

1. Captain W. A. Gladwin ... (Railways—North Russia).

MILITARY ORDER OF AVIZ (PORTUGAL), 3rd Class.—1.

1. Major E. F. Rendell ... (A.A. searchlights—France).

ORDER OF EL MAHDA (HEDJAZ).—1.

1. Captain T. Henderson ... (R.F.C.—Arabia).

BROUGHT TO THE NOTICE OF THE SECRETARY OF STATE FOR VALUABLE SERVICES IN CONNECTION WITH THE WAR. (AT HOME.)

Officers—26.

1. Lt.-Col. E. Robinson.
2. Lt.-Col. A. K. Tasker.
3. Major O. M. Short.
4. Major N. H. Firmin.
5. Major F. D. Pyne.
6. Captain E. H. E. Woodward.
7. Captain K. S. M. Scott.
8. Captain R. Sharp.
9. Captain J. F. S. Hunter.
10. Captain A. B. Williamson.
11. Captain H. Sherlock.
12. Captain G. I. N. Deane.
13. Captain D. A. Williamson.
14. Captain W. H. James.
15. Captain T. S. Rea.
16. Lieut. K. A. Mountain.
17. Lieut. A. Howell.
18. Lieut. B. H. Leeson.
19. Lieut. E. B. Preston.
21. Lieut. C. W. Rood.
    Five others. Names not known.

Other Ranks—48.

1. C.S.M. A. H. Chard.
2. Sjt. (a/C.S.M.) J. Chambers.
3. Sjt. (a/C.S.M.) J. F. Melville.
4. Sjt. (a/C.S.M.) J. A. Jones.
5. Corpl. (a/C.Q.M.S.) W. Dean.
6. Corpl. W. Atchinson.
7. 2nd Corpl. (a/Sjt.) W. Fielding.
8. 2nd Corpl. (a/C.Q.M.S.) J. J. Innes.
9. Spr. (a/C.S.M.) G. Ross.
10. Spr. (a/C.Q.M.S.) E. Andrews.
11. Spr. (a/C.Q.M.S.) W. R. Davison.
12. Spr. (a/C.Q.M.S.) G. J. Fortens.
13. Spr. (a/S/Sjt.) T. W. Pattenden.
14. Spr. (a/S/Sjt.) H. Shaw.
15. Spr. (a/Sjt.) H. C. Cartwright.
16. Spr. (a/Sjt.) F. B. Hammond.
17. Spr. (a/Sjt.) H. C. Jackson.
18. Spr. (a/Sjt.) C. J. Keep.

G

Other Ranks—continued.

19. Spr. (a/Sjt.) J. Carter.
20. Spr. (a/Cpl.) A. Pratt.
21. Spr. T. A. Churcher.
22. Spr. G. W. Nash.
23. Spr. A. Petrie.
24. Spr. W. Potter.
    Twenty-four others. Names not known.

––––––––––

NOTE " D."

A certain number of officers[1] continued after the end of the War to be employed on military services other than those directly associated with the Tyne Electrical Engineers. Among these were :—

Captain T. HENDERSON, who was at that time seconded to the R.A.F., was selected, early in 1919, to organise an aerial relief expedition for the British North Russian Force which was isolated on account of the closing of the Baltic by ice. For this purpose he was posted to command No. 58 Squadron, R.A.F., which was equipped with Handley-Page twin-engined aircraft. This Squadron, however, was very shortly ordered to proceed to Egypt on the first long-distance formation flight to be attempted. Captain Henderson laid out the route and, after considerable difficulties involving two fatalities, succeeded in reaching Heliopolis with nine out of the original ten machines. The route taken was mainly that which was later adopted as the official air route to Egypt. For this enterprise, Captain Henderson was complimented and awarded the Air Force Cross. His next post was in command of the Central Air Communication Squadron which was originated for the purpose of surveying and preparing a new air route to India. Aerodromes were established all down the Persian Gulf and a route was established across the Great Northern Desert between Baghdad and Damascus. In 1920 Captain Henderson returned home sick, and on the 31st July 1923 relinquished his commission in the Tyne Electrical Engineers and was commissioned flight-lieutenant, Royal Air Force, being posted to the War Office as G.S.O. 3 to the Department of Intelligence (M.I. 4). Here he acted in the capacity of air liaison officer and was largely employed with the geographical section, being responsible for the production of the original models for air maps and charts and assisting to design the conventional signs, colouring, type and style of projection to be used in the new maps. This appointment he held for two years and six months and returned to flying duties in 1926 being posted overseas to No. 70 Squadron at Hinaidi, Baghdad. While in Iraq, Flight-Lieutenant Henderson was given command of a section of

––––––––––

[1] It will be observed that certain officers were employed in Mesopotamia after transfer to the Reserve (see Appendix II). This was due to the fact that they were not employed as " Officers " but as " Gazetted Officials."

armoured cars and, while operating on the Kurdisk Frontier, he contracted double-pneumonia and was brought back to Baghdad by aerial ambulance. After convalescing he returned to England in 1927 and resigned his commission in 1928 being medically boarded as unfit for further flying. (See also Chapter III, note " B.")

Major W. SHARP. In February 1919 the administration of the Palestine Railway, on which Major Sharp had been employed as second-in-command of the operating division, was altered, and this officer was then sent to Constantinople for duty on the Anatolian Railway which was still under Allied military control. For the next two years Major Sharp acted as port superintendent at Haidar Pasha on the Asiatic shore of the sea of Marmora, and as Assistant Chief Engineer for the controlled line. He was locally demobilized on the 30th November 1921 at Constantinople, but continued to serve with the civil administration of the railway until the 30th April 1922. (See also chapter III, note " B.")

Captain R. PHILLIPS served with the 35th (Fortress) Company R.E. at Pembroke Dock from the 5th July 1919 until the 30th April 1920.

Captain F. B. C. SUTTHERY completed his service with the Royal Marines at the end of January 1919 and was restored to the establishment of the Tyne Electrical Engineers on the 1st February. He was then seconded to the Royal Engineers on the 17th February 1919 and took up the post of Chief Electrical Engineer, Southern Command, in succession to Major F. D. Pyne. (See chapter IV.) The duties of this post covered an extensive area and included the supervision of all the power stations (which were mostly operated by civilians) in the Command. Captain Sutthery was promoted acting-major in March and was demobilized on the 15th November 1919. (See also chapter III, note " B.")

Captain W. A. GLADWIN, after leaving the A.A. Searchlight and Sound Locator School at Ryde in April 1919, joined the North Russian Expeditionary Force and was first employed as Officer i/c Electric Lighting, Murmansk. He was very shortly transferred to the transportation branch of the Royal Engineers, where his valuable mechanical engineering experience resulted in his appointment as officer i/c internal combustion engines, Syren Forces. In this post, he took over all duties connected with the pertol-driven rail-cars, lorries, workshops, stores and equipment. Working in conjunction with the Russian railway troops, this officer ultimately became responsible for the equipment of over seven hundred miles of railway, mostly operated by I.C.-engined vehicles such as Caledon lorries fitted with 40 H.P. Dorman engines. The subordinate personnel consisted largely of Russian tradesmen with a few Bolshevik prisoners and a leavening of R.A.S.C., M.T. drivers. For his services in this connection Captain Gladwin was mentioned in the dispatches of the General Officer Commanding-in-Chief (General Rawlinson) and was awarded, by the Provisional Government of Russia, the Order of St. Stanislas, 3rd Class, with crossed swords and bow. Following the evacuation of North Russia by the British forces in

October, Captain Gladwin was transferred to Mesopotamia where, on the 6th December 1919, he joined the Inland Water Transport (I.W.T.) in the post of assistant to the A.D.I.W.T. (Motors). On the 1st May 1920 he was placed in charge of the motor section of the Basrah dockyards where he remained until his demobilization in 1921.

Lieutenant A. G. DICKSON, on leaving No. 11 A.A. Searchlight Section on the Rhine in September 1919, was seconded to the Royal Engineers for duty with the E. & M. companies in Mesopotamia. In March 1920 he was sent up to North-West Persia with a coast defence searchlight detachment for service with " Norperforce " at Enzeli Harbour on the Caspian Sea. The personnel allotted for this purpose consisted of one British N.C.O. and a mixed party of Indians. The detachment was equipped with two 90 cms. searchlights which were in operation nightly from the 29th March to the 24th June, but the only attack made by the Bolsheviks was one morning at dawn so the lights were not brought into action. Following this duty, Lieutenant Dickson was officer i/c armoured searchlight trains on the Baghdad–Sherghat and Baghdad–Hillah lines from the 25th June to the 19th December 1920 during the Arab rebellion. For this service acetylene lights were at first provided but these were later converted to electric. The trains patrolled the lines every night and the lights proved very effective in scaring away Arabs who were engaged in sniping at block-houses. The lights also proved useful on one occasion in illuminating a wrecked train and sweeping the desert round about while the breakdown gang was clearing the line. The Arabs were close at hand and would undoubtedly have made the work very difficult had it not been for the searchlights.

Lieutenant E. H. WHITE, while serving at Cologne as Officer i/c Searchlights, Northern Zone, applied for service in Mesopotamia and embarked on the 31st October 1919 arriving in Baghdad on the 27th December. From there he was posted to Quiarah to superintend the working of the oil wells situated at that place and the manufacture, from the crude oil, of petrol and paraffin for mechanical transport in the Mosul—Sherghat area. After three months Lieutenant White handed over these duties and returned to Baghdad, where he was employed in the design and construction of a new pumping station on the banks of the Tigris and in laying the pipe lines into the town. While this work was in progress, the Arab rebellion of 1920 broke out and Lieutenant White was engaged, for a while, in the construction of field defences. At the end of 1920 this officer was posted as D.O.R.E. " Norperforce " lines of communication and proceeded to Kermanshah. His duties here included the maintenance, throughout 250 miles of road from railhead at Quirito, of stone-crushing plant and petrol-electric road rollers, and the supervision of various electric lighting, water-supply and ice plants. After the Bolshevik rebellion in the spring of 1921, the North Persian Force was withdrawn and all the material had to be dismantled and either sold locally or returned to Baghdad. Lieutenant White then prepared designs and estimates for the supply of water to the town of Mosul. These plans were subsequently executed by the civil authorities.

Lieutenant White next became D.O.R.E. Nasiryah until this area was handed over to Arab troops, and then returned to Basrah where he had under his charge all the electric lighting and power supplies for the town and also for the R.A.F. headquarters, about 15 miles distant. Lieutenant White finally returned to England in November 1923 after four years service in the East.

Lieutenant P. F. FAWCETT returned from Germany with Nos. 8 and 36 A.A. Searchlight Sections and was immediately posted to No. 1 Mobile Searchlight Group R.E. which was one of two groups then being formed at Hilsea for service in Ireland. A proportion of other ranks from the Tyne Electrical Engineers also found their way into these groups, which sailed from Holyhead and Liverpool on the 9th April 1920. The establishment of each group was 3 officers and 56 other ranks and the equipment consisted of 6 petrol-electric lorries with 60 cms. searchlights. The officers commanding the two groups were Captain Mathews and Captain Rogers, both of the London Electrical Engineers, and the personnel consisted of a motley assortment of Regulars and Territorials. No. 1 Group proceeded to Dublin and No. 2 to Cork. On arrival in Dublin No. 1 Group was divided into two sections, No. 1, under Lieutenant Fawcett, being posted to Dublin Castle and No. 2, under Lieutenant Chapman R.E., to the Curragh. In June Captain Mathews returned to England, and the Group was then commanded by Lieutenant Fawcett until his demobilization in December 1920. The sections were employed for escorting infantry patrols in Dublin, Wicklow and Meagh. The projectors were mounted on the lorries, each light having a machine-gun attached to it. The lights were used in this way for locating Sinn Feiners drilling in the country districts and also for illuminating the streets when raids were taking place. One of the lorries of No. 1 Section was especially equipped with six portable hand-lamps which proved valuable for illuminating cellars and the interiors of houses during raids. The detachments were in operation almost nightly and were frequently fired upon by the Sinn Feiners. The O.C. of the Group had many other duties to perform, such as officiating as adjutant to the C.R.E., acting as E. & M. officer for the garrison and technical adviser to Dublin District Headquarters, and supervising the oxy-acetylene searchlights installed for defence purposes at the Vice-Regal Lodge and Chief Secretaries' Lodge, Dublin. The mobile group also, on occasions, took charge of the power station and docks, assumed control of the unloading of all transport arriving at Dublin and Kingstown and, in one instance, carried out an emergency lighting scheme for Dublin Castle run off P.E. lorries and Lister generating sets. The Group remained in Ireland for some time after the return of Lieutenant Fawcett to England and the Territorial personnel was gradually all replaced by Regulars. The Group eventually removed to Blackdown, near Aldershot, and formed the nucleus of the 1st A.A. Searchlight Battalion, Royal Engineers.

Lieutenant W. S. WATTS was serving with the 33rd (Fortress) Coy. R.E. stationed at Cork and was kidnapped on the 15th November 1920. He is believed to have been murdered by the Sinn Feiners.

CHAPTER V.

1914—1919. THE GREAT WAR.[1]
NO. 1 (LONDON AND TYNE) E. & M. COMPANY.

Formation — Functions — Work in H.Q. and First and Second Army areas — Water supply for first battle of the Somme — Fifth Army workshops — Increase in E. and M. establishment — The Company split up — Duties of the new Companies — Work in preparation for the third battle of Ypres — The battle of Messines — Passchendaele — Attached personnel — Work in forward areas — The second battle of the Somme — " Carey's Force " — The battle of the Lys — The Companies re-unite — Final British advance — Demobilization — The Army of Occupation at Cologne — Disposals board — Disbandment.

I N 1914 the British Army, equipped for mobile warfare, was provided with no units to deal with the more permanent engineering services. The work of operating and maintaining machinery required for electric lighting, water-supply and workshops in the field was, at first, carried out by the field units of the divisions and by certain fortress companies converted into works companies for service on the lines of communications.

The lack of specialist units was not immediately felt owing to the small size of the original expeditionary force but, by 1915, the amount of machinery in the field had so increased that it was decided to form special R.E. companies to deal with it. The formation of one of these companies for France was approved in September 1915 and a second for Mesopotamia in 1916. The unit for France was actually formed on the 16th November 1915 from personnel of the Tyne and London Electrical Engineers and was designated No. 1 (London and Tyne) Electrical and Mechanical Company R.E. (T.F.).

The Tyne Electrical Engineers' complement for this Company was provided in part from the Headquarters at Clifford's Fort and in part from the Depot at Haslar. The company assembled at the Headquarters of the London Electrical Engineers, Regency Street, Westminster, on the 20th November, the men being billetted in the neighbourhood. This being the first company of the kind no " mobilisation stores table " existed and the War Office asked the Company for suggestions. The O.C. and other officers drew up an extensive table, largely based on the " vocabulary of stores," but including a good deal of other material. It was approved exactly as submitted and purchase of the non-vocabulary items direct by the Company was authorised. This accounts for the somewhat lavish equipment of small tools and stores which the Company was able to take with it to France.

---

[1] The maps at the end of the book will be found useful when reading this chapter.

Arms were drawn at the Tower of London, the whole Company being marched there from Westminster for the purpose.

The Company, 4 officers and 122 other ranks strong, under the command of Major A. E. Levin of the London Electrical Engineers, embarked at Southampton and landed at Havre on the 15th December 1915, and marched from the docks to the docks rest camp where a few days were spent under canvas in dismal surroundings and cold wet weather.

The Company was at this time composed as follows :—

Officer Commanding	...	Major A. E. Levin, A.M.I.E.E. (L.E.E.)
Second-in-Command	...	Captain G. L. Russell, A.M.I.E.E. (T.E.E.)
Section Officers	...	Captain C. Tufnell, A.M.I.E.E. (L.E.E.)
		Lieutenant F. H. Bowers (T.E.E.)
Coy.-Sjt.-Major	...	Bowmaker, E. V. (L.E.E.)[1]
Coy-Qu.-Mr.-Sjt.	...	Mewis, N. C. (L.E.E.)
Mechanists	...	Hews, C. J. (L.E.E.)[2]
		Watson, W. (L.E.E.)[3]
		Burdis, A. S. (T.E.E.)[4]
		Ruderman, L. (T.E.E.)
Section Serjeants	...	Howe, T. G. (L.E.E.)
		Hurrell, M. (L.E.E.)[5]
		Forsyth, C. H. (T.E.E.)
		Dunn, J. S. (T.E.E.)
Total R.E. other ranks	116	
A.S.C. drivers attached	6	
Total	122	

The transport with which the Company was equipped consisted at the outset of the following vehicles :—

1 Thorneycroft workshop lorry which was fitted with an electric generator supplying power for various portable tools.
1 Commer lorry.
1 Packard store lorry.
1 Napier van.

Lieutenant Bowers returned to England almost immediately and was replaced by Second-Lieutenant J. O. Baird, Tyne Electrical Engineers.

On receipt of orders as to the destination of the Company, the transport proceeded by road under Staff-Serjeant Ruderman accompanied by an interpreter. The personnel entrained at Havre and the first night's journey was to Rouen. The whole of the next day was spent in the goods yard of the railway station, only the officers being allowed to enter the town. The next night's

---

[1] Subsequently awarded the M.S.M.
[2] Later commissioned in the L.E.E. and posted to the E. & M. Coy.
[3] Later commissioned.
[4] Later commissioned in the T.E.E.
[5] Later commissioned in the R.A.S.C. (M.T.)

journey brought the Company via Calais to St. Omer by about midday, but Arques, about 3 miles away, was not reached till dusk. No arrangements appeared to have been made but the O.C. saw the Maire of Arques and arranged for temporary billetting in a disused glass factory. In these rather dismal surroundings the Company spent its first Christmas. Shortly afterwards a more satisfactory billet was arranged in the " Patronage "—a church school and orphanage which were unoccupied. The premises included a yard, convenient for parking the transport and mobile workshop, and buildings for offices, stores and workshops. This constituted the headquarters of the Company from the end of December 1915 till the 2nd December 1917. The workshop in the building contained no machine tools and was used as a carpenters' and fitting shop, the machine tools being on the workshop lorry.

On arrival, the Company was attached to General Headquarters, British Expeditionary Force, and carried out electrical and mechanical work in the H.Q. area, and, when requests for assistance came from the headquarters of armies, detachments were sent from the Company.

The Company was immediately under the C.R.E., G.H.Q. area, for discipline etc. and under the Engineer-in-Chief (Major-General G. H. Fowke,[1] shortly succeeded by Major-General Spring-Rice) for work. The O.C. was, however, authorised to correspond direct with the Chief Engineers of Armies for work in the different areas.

The functions of the Company in respect of electrical and mechanical plant had been defined on its formation as follows :—

   i.   Instalment.
   ii.   Training of personnel for the running of such plant when installed.
   iii.   Inspection and advice to ensure full value being obtained from any plant available.
   iv.   Preparation of projects for installations.
   v.   Minor repairs.

It will be seen that the permanent detachment of Company personnel for the management of plant, which later became the rule, was not at this time contemplated.

In the early days of 1916 the Company worked chiefly on such undertakings as the electric lighting of hospitals, casualty clearing stations, etc., in the Army areas, where there were a number of civil electric power circuits and in which a considerable amount of machinery had already been installed. The average numbers available for these works appear to have been between 80 and 90, the remainder of the Company being employed permanently in or near headquarters at Arques, near St. Omer.

---

[1] Subsequently Lieut.-General Sir George H. Fowke, K.C.B., K.C.M.G.

The part of the country in which the Company was now operating being plentifully supplied with canals, opportunity was taken to augment its mobile equipment by the provision of a large barge equipped by the Inland Water Transport as a workshop. This was provided with an internal combustion engine and proved extremely valuable, as, while retaining a considerable degree of mobility, it was able to carry heavier machine tools than was possible in the workshop lorry.

In February 1916 the O.C. was instructed to report on the system of elecrified wire entanglements on the French front between Bully Grenay and Lièvin. Although the system had been well worked out technically, it was decided by G.H.Q. that it was too much suited to passive defence, so that the taking over of this piece of electrical work was not added to the activities of the Company.

Among other works undertaken was the erection of a large printing establishment for the advanced printing and photographic section of the Army Printing and Stationery Service at Terdeghem near Steenvoorde.

The first trench locomotive was also made in the workshops of the Company at Arques under the direction of Major Levin. It consisted of a Vulcan engine mounted on an ironwork chassis constructed by L/Cpl. R. Charman.[1] The mechanical work was superintended by L/Cpl. Tregear. This locomotive was later used on a section of the front near Laventie. Criticisms, compiled after some experience in use, ranged from complaint that it was too heavy to complaint that it was not completely armoured.

The water supply at Arques itself was at times a matter of some anxiety to the Company. A deepwell pump driven by a town-gas engine with small storage in a water tower was sufficient for the civilian population, but was rather heavily taxed for the two hospitals and the troops billeted in the town. The engine and pump required at times a good deal of nursing.

Work in connection with water supply had not yet, however, assumed the proportions which it was destined to do later. Early in March, in connection with the preparations for the Somme offensive, an army circular was issued calling for names of officers with experience in pumping machinery and general mechanical work. As a result of this, Lieutenant Baird was ordered to report to the Chief Engineer of the Third Army which was then holding a long front in the Somme area. To this officer was assigned the particular duty of studying the forward areas with a view to selecting suitable sites for installing pumping machinery.

Later in the month the Fourth Army took over from the Third Army the front from the River Somme in the south to the Gommecourt salient, about 10 miles south of Arras, and at the same time orders were issued that work was not

[1] Subsequently head foreman blacksmith to Messrs. Harland & Woolfe at North Woolwich.

to be confined to preparations for a large concentration only but was to include all the necessary arrangements for a general advance. Such an advance would take the Army across a practically waterless plateau and the organisation for supplying troops and horses with water during the advance at once assumed great importance.

Suitable sites were accordingly located for machinery, both to supply water to a large concentration of troops and horses and to pump water forward, from the positions then held, over the high ground across which the main attack was to take place.

In May supplies of pumping machinery of all types, both steam and I.C., began to arrive in the R.E. park, but there was, at that time, no organization in the Army to deal with this plant. A Section of No. 1 (L. & T.) E. & M. Company was, therefore, sent forward for this purpose and placed under Lieutenant Baird.

There had previously been sent out from England a number of men, furnished by the Tyne Electrical Engineers, for the operation of oxy-acetylene searchlights with troops in forward areas, and these had been attached to field companies on a scale of 13 men per company. The oxy-acetylene lights had, however, proved a failure and the men remained for general duties with the companies to which they had been attached. Seeing some of these men working in the Fourth Army area, Lieutenant Baird succeeded in getting them attached to his section, thus bringing the strength of the whole detachment up to about 80 other ranks.

During the preparatory period prior to the 1st July, a very large number of power installations were installed with rising mains up to five and more miles in length with more or less elaborate distribution systems. About 20 miles of mains were laid in all and over 100 power pumps installed. A great deal of the work of installing was done by fortress and army troops companies R.E., but, owing to the movements of formations, this was not very satisfactory, as incomplete work had sometimes to be handed over to incoming formations with consequent loss of efficiency.

Actually, the advance took place by stages, and, although several of the advanced lines were able to be pushed forward rapidly and forward water points were opened, the general scheme of carrying water to the divisional water-carts did not materialize.

For the part he played in this important work Lieutenant Baird was awarded the Military Cross. It was said of him that throughout the War " his character, energy and great technical ability, coupled with his extraordinary ingenuity and skill in improvisation, made him the ideal E. & M. officer, and " later, when the opportunity arose, " company commander."

After the formation of the Fifth Army during the Somme battle, this section, which had been chiefly occupied with the repair and maintenance of the water-supply plant in the area, became known as the Fifth Army R.E. workshops and the headquarters was established at Varennes. This title was in no way related to the Army Workshop Companies R.E. which were formed later and were derived from the Bethune Bomb Factory of 1914, afterwards known as the Army Workshop Company. The function of these Workshop companies was production ; that of the E. & M. companies installation and repair of machinery.

During all this period the main body of the No. 1 (L. & T.) E. & M. Company continued on general electrical and mechanical work in the Headquarters area, carrying out about May the wiring of the buildings taken over for G.H.Q. at Montreuil in preparation for the move from St. Omer.

During 1915 and 1916 mining and countermining along the front had been developing on an unprecedented scale, and, in the course of the first battle of the Somme, several failures had occurred in the use of dynamo exploders for firing simultaneously numbers of detonators distributed throughout the large charges which were then becoming customary.[1] The Company was accordingly instructed to make experiments to determine the best methods of firing a number of detonators in series. The obvious expedient of connecting two service exploders in series was found to be unsatisfactory and extensive experiments were carried out under the direction of S/Serjeant (later Lieutenant) Hews, in the course of which he made, with the resources available in the Company workshops, an oscillograph by means of which photographic records were obtained of the performance of the exploder at the moment of firing.

While the Company was so engaged, Serjeant Forsyth was awarded the Royal Humane Society's certificate and was mentioned in army orders for saving the life of a French soldier, Pierre Berthelot, who had got into difficulties when bathing in the river Lys. Serjeant Forsyth, who was walking along the bank, seeing him sink, dived in and succeeded in rescuing him.

In 1917 it was decided to build a laundry for the Second Army at Arques. The machinery was bought from an asylum at Armentieres. Six washing machines were at the asylum, but when the time came for removal only five were found. After some investigation it was found that an Australian division had passed that way . . . ! (The machine was subsequently recovered.) A fine horizontal steam engine for driving the machinery was obtained from the asylum, but, as the asylum was fairly near the line and in sight of the enemy, it was decided not to remove the heavy and cumbersome elephant boilers, two new vertical boilers being provided for supplying steam to the engine and drying rooms.

---

[1] The great mine of a million pounds of explosives, fired at Messines on 7th June 1917, was fired by connecting the detonators across the terminals of a 220 volt electric lighting set.

From the experience gained in the first Somme offensive, it was realised that, owing to the tremendous concentration of men and animals, no great offensive operations could in future be undertaken without adequate organisation for dealing with water-supply. Accordingly, in November 1916, the Field-Marshal Commanding-in-Chief in France recommended the formation of five E. & M. Companies, one for each Army.

The establishment recommended was :—6 officers and 259 other ranks, a considerable increase over the previous one. In addition, a well-boring section was attached to each company. In order to provide in part for these increases in personnel, certain reductions in units already in France were recommended, with the result that the total number of officers and other ranks to be found from England was reduced to 12 officers and 531 other ranks, but none of these was provided by the Tyne Electrical Engineers.

The proposals were approved and the numbers 350—354 were allotted to the five E. & M. companies so formed.[1] On the 1st February 1917 the original company was practically divided, half the personnel remaining with the Company at Arques and forming the nucleus of the 351st E. & M. Company for the Second Army, while the other half, most of whom had been in the Fifth Army under the title of Fifth Army Workshops, was absorbed into the 354th E. & M. Company, which formed at Varennes in the Fifth Army area.

The 351st Company remained under the command of Major Levin, while Captain H. H. Bateman, R.E.[2], took command of the 354th Company with Captain Baird as his second-in-command. Of the other officers with the original company, Captain Russell (T.E.E.) took over the post of Machinery Officer, G.H.Q., vacated by Captain Bateman and his place was taken in the 351st Company by Captain W. W. E. French, R.E. (New Army). Captain Russell subsequently went to the War Office where he did valuable work in sending out the right type of machinery for the use of the E. & M. companies. Captain Tufnell (L.E.E.) was, for a short time after formation, with the 352nd Company, Third Army, and then took over from Captain Russell at G.H.Q. where he remained to the end of the War under the Director of R.E. Stores.

A few of the original London-and-Tyne personnel found their way into the other E. & M. companies, but the bulk, as has been stated above, were divided

---

[1] The following was the distribution of the E. & M. Coys. on formation :—

First Army	... ...	350 Company.	Fourth Army ...	353 Company.
Second Army	... ...	351 Company.	Fifth Army ...	354 Company.
Third Army	...	352 Company.		

[2] Born 3/7/88. 2nd-Lieut. 29/7/08, Lieut. 19/8/10, Captain 30/10/14, Bt.-Major 3/6/19, Major 26/6/25, Bt. Lieut.-Col. 1/7/29, Lt.-Col. 18/1/33. D.S.O., M.C.

between the 351st and 354th Companies[1] and formed the stiffening for both. C.S.M. Bowmaker and C.Q.M.S. Mewis remained with the 351st Company, while C.S.M. Scott[2] from the Tyne Electrical Engineers, a man of outstanding personality and ability, was appointed to the 354th Company and remained with it until after the armistice.

S/Sjt. Hews, who had been commissioned in the London Electrical Engineers in October 1916 and posted back to the E. & M. Company, now went to the 351st Company. Serjeant Howe was appointed S/Sjt. mechanist in his place and remained with the company in charge of the headquarter workshops to the end of the War. S/Sjts. Watson and Burdis were also commissioned and the latter was now employed in anti-aircraft searchlight duties in the home defences. S/Sjt. Ruderman, who alone remained of the original mechanists, now went to the 351st Company and continued there until his demobilisation in 1919. Sjt. Hurrel, one of the section serjeants in No. 1 (L. & T.) E. & M. Company, went as staff serjeant (electrician) to the 354th Company on its formation and was commissioned in the R.A.S.C. (M.T.) about June 1918. Serjeant J. S. Dunn (T.E.E.) remained with the 351st Company in which he was subsequently promoted C.Q.M.S. Serjeant C. H. Forsyth (T.E.E.) was one of two men of the original company who were posted to the 350th Company on its formation and with this unit he remained until the end of the War. The technical ability of the reinforcements was for the most part considerably inferior to that of the original company. One man who came up, described as a " fitter," is said to have had more experience of tailoring than of engineering !

The new E. & M. companies were Army companies and worked under the orders of the Chief-Engineer of Armies but in liason with Corps water-supply officers and Army Troops Companies R.E. Owing to the inconveniences caused by handing over works and plants on the moves of formations, E. & M. companies came to be regarded as belonging to the various areas and did not long remain attached to a given Army. Although they sometimes did move with an Army, these cases were exceptional, and, as the moves of Armies (or of Army Headquarters) were frequent, the tracing of the changes in attachment

---

[1] 351st E. & M. Company. On formation :—Major A. E. Levin (L.E.E.).
                                        Captain W. W. E. French (New Army).
                                        Lieutenant H. F. A. Kinder (New Army).
                                        Lieutenant C. J. Hews (L.E.E.).
                                        C.S.M. Bowmaker (L.E.E.).
                                        C.Q.M.S. Mewis (L.E.E.).
                Posted later :— Lieutenant J. Heyworth-Davis (New Army).
                                          Lieutenant T. S. Hall (New Army).
 354th E. & M. Company. On formation :—Captain H. H. Bateman (R.E.).
                                          Captain J. O. Baird (T.E.E.).
                                          Lieutenant A. R. Evans (New Army).
                                          Lieutenant L. J. Harris (New Army).
                                          Lieutenant H. C. Day (New Army).
                                          C.S.M. J. Scott (T.E.E.).
                                          C.Q.M.S. Snow (R.E.) (from 25 A.T. Coy.).
                Posted later :— Captain Ross (New Army).
                                          Captain E. I. W. Jardine (Canadian Infantry).
                                          Captain N. W. Woodward (Canadian Infantry).
                                          Captain M. N. Bowmer (New Army).
                                          Lieutenant Pickersgill (New Army).
                                          Lieutenant Bower-Bower (New Army).
                                          Lieutenant Deane.

[2] Subsequently awarded the Medal of the O.B.E. and the M.S.M.

of the various companies becomes somewhat complicated. They had numerous other duties connected with electricity, machinery and workshops and were, from the first, very much overburdened, but military considerations and the shortage of man-power prevented the provision of sufficient personnel.

The work of these units often covered a very large area as the front held by an Army would be anything from three to twelve miles wide and the depth twelve to twenty miles. An idea of the extent of their work can be formed from the fact that the 354th Company, on its formation, immediately took over the running of some 80 pumping plants and about 20 electric light installations, mostly in casualty clearing stations.

In order to cope better with these widely distributed works, the 351st Company, with headquarters at Arques, threw out advanced sections to Abeele (under Lieutenant Kinder) and Armentieres (under Lieutenant Hews) with a workshop lorry and the workshop barge respectively. The barge was later withdrawn to Estaires.

The duties of these companies were now laid down roughly as being the " installation, maintenance *and* operation of all R.E. machinery in Army areas, which included that of water-supply. electric lighting, workshops, laundries, bakeries and forestry units " and also drainage installations in low-lying districts which had previously been run by land drainage companies of specially recruited men from the Fen district. The 350th Company, in particular, was formed largely from personnel of the land drainage companies and included two of the Tyne Electrical Engineers' members of the original E. & M. Company.

The E. & M. companies were also made the sole channel through which the Armies could obtain spare parts and each E. & M. company was responsible for keeping a register and running records of all machinery in its area. In Army areas where offensives were projected, the pumping installations for water-supply formed the greatest and most important part of the companies' work. The normal distribution of work was, roughly, as under :—

Water-supply installation and running			...	...	50%
Water-supply repairs ...	...	...	...	...	25%
Electricity	...	...	...	...	...	15%
Miscellaneous	...	...	...	...	...	10%

For technical details of the works carried out by these companies the reader must be referred to " The Work of the R.E. in the European War " Volumes " Miscellaneous," Section VI. (Machinery, Electricity, and Workshops), and " Water-Supply, France." Most of the illustrations in the latter volume represent works with which both the 351st and 354th Companies were associated at some time and shew, in some cases, officers and other ranks of the Tyne Electrical Engineers at work.

The 354th Company was responsible for two technical achievements of outstanding importance in water-supply work. The first of these was the development of the air-lift pump for military purposes. Whereas the maximum quantity raised from a 6″ bore-hole by an ordinary pump had been about 3,000 g.p.h. it was found possible with the air-lift pump to raise from 8,000—10,000 g.p.h. in suitably chosen positions. Two bore-holes at Aveluy necessitated the installation of two large petrol-driven compressor sets, and, during the German withdrawal from the Hindenburg Line in March 1917, a section of the Company was sent to Bapaume to instal the air compressors to work bore-holes drilled by the recently formed No. 3 Water-Boring Section (Captain G. Hindson). The bore-holes put down in this way at Sapignies, Behagnies, Vélu, Haplincourt, Frémincourt (2 bores), Vraucourt and Achiet-le-Grand gave sufficient water to provide the whole of the advanced portion of the Fifth Army, including horses.

As an example, the hole at Sapignies was started on the 25th March with a view to supplying a cavalry division. "On the 2nd April large numbers of horses were actually watering at the water-point which had been constructed while the hole was being driven. This spot is nearly at the summit of the Bapaume Plateau and many miles from any surface water. This and other bores saved the situation—five bores put down during March and April delivered a total of some 400,000 gallons per day—these were all high level bores, the average depth being 100 metres." (*Work of R.E.*, Vol : " Water Supply—France," p. 71.)

The second of the technical achievements of this Company was the development of the belt pump. The Company was detailed to carry out experiments in order to evolve a good type and this was done, thirty being manufactured in the company shops and operated with success. The pattern adopted consisted of a canvas belt running on a pulley and dipping into the water in the well. The water was brought up on the belt, and thrown off by centrifugal force into a collecting box on reaching the top pulley. This type of plant was evolved from a belt pump found in a French chateau.

In certain advanced and battle areas reliable and good water-supply was not available, and, with the object of providing the troops in such areas with water in the shortest possible time, the 351st Company operated a mobile system consisting of lorries carrying collapsible framework and waterproof sheeting for purifying and settlement-tanks together with the necessary pumps.

The severe frosts of early 1917 resulted in the cracking of large numbers of motor and oil engine cylinders the repair of which was carried out in the workshops of the E. & M. companies.

The electrical sections installed and ran electric lighting sets in dugouts, brigade headquarters, dressing stations and casualty clearing stations. Other work carried out from time to time included the repair of X-ray plant and making surgical tools for boring into the human skull.

On the 1st July 1917 the 354th Company moved with the Fifth Army to Flanders and established headquarters at Proven, where the following work was carried out in preparation for the Ypres offensive :—[1]

 (a) Building new workshops and stores (see photograph).

 (b) Installing and running electric light for 15 casualty clearing stations.

 (c) Installing 6 mechanical filters.

 (d) Installing 20 new pumping stations.

 (e) Collecting and testing all machinery prior to issue to three new Corps workshops.

 (f) Collecting and testing large quantities of new pumping plant.

 (g) Taking over and running 30 existing installations.

 (h) The erection of a factory for the production of 1,000 S.O.S. rockets per day, and providing a stock of 10,000 before July 31st. (This necessitated experimenting, training unskilled personnel and obtaining various components from civil firms.)

All the above work was completed within two months, in addition to the installation of printing works and the lighting of front-line dug-outs. At the same time the Company set forward an advanced section to Ypres where it was accommodated in dug-outs near the swimming-bath.

Meanwhile, in preparation for the battle of Messines (fought from 7th—14th June), the 351st E. & M. Company had erected pumping stations capable of supplying nearly a million gallons of treated drinking water and five hundred thousand gallons of untreated water for animals per day. This included the installation of plant in connection with existing storage supplies in Dickebusch and Zillebeke Lakes and in the moat at Ypres. At the same time, this Company began work on a large installation at Haringhe which has been described as " probably the most ambitious scheme of a semi-permanent nature for the supply of pure water to a large area " undertaken on the Western Front. The installation, and later the operation of the plant here, was carried on by the 354th Company when the preparations for the Ypres and the first Passchendaele attacks were taken over from the Second Army by the Fifth Army at the end of June 1917. These preparations included arrangements for carrying forward pipe lines after the advance, and the account of these operations states that " by extraordinary efforts one of the lines from Ypres was got as far as Wieltje before the attack, but was almost entirely destroyed during the first day's fighting." Another line was completely destroyed during the night preceding the attack and the official account continues " the intense and continuous shelling made it impossible, both during the preparatory period and during the battle, to maintain water-cart points or pipe lines within about 5,000 yards of the line."

---

[1] The third battle of Ypres opened on the 31st July and concluded on 10th November, 1917.

A temporary workshop in France put up in a few days by the 354th E. & M. Company R.E. for dealing with water-supply machinery during the last battle of Ypres, July 1917.

The E. & M. companies were responsible for laying and maintaining the suction pipes to the pumps and for the pumps themselves, but were not usually responsible for the delivery lines, as in many cases during heavy fighting it took a whole R.E. company to maintain a single pipe line.

The strength of the companies by themselves proved totally inadequate to cope with the enormous amount of work required, and a great deal of the operational work was carried out by men selected from other branches of the service on account of their civilian experience as mechanics, electricians, or engine drivers. It also became necessary to train as engine drivers men with no previous experience. A company course of seven days' duration produced an engine driver who could be given charge of a steam (Merryweather) or internal combustion engined pump at a detached station, provided he did what he was told and reported all defects at once without attempting to repair them himself. As many as 3 officers and 300 other ranks would be attached in this manner to a single company.

Many of the installations, particularly pumps for dealing with trench drainage, were installed in dug-outs near the front line, where conditions of operation were frequently very trying. In some places, where the fall of the land was from the enemy towards the British positions, it was necessary to instal plant capable of dealing not only with the drainage of our own trenches, but also that from the German system, which was being pumped virtually out of their lines into ours. In this way a veritable " battle of the pumps " developed in some areas and any breakdown in the machinery soon led to complete flooding and considerable difficulty in subsequently recovering submerged plant. As these installations were subject to shell fire, damage from this cause was not infrequent and it demanded a high standard of personal courage and technical efficiency on the part of the men posted, frequently alone, in charge of these engines.

During heavy fighting on the night of the 19/20th September 1917 for example, the exhaust pipe of an I.C. engine working electric lighting in a front line dug-out was hit or jarred by shell-fire, so that the C.I. elbow on the engine itself was broken. Corporal Bromley (L.E.E.) managed to wedge up the broken part and kept the set running at the cost of a severe gassing from the exhaust, while S/Serjeant Ruderman (T.E.E.) made his way up from the Abeele detachment through heavy shell-fire with a replacement. For this act Corporal Bromley received the D.C.M.

The winter of 1917-18 caused a great demand for bricks for horse standings in the Ypres salient. The 354th Company put into working order and operated (until the armistice) the machinery of a large brickworks at Houlle. This Company also provided steam-heating for several casualty clearing stations and evolved and manufactured various appliances such as radiant heat baths.

In November 1917 the newly formed Army Tramway Companies R.E.[1] were provided with Ford tractors for operating the 60 cms. track, and, owing to the lack of trained mechanics in these new units, the 354th Company set up small repair shops at forward tram bases and undertook all tractor repairs in what was then the Second Army area. More serious repairs were dealt with at the Company headquarters' shops. In connection with this work a design was prepared and executed for bringing the gearing of the tractors to give a reasonable speed on direct drive and in reverse. The experimental tractor proved very successful and a great advance over any standard Ford tractor in the area.

In the same month the Second Army Headquarters was transferred to Italy, together with XI and XIV British Corps and French troops, to assist in stemming the Austrian break-through on the Isonzo, and the 351st Company was transferred to the Fourth Army. On the 3rd December this company moved to Ouderdom, the two advanced detachments being withdrawn, and set up at that place considerable workshops with electrical drive, tanks for testing pumping machinery, etc., and large stores for stocks of machinery.

Early in 1918, the Second Army Headquarters and XI Corps returned from Italy and took over approximately its former area on the western front, at the same time re-absorbing the 351st Company and taking over the 354th Company.

The second battle of the Somme commenced on 21st March 1918 and lasted until the early days of April, during which period the Fifth Army,[2] holding a front of 40 miles, was driven back almost to the gates of Amiens. There were a number of Tyne Electrical Engineers in the 353rd Company which was seriously involved in this German advance, becoming part of the so-called "Carey's Force," "a curious assortment of details made up of entrenching troops, American sappers, and staffs of various army schools, re-inforcing units and stragglers .... including 500 cases out of hospital"[3] which had been hastily collected to fill a vital gap in the line. After they had been relieved on the 30th March, this Company was occupied in part with heavy bridging work from April to July. The "Work of the R.E. in the European War," Volume "Bridging," p. 34, states, "the most important work done during this period was probably that of the Fourth Army east of Amiens, which was carried out under considerable difficulties. Troops engaged were the 216th and 574th Army Troops Companies, the 353rd E. & M. Company and two companies of American Engineers."

Towards the end of March, on account of the German offensive gains to the south, orders were issued confidentially to the E. & M. companies to reconnoitre new positions for workshops and headquarters further to the rear. Several positions were proposed but not approved.

---

[1] Re-named "Foreway" Companies in March, 1918.
[2] Subsequent to the battle of Cambrai (ended 30th November, 1917) the Fifth Army moved south to the Somme but the 354th Coy. remained in the Ypres salient.
[3] Conan Doyle "The British Campaigns in Europe."

On April 9th, in the morning, there was somewhat heavy shelling of the 351st Company's camp at Ouderdom and one lorry was disabled, the driver being wounded. Accordingly, orders were issued for all the workshop machinery, other than the workshop lorries, and all the stock of pumps, electrical plant etc., to be sent to Calais. This work was immediately set in hand by both the 351st and 354th Companies.

On April 12th, after all machinery had been loaded on the railway at the adjacent sidings, the 351st Company withdrew to a farm near Steenvoorde, coming under the orders of the IX Corps. This date actually marked the opening of the battle of the Lys during which the Germans captured Mt. Kemmel, and the whole of the position of the Second Army round Ypres was gravely threatened.

On the morning of the 14th April, the 351st Company received orders to remove, that night, all pumping plants round Zillebeke Lake, the electric lighting sets at Vierstraat dressing station, and to remove or destroy the electric lighting sets in forward dug-outs. All pumping sets were removed by the morning and delivered to the R.E. park at Busseboom. On reporting completion, further orders were received for the removal of the two installations from Dickebusch sluice gates and this was done forthwith.

The 354th Company were also similarly engaged about this date in recovering pumping machinery which they had previously installed in the moat at Ypres and at other places, although they had not received their orders so early as the other company. The German advance was so rapid that the parties dispatched to recover the plant found it in some cases already in the hands of the enemy, but losses were, fortunately, confined to a few cooling tanks at, or near, Westoutre and a few demolished pumps which were subsequently recovered.

There were also three other pumping stations at Dickebusch in addition to those removed by the 351st Company. These had been entrusted to an army troops company to destroy, but it was moved away on other duty, leaving L/Corporal Hall, of the 354th Company, to complete the demolition alone. This he was able to do successfully, but not until the enemy was already advancing round the corner of the lake. The 354th Company also mined most of the roads round Ypres by boring under the foundations with earth augers and sliding in place 6″ diameter water pipes filled with ammonal.

A week later, the area to the south of Ypres in which the 351st Company had been working was handed over to the French Army, and the Company left Steenvoorde and occupied the then disused army sawmills at Abeele until the 27th April when they moved to Proven, coming under the XXI Corps and sharing the accommodation of the 354th Company, who had been at that place since July of the previous year.

Violent enemy attacks had continued during April, and, although Mount Kemmel had been lost on the 25th, the enemy was slowly beaten to a standstill, and his defeat on a front of nine miles between St. Jans Cappel and Zillebeke lake on the 29th marked the end of the battle of the Lys.

On the morning of this date the camp at Proven was shelled, Lieutenant Campbell (attached) and 3 other ranks being wounded and 2 other ranks killed, all of the 354th Company. The companies escaped very lightly (though two 12″ A.P. shells fell just clear of the huts where the officers of both companies were sleeping) and were moved to a fresh camp on an adjoining site a short distance to the north.

During the move a driver attached to the 351st Company and one sapper were hit by shell splinters and each lost a leg.

This constitutes probably the only occasion during operations when two E. & M. companies occupied the same camp, and here they remained until, on the 2nd May, the 351st Company, being again taken over by the Second Army, moved away to Lumbres.

As soon as the front became stabilised it was possible to bring forward again the heavier machinery and the headquarters of the 354th Company were moved to Houlle where the workshops were re-established. A section however remained at Proven, where their anti-aircraft machine gunners (under Lieutenant N. A. R. Evans) were successful in bringing down a large enemy bombing aeroplane which gained the Company the commendation of the Army Commander, General Plumer. This A.A. machine gun detachment included at least one of the original London & Tyne men, Corporal Brown.

In the following month, June, the 351st Company again removed, this time to Rouen, whence, in the second half of July, it was transferred to the re-formed Fifth Army and re-installed its heavy plant at Dennebroeucq. At the same time the advanced section of the 354th Company, which had been at Ypres, was moved to Steenvoorde.

When the final British advance began on the 8th August 1918 it was found that the Germans had removed or destroyed the motors and centrifugal pumps which supplied their extensive systems of rising mains.

The 351st Company undertook the re-instatement of the Tournai water and electricity supply. The former involved the repair of the machinery at the Tournai pumping station, the reconditioning of the high level water reservoir outside the town and repair of the water mains which had been damaged by land mines. While this work was proceeding the town was supplied by a special service of water lorries. The overhead electricity mains between Tournai and the power station, situated some miles outside the town, were repaired and the power supply resumed within a few days after this duty had been assigned to the Company. A very large amount of machinery and technical plant, especially electrical, was captured and taken over by the E. & M. Companies.

On the 13th August 1918 Major Levin was appointed electric light and machinery officer at G.H.Q. Italy and Major Baird took over the command of the 351st Company at Dennebroeucq, being replaced in the 354th Company by Captain M. N. Bowmer. When the British Army re-occupied Lille in September, both the town water-supply and what remained of the electric supply system were operated by the 351st Company.

Meanwhile, the 354th Company was engaged in following up the advance of the Second Army with pumping and electric light installations, carrying out 32 separate installations of plant and operating 120 at the same time. In October difficulties arose owing to the length of the communications from the headquarters at Houlle and also from the advanced sections at Proven and Steenvoorde, and accordingly the company headquarters and workshops were moved to Roncq near Roubaix, upwards of 200 tons of stores and machinery being moved by road in captured German trailers. Even then it was necessary to depart from the established principle of administering the men direct and it became necessary to attach individuals for rations, work and discipline to the R.E. units of the advancing Army. On the signing of the armistice, on November 11th, the Second Army marched into Germany and the 354th Company became once more attached to the Fifth Army.

After the armistice, when Leslie Henson was appearing in the Revue 'Alladin' at the Army Theatre, Lille, the 351st Company provided the stage manager, in the person of Sapper S. Sleap who had had experience in this class of work in London before the War. At the same time, personnel of the A.A. searchlight sections, Fifth Army, under Captain R. O. Porter, Tyne Electrical Engineers, were providing the electric light for the theatre.

As soon as possible after the armistice, the work of demobilization commenced, about 15 per cent. of the men of the E. & M. companies proceeding on leave to England where they were disembodied and returned to their civil occupations.

On the 30th January 1919 the bulk of the 354th Company moved into Germany and rejoined the Second Army at Cologne, where the Headquarters of the British Army of Occupation had been established, but the section at Steenvoorde remained there under Captain N. W. Woodward, and certain details, under Lieutenant Bower-Bower, were left at Houlle looking after the old workshops and running pumps, etc. At the same time, Major Bateman handed over the Company to Major Bowmer. Later, on the demobilization of the latter officer, the Company was commanded by Major B. M. Owen, M.C. who had been the original commander of the 353rd E. & M. Company with the Fourth Army.

No plant was taken up to the Rhine. The Company took a week to move from Roncq to Cologne, but, as they had previously contrived to instal stoves in the railway trucks and played football during the frequent halts, the journey was not without its enjoyment.

There was very little real E. & M. work to be done on the Rhine as all requirements were met by the Germans on requisition.

In April, Captain E. V. Baldwin, Tyne Electrical Engineers, joined the 351st Company at its headquarters at Annappes, near Lille, and took over the Company from Major Baird on his demobilization on the 20th April.

In addition to the work mentioned above, the Company was now engaged in maintaining pumps and electrical plant for hospitals and demobilization camps in a considerable area round Lille, and also in salving surplus plants and handing them over to the Disposals Board.

Demobilization was also proceeding apace and there were now very few of the original men left, but C.S.M. Bowmaker, C.Q.M.S. Dunn and Staff Sergeant Ruderman still remained. As men became demobilized they were replaced by re-enlisted R.E. of poor technical ability who were mostly engaged on salvage and repair work.

On the 30th July all the plant at Annappes was handed over to the Disposals Board and the Company moved to Renescure, near St. Omar, where there was a large machinery park, and, on the 19th September, the Company moved to Peselhoek near Poperinghe where, at approximately half strength, it continued with similar work. On the 1st November the 351st Company moved finally to St. Pol, where the office staff was accommodated with the headquarters of the Graves Registration and Enquiry Department and of what was then known as Forward Districts. Here the 350th, 353rd and remnants of the 354th E. & M. Companies came under the charge of Captain Baldwin and were disbanded by him. All plant and stores were finally handed over to the Disposals Board and the personnel dispersed by the 24th December. Captain Baldwin was then transferred to the staff of the C.R.E. Forward Districts under the Chief Engineer, British Troops in France, where he carried out the duties of E. & M. officer until demobilized in February 1920. There remained only the E. & M. section with the British Army of the Rhine, which retained its identity as an offshoot of the 354th E. & M. Company, first at Cologne and later at Wiesbaden, until the British troops finally evacuated Germany at the end of the year 1929.

TABLE A.

Electrical and Mechanical Company, R.E., France.
War Establishment.

(Extracted from " the Work of the R.E. in the European War."
Vol : " Miscellaneous," p. 292).

(i) Personnel, etc.

Detail.	Officers.	W.Os.	Staff Sjts. and Sjts.	Rank and File.	Total.	Bicycles.	Motor Bicycles.
Major	1	–	–	–	1	—	—
Captain	1	–	–	–	1	—	1(b)
Subalterns	4	–	–	–	4	—	4(b)
Coy.-Sjt.-Major	–	1	–	–	1	—	—
Coy.-Qr.-Mr.-Sjt. ...	–	–	1	–	1	—	—
Mechanists (Mach.) ...	–	–	2	–	2 }	—	2
Mechanists (Elec.) ...	–	–	2	–	2 }		
Serjeants...	–	–	9(a)	–	9 }		
Corporals	–	–	–	8	8 }	12	—
2nd Corporals	–	–	–	8	8 }		
Sappers and pioneers ...	–	–	–	228(c)	228 }		
Total (excluding attached) ...	6	1	14	244	265	12	7
Attached :—							
Batmen(d)	–	–	–	3	3	—	—
For general duties(d)...	–	–	–	6	6	—	—
Drivers A.S.C., M.C....	–	–	–	5	5	—	—
Total (including attached) ...	6	1	14	258	279	12	7

(a) 1 Serjeant for pay duties. (b) 2 with side-cars. (c) Includes 12 Lance-Corporals and 2 Pioneers for water duties. (d) Of category lower than " A."

(ii) Transport.

Detail.	Vehicles.	Drivers.	Remarks.
Box car	1	1	
Workshop lorries	2	2	
Lorry, 3-ton, for tools ...	1	2	(e)Provided by R.A.S.C.
	4	5(e)	

Distribution of Serjeants and Rank and File by Trades.

Mechanists (Mach.)	2	Mechanists (Elect.)	2
Engine drivers (Steam and I.C.) ...	150	Fitters and turners	40
Blacksmiths	5	Whitesmiths	4
Moulders	3	Pattern makers	2
Plumbers	2	Electricians (field and fortress) ...	30
Instrument repairers	2	Draughtsmen (Mech.)	2
Clerks	2	Other Trades (incl. storesmen, etc.)	10

Total :—256.

CHAPTER VI.

1915—1919. THE GREAT WAR.
THE ANTI-AIRCRAFT SEARCHLIGHT UNITS AT HOME.

A.A. Organisation prior to the Great War — the Admiralty takes charge — First Zeppelin raids — Searchlight companies formed for London — The War Office once more assumes responsibility — Mobile searchlight companies formed — The Zeppelin L.15 shot down — Aeroplane defences — Aeroplane searchlight units — Re-organisation of London defences — Development of defences in North and Midlands — Establishments and technical equipment — Companies formed for Tyneside — And at Leeds, Hull and Sheffield — Transfer of Tyne Electrical Engineers from London to Coventry — Further Aeroplane searchlight units formed — Further mobile companies — The Cuffley, Billericay and Potter's Bar Zeppelins — Searchlight company for Birmingham — Aeroplane searchlight companies for York and Leeds — Commercial gas for engines — L.34 brought down off Hartlepool — Extension of searchlight barrage — Company for Gretna — Aeroplane searchlight companies converted — Company for Nottingham — Distribution of officers and units in April, 1917 — Progress in organisation and tactics — Aeroplane raids — Formation of L.A.D.A. — The A.A. defences re-organised, January, 1918 — Establishments — Appointments to G.H.Q. H.F. — Formation of Northern air defences — Dover A.A.D.C. — Superiority of the defence over the attack — Developments in equipment and organisation — Height-finding and track-plotting — Distribution of officers and units, July, 1918 — Further re-organisation — The armistice — Disbandment — Observations on Home service.

PRIOR to the outbreak of the Great War in 1914 the air forces of the world powers were so small, and the state of technical development so imperfect, that the uses to which aircraft might be put in the event of war had hardly begun to be considered. In 1913, nevertheless, the text-book of " Electric Lights for Coast Defence," published privately by the electric lighting school at Stokes Bay, had drawn attention to the possibility of attack by airships or aeroplanes " with a view to obtaining information and damaging ships, docks, etc. . . ." In this same year the responsibility of the War Office for the protection of the defended ports in the British Isles was extended to include defence against air attack, and two 6″ howitzers were mounted for the guarding of the naval magazines at Chattenden and Lodge Hill.

By 1914 a few Q.F. guns were on order but, beyond this, little progress had been made in the provision of air defences before the War broke out. It then became apparent that the formidable German fleet of airships would be used to deliver attacks on this country, and we have already observed in what manner the defensive arrangements were developed, not only at the naval bases of paramount importance such as Portsmouth, but also, to a lesser extent, at those commercial ports, such as the Tyne, which had previously been considered of sufficient national importance to demand defence from sea attack.

It is not the purpose of this account to trace in detail the various enemy raids on these shores during the course of the War, and the reader who seeks further information must be referred to the several books which have been published dealing exclusively with this subject.[1] Some reference to the progress of the attack is, however, necessary in order to follow the development of the defensive organisation in which the Tyne Electrical Engineers were destined to play so important a role.

The year 1914 passed with only two minor enemy daylight raids by aeroplane on Dover and the mouth of the Thames respectively, but these were sufficient to demonstrate that the army, fully pre-occupied with the onerous task of providing men and material for the expeditionary forces overseas, was not at that time capable of effectively discharging its responsibility of defending the British Isles from air attack. Accordingly, on the 3rd September, 1914, the War Office handed over the duties of home anti-aircraft defence, with the exception of that of the defended ports, to the Admiralty, and there then sprang into existence a force for the defence of London composed of members of the Royal Naval Volunteer Reserve and of special constables. A meagre armament was collected, the guns being manned in part by the R.N.V.R. and in part by Royal Marines, while the searchlights were at first operated by the special constables.

Zeppelin airship raids began early in 1915, and we have already traced the development of air defences on Tyneside as a result of the raid of the 14th April, culminating in the formation of the (Special) Tyne Searchlight Company to assist in the protection of that locality. (See Chapter III). The first air raid on the capital took place, after a prelude of three raids on the Thames estuary, on the 31st May, forty-one people being killed or wounded, and damage to the value of £18,000 being done in the East end of London. Further raids were made on Kent and Yorkshire on the 4th June, and, on the 6th June very serious damage was occasioned in Hull, sixty-four people being killed. There followed the disastrous raid of the 15th June on Tyneside which has already been described.

As a result of these attacks the Admiralty hurried on with its measures of defence, and, on the 12th September, Admiral Sir Percy Scott was appointed to command the gun defences of London, which included the famous R.N.V.R. Mobile Brigade under Commander Rawlinson and a (Special) Company of the London Electrical Engineers. During 1915 the coast lights were extinguished, and lighting restrictions, which had first been imposed in August, 1914, were made more stringent.

---

[1] e.g. " The Defence of London. 1915–1918." Rawlinson. London 1924.
 " The German Air Raids on Great Britain. 1914–1918." Morris. London N.D.
 " Air Defence." Ashmore. London 1929.

Air raids continued throughout the ensuing months, and, by December, it was decided that the army was once more in a position to take a hand in the defence of London and of other vulnerable points in addition to the naval ports. Accordingly, to man the searchlights, anti-aircraft searchlight companies were raised as part of the Territorial Force. The greater number were provided by the London and Tyne Electrical Engineers, the remainder by various fortress units such as the Hampshire, Kent, and Essex Fortress R.E. (T.F.).

The first of these companies to be formed, if we exclude the nuclei of the two (Special) Searchlight Companies of the London and Tyne Electrical Engineers, were No. 1 (London) and No. 2 (Tyne) A.A. Companies R.E. (T.F.), the latter being formed at Haslar during the first week in December under the command of Captain C. M. Forster. The other officers appointed to this company on its formation were Lieutenant F. B. C. Sutthery and Second-Lieutenant H. S. Ripley and the company-serjeant-major was C.S.M. R. Rood. Captain Forster arrived in London on the 13th December and was informed that the Company was required to establish a ring of A.A. lights, 10 or 12 miles in radius, round the outskirts of London. It was considered extremely important to have this circle completed as far as possible before the next full moon when it was expected that the Zeppelins would return to the attack. Such indeed was the urgency that the normal military channels of administration were short-circuited and Captain Forster was instructed to deal direct with the appropriate branch of the War Office, while the whole of the R.E. Stores in the areas concerned were put at his disposal and the C.R.Es at Woolwich and Hounslow were instructed to give him every assistance in their power. Captain Forster was also empowered to select the most suitable sites for the lights and to commandeer the most convenient billets for his detachments. The actual words of Captain R. M. Crosse, the responsible officer at the War Office, were " Take a duke's castle or his stables, whichever you consider the more suitable ! "

Electric lighting plants of various types were collected and, as each station was ready for occupation, detachments of men were ordered up from Haslar to man them. These detachments began to arrive on the 18th December and, by Christmas Day, nineteen or twenty searchlight stations were in action. Electric supply was obtained from all manner of sources ; motor-generators run from the public electric supply systems, various types of generating plant, and even motors and dynamos fixed in tramcars with projectors on the top decks. For the driving of these tramcars on the road, men were specially recruited on Tyneside from the Tynemouth and South Shields tramways.

The headquarters of the Company was established at 50 Parliament Street, and the administration of all the searchlight stations, scattered as they were all round the outskirts of London, introduced many unusual difficulties and involved a great deal of travelling on the part of the officers in the course of their daily duties.

During the winter, No. 3 (Yorkshire) A.A. Company was raised from personnel of the North- and East-Riding Fortress R.E. (T.F.) and No. 5 (Lancashire) A.A. Company by the Lancashire Fortress R.E. (T.F.). Nos. 4, 6, 7 and 8 A.A. Companies were gradually mobilised by the London Electrical Engineers during the months which followed.

At the beginning of 1916 the ground defences underwent considerable re-organisation and on the 16th February the War Office took over the defence of London. By April their responsibility had extended to include the defence of the entire Country against air attack.[1] The Home Defence Branch at the War Office transferred to the Horse Guards, there to form part of the staff of Sir John French, Commander-in-Chief, Home Forces. The guns and their attached searchlights then came under the command of Colonel M. St. L. Simon, and part of the staff of the Commander-in-Chief was organised as an anti-aircraft section responsible for the home defence intelligence and operations. On the 5th May Captain Forster was appointed to the post of G.S.O.A.A.3. at G.H.Q. Home Forces, and handed over the command of No. 2 (Tyne) A.A. Company to Captain D. A. Williamson who arrived from Haslar. The duties of the G.S.O.A.A.3. included all matters dealing with the design and provision of anti-aircraft searchlights and accessories, their distribution among the various defended localities, the establishments of personnel for their operation and the organisation of the R.E. searchlight companies.

Meanwhile the defences were being considerably extended, one of the earliest actions of the new organisation being the formation of mobile anti-aircraft brigades, intended to provide defences for threatened localities pending the provision of fixed defences. For these, mobile anti-aircraft brigade searchlight companies came into existence. One of the first to be created was No. 9 (Tyne) Mobile Searchlight Company R.E. (T.F.), which formed at Haslar and proceeded, under the command of Major N. H. Firmin, to London on the 11th March, for the purpose of completing mobilisation. Each mobile brigade consisted of three batteries, each of six 13 pdr. A.A. guns, and one searchlight company of three sections, each of four lights. The establishment of this Company on formation was 7 officers and 138 other ranks including 32 A.S.C. drivers attached. The officers appointed to this Company at the outset were :—
Major N. H. Firmin, Lieut. J. R. Abbott, Lieut. T. T. Tucker, 2nd Lieut. J. Lawther, 2nd Lieut. C. G. Huntley, 2nd Lieut. J. A. Power, 2nd Lieut. J. Robinson.

There was no company-serjeant-major on the establishment, the senior N.C.O. being C.Q.M.S. J. Peat. There were two mechanists, Staff-Serjeants G. T. B. Scullard and F. Thompson, and Serjeant J. Welsh was the N.C.O. i/c telephones.[2]

---

[1] The Admiralty remained responsible for dealing with all hostile aircraft attempting to reach this Country, while the Army undertook to deal with all such craft which succeeded in doing so.

[2] Staff-Serjeants Scullard and Thompson and Serjeant Welsh were all subsequently commissioned in the Tyne Electrical Engineers.

During mobilisation, the headquarters of the Company was established first at the headquarters of the London Electrical Engineers at No. 46 Regency Street (from which many Tyne Electrical Engineers' units were destined subsequently to set forth), and later at 6 Earl Street, Westminster, where a convenient brewery opposite was used as a garage, store and workshop.

The equipment of the Company was drawn from Woolwich Arsenal and consisted of twelve 60 cms. projectors with various types of generating sets carried on 3-ton Daimler lorries. For his personal service, the O.C. was furnished with a Ford box-van, and six Fiat 30-cwts. lorries were issued for company baggage and stores. The vehicles all came from the depot at Kempton Park and were driven by A.S.C. drivers under Corporal W. Gladwin.[1] There were also six motor-cycles for the use of the other officers. This Company was the first to adopt the triangular transport sign which is now borne on all vehicles of the Tyne Electrical Engineers.

The mobilisation store table of the Company proved to be very incomplete: for example, the table contained a plentiful supply of twist drills but no form of brace or drilling machine in which to use them. By exercising much persuasion the equipment was finally more or less completed. The function of No. 9 (Tyne) Company was at first that of a " stop-gap " in the various A.A. defences springing up all over the United Kingdom, and, as soon as the individual detachments were complete with equipment, they were sent out, under the direction of G.H.Q. Home Forces, to occupy temporary stations in such widely scattered localities as Dartford, Gravesend, Ipswich, Syston (Leicestershire), Whitley Bay, Barrow-in-Furness, Southminster, Deal, Folkestone, Brentwood, Herne Bay, Barking, Welling, High Wickham, Scunthorpe and Stowmarket.

While the Company was thus split up, the detachment at Darenth, near Dartford, had the good fortune to be in action when the first Zeppelin to be brought down in this Country, the L.15 (Kapitanleutnant Breithaupt), was shot down in the Thames estuary on the night of the 31st March. This detachment[2] had the additional distinction of receiving the gold medals promised by the Lord Mayor of London to any members of the services who should be instrumental in bringing down the first enemy airship over England.

---

[1] Corporal Gladwin was, before the War, the works manager of the Benz Motors at Frankfurt and was interned at Ruhleben until the latter end of 1915. In order to obtain his exchange he ate a quantity of tobacco which affected his heart, and he was eventually exchanged for an incapacitated German prisoner in British hands. He returned to England, and, having quickly recovered from his disability, enlisted in the A.S.C. He was subsequently commissioned in the Tyne Electrical Engineers.

[2] The personnel of this detachment was as follows :—

Serjeant J. Donaldson.	Sapper J. King.	Sapper W. Milburn.
Corporal P. Bolt.	Sapper F. Allan.	Sapper A. L. Hastings.
L/Corpl. S. Armstrong.	Sapper R. Elliott.	Driver Colman A.S.C.

GOLD MEDAL.

Presented by the Lord Mayor of London to members of the
Army and Navy who were instrumental in bringing down
the L.15 on the night of the 31st March 1916.

By this date, there were also in existence certain un-numbered searchlight sections of the London Electrical Engineers, which had been formed for co-operation solely with the R.F.C. squadrons of defending aircraft. Up to April, the aircraft for home defence purposes had been furnished through the existing training organisation, but experience had already shewn that, to deal effectively with the problem, it would be necessary to have separate operational units for home defence co-operation. Thus, on the 15th April, No. 39 Squadron R.F.C. was formed at Hownslow especially for this purpose with flights at Hainault and Sutton's Farms, Hornchurch. The formation of further home defence squadrons was set in hand[1] and searchlight sections were allotted to these squadrons. These ' aeroplane lights,' as they were called, were trained to work in close co-operation with the aircraft and were quite distinct from the ' gun lights ' which worked in conjunction with the Artillery. By the end of April, there were six ' anti-aircraft aeroplane squadron searchlight sections ' of London Electrical Engineers, while a company known as No. 33 (Tyne) A.A. Company[2] had been formed of Tyne Electrical Engineers for co-operation with the aeroplane squadrons in the north.

This Company was mobilised at Earl Street, London, on the 26th April, under the command of Captain J. F. S. Hunter. The establishment allowed six subalterns, on the somewhat lavish scale of one for each of the six lights. Four of these officers were the following :—Second-Lieutenants C. B. Williamson, C. H. R. E. Raven, E. H. White, and W. Fox. The equipment consisted of six 60 cms. projectors with Boulton & Paul and Gardner engines carried in lorries. At first two of the lights were at Hull and four round Cramlington, Northumberland. Later, the lights from Hull were moved north and the Company then occupied positions in Northumberland and Durham and co-operated with No. 36 Squadron R.F.C., the headquarters of which was at Cramlington. A flight of this Squadron was stationed at Seaton Carew, near Hartlepool, and three lights, under Lieutenant C. B. Williamson, were disposed at Hutton Henry, Elwick and Greatham to work with it.

During April a plan was introduced to supplement the naval air stations on the coast with a great barrage line of aeroplane patrols and searchlights, extending from Northumberland to Sussex and about 25 miles inland from the coast, and, at the beginning of May, the block of numbers 20 to 29 was allotted for aeroplane searchlight units to serve this barrage.

The existing sections of the London Electrical Engineers were forthwith re-named Nos. 20 and 21 (London) Aeroplane Squadron Searchlight Sections R.E. (T.F.), and at the same time, a new section, which was already in formation at Haslar, received the title No. 22 (Tyne) Aeroplane Squadron Searchlight

---

[1] By the end of June there were ten squadrons of the R.F.C. organised for this barrage line, and they were then formed into a home defence wing whose sphere of responsibility extended to the Wash. Further squadrons north of the Wash worked through the appropriate garrison commanders, such as the Humber, Tees and Tyne.

[2] This title, which only lasted a short while, was presumably in anticipation of the great scheme of defence which came into being a month or two later.

Section R.E. (T.F.). This latter unit, under the command of Captain F. H. Bowers, went to London in the first days of May and completed mobilisation at the Earl Street headquarters. The establishment on its formation was six officers and 55 other ranks R.E. with 16 A.S.C. drivers attached. The officers were Captain F. H. Bowers, Lieutenants E. F. Rendell and R. P. Winter and Second-Lieutenants T. W. Crawford, J. B. Murray and C. Graham. The equipment consisted of six 60 cms. Johnson and Phillips searchlights with Gardiner and Astor generating sets carried in Daimler and A.E.C. lorries. The O.C. was provided with a motor car, and there were three 30 cwts. lorries for baggage and stores and five motor cycles. The establishment of this Section included no C.S.M. or C.Q.M.S. but contained two mechanists and four serjeants.

The headquarters and four of the lights proceeded at once to Sutton's Farm, Hornchurch, where they functioned in co-operation with No. 39 Home Defence Squadron R.F.C., but, as a result of the severe bombing which the city of Hull had recently experienced, two lights of this Section, under Lieutenant Winter and Second-Lieutenant Murray, were rushed to Beverley, Yorks., to supplement the defences of that place. This detachment subsequently remained at Beverley until July and worked in co-operation with No. 52 Squadron R.F.C., but there was no enemy action in their area during this period. Night flying was still a precarious undertaking and the dangers of landing by night were at that time proving a severe handicap to the defence. While at this station, the detachment of No. 22 Section was engaged in experiments in ground lighting with a view to minimising these dangers but the use of searchlights for this purpose was found to be unsatisfactory.

During May the anti-aircraft defences of London were re-organised into four sections or 'controls' viz. :—N.E., N.W., S.E. and S.W., each under a separate A.A. Defence Commander, and the presence of certain of the detachments of No. 2 (Tyne) Company in each of the 'controls' aggravated seriously the existing administrative difficulties. It was therefore decided that this Company should be concentrated in the N.W. area, and the light stations previously occupied by them in the other areas were then handed over to the companies of the London Electrical Engineers detailed to each particular 'control.' The headquarters of No. 2 (Tyne) Company, now under Captain D. A. Williamson, was thereupon moved from Parliament Street to Wembley Hill, where the headquarters of the N.W. area was being formed, and the Company then took over a number of stations in that area which had previously been manned by the R.N.V.R. These stations were found to be modelled on a ship and fitted out in a manner true to the traditions of the Royal Navy and, while the equipment and accommodation was of the best, many of the articles of store, such for example as naval cutlasses, appeared to be of but little use to the army in anti-aircraft defence. The quarters were fitted with collapsible bunks and equipped with everything that could be desired in the way of cooking facilities, utensils and furniture; alongside were various storerooms and a room containing the engine and generator, while the whole

60 CMS. SEARCHLIGHT PROJECTOR.

Of the type with which all Home Defence A.A. Companies
were equipped until 1918.

This photograph is of the Botany Bay Station of No. 2
(Tyne) Searchlight Company in the N.W. Control of the
London Defences, May 1916.

was covered by a flat roof upon which the projector was erected. A wooden palisading enclosed the entire station, which seemed better fitted to resist attack from the land than from the air.

Further light stations were erected by the Office of Works and these were handed over when ready for occupation.[1] No receipts were given or asked for and no vouchers passed. Full inventories of all stores which were in the station when the Company marched in were taken, but the question of writing up the ledger became very difficult because very few of the articles that were on the stations appeared in the 'vocabulary of stores.' An appeal for assistance was made to Haslar Barracks with the result that, unofficially, the quartermaster, Lieutenant A. J. Sergeant, spent three or four days with the Company and put everything on the right lines, and in due course all stores and equipment were put on ledger charge.

The plant was a heterogenous collection and hardly two stations had similar equipment. Motor generators taking power from the electric mains and from the tramway systems, Gardner, Parsons, Keighley and Crossley petrol engines, and Blackstone, Hornsby and Tangye oil engines all gave efficient and reliable service. The Dollis Hill light was of particular interest as it consisted of twin 60 cms. projectors mounted after the fashion sometimes met with in coast defences. The Company had now only two tramcar lights in its area, at Finchley and Hendon, and these proceeded nightly to their appointed places.

All stations were in communication by telephone with the A.A.D. Control Headquarters at Wembley Hill. These headquarters consisted of a high tower with a flat roof to which the A.A.D.C. proceeded on the occasion of a raid. He then shouted down a speaking tube to his telephone operators in the room below any orders he cared to endeavour to transmit. In the general confusion it appeared that very few orders of actual value ever reached their intended destination ! The R.E. company commander was never given any orders and he was not allowed to use these telephones at night on account of the mixed intercommunication between the guns and lights. The standing order for the lights was, ' illuminate the target,' and it cannot be denied that this object was attained most successfully when the opportunity arose. All detachments lived in huts on their searchlight sites, and, while the army huts did not attain the luxury of the naval quarters, they were quite comfortable and convenient. The Company was on ration allowance, and this worked admirably both from the men's point of view and from that of the Company headquarters ; rations in kind would have been an impossible proposition in view of the large area covered and the lack of transport.

---

[1] The searchlight stations in the N.W. area manned by No. 2 (Tyne) A.A. Co. R.E. were as follows :—

Dollis Hill.	Parliament Hill.	Hendon.	Finchley.
Welsh Harp.	Kenton.	Mill Hill.	Harrow.
Pinner.	Ealing.	Hatfield.	Botany Bay.
Uxbridge.	Harpendon.	Acton.	Watford.
Barnet.			Rickmansworth.

127

The only transport which No. 2 (Tyne) Company had at this time amounted to three motor cycles. Arrangements were made with the petrol and oil companies to distribute all fuel to the sites, and when any stores were required to be moved the transport had to be requisitioned from the A.S.C. (M.T.). The mechanists were each provided with a book of warrants to enable them to use the 'District' and other railways for conveyance. One transport privilege was bestowed, however, consisting of what was known as a "green pass." This pass allowed the holder, when on duty, to use full headlights on a car or motor cycle and it enabled the transport, in spite of frequent stoppages by irate police constables, to move at night with speed and a certain degree of safety.

In June a new establishment for the Company was evolved. That for officers remained the same in number but the O.C. was henceforth to be a major. There was no provision for a C.S.M., consequently C.S.M. R. Rood returned to Haslar. In place of the C.S.M., company headquarters included a C.Q.M.S. and a pay serjeant, Lance-Corporal G. Gill being promoted C.Q.M.S. and Sapper A. Gilhespy pay serjeant.

It was in this month that the establishments of all Territorial Force units were doubled to enable them to provide for the many additional duties which they were by then undertaking, and this furnishes a fitting opportunity to review the development of the anti-aircraft searchlight organisation for home defence up to this time. Apart from the A.A. searchlights installed in the various naval ports which were operated by the units manning the coast defence searchlights, there were then in existence the following army anti-aircraft searchlight units in Great Britain :—

    (i)   (Special) Tyne Searchlight Company—Tyne Electrical Engineers.
   (ii)   (Special) London Searchlight Company—London Electrical Engineers.
  (iii)   No. 1 (London) A.A. Company R.E. (T.F.)—London Electrical Engineers.
  (iv)   No. 2 (Tyne) A.A. Company R.E. (T.F.)—Tyne Electrical Engineers.
   (v)   No. 3 (Yorkshire) A.A. Company R.E. (T.F.)—East- and North-Riding Fortress R.E. (T.F.).
  (vi)   No. 4 (London) A.A. Company R.E. (T.F.)—London Electrical Engineers.
  (vii)   No. 5 (Lancashire) A.A. Company R.E. (T.F.)—Lancashire Fortress R.E. (T.F.).
 (viii)   No. 6 (London) A.A. Company R.E. (T.F.) ⎫
  (ix)   No. 7 (London) A.A. Company R.E. (T.F.) ⎬ London Electrical Engineers.
   (x)   No. 8 (London) A.A. Company R.E. (T.F.) ⎭
  (xi)   No. 9 (Tyne) Mobile Searchlight Company R.E. (T.F.)—Tyne Electrical Engineers.
 (xii)   No. 20 (London) Aeroplane Squadron Searchlight Section R.E. (T.F.)—London Electrical Engineers.

(xiii) No. 21 (London) Aeroplane Squadron Searchlight Section R.E. (T.F.)—London Electrical Engineers.

(xiv) No. 22 (Tyne) Aeroplane Squadron Searchlight Section R.E. (T.F.)—Tyne Electrical Engineers.

(xv) No. 33 (Tyne) A.A. Company R.E. (T.F.)—Tyne Electrical Engineers.

Immediately subsequent to the doubling of the Territorial Force establishments the title " (Special) Searchlight Company " disappeared, and, in the ensuing re-shuffle of the units, the (Special) Tyne Searchlight Company at Newcastle became, as has already been mentioned in an earlier chapter, No. 4 (A.A.) Company, Tyne Electrical Engineers.

Up till August, nearly all the airship raids were made on Scotland and the North of England, and the re-organisation of the defences of the Capital was permitted to proceed more or less unmolested. As trained personnel of the London Electrical Engineers became available, it was found possible gradually to relieve the personnel of No. 2 (Tyne) A.A. Company in the N.W. Control and thus to liberate them for service in the Midlands, where the defences were being developed with all possible energy.

The general scheme was to furnish the great industrial and manufacturing towns of the North and Midlands with anti-aircraft gun defences quite separate from the aeroplane barrage which has already been referred to, and for this purpose the formation of a series of A.A. 'Gun Lights' Companies R.E. was planned. These companies were allotted to such localities as Bradford, Leeds, Birmingham, Hull and Sheffield, and were numbered consecutively, commencing with No. 34 in the north.[1] With the meagre resources of men and material available, it was impossible to form these companies completely at the outset, but drafts of men were sent into the different areas as soon as they were trained and as soon as plant was ready for them. The organisation of this system thus took a considerable time to complete, and during the earlier part of 1916 the North and Midlands remained practically defenceless.

The establishment of the companies now being formed allowed a C.Q.M.S., one serjeant and two mechanists in company headquarters, and each searchlight detachment consisted of a serjeant or a corporal, one second-corporal or lance-corporal and six sappers of whom two were engine drivers and four electricians. The companies were not all alike, the number of lights, and consequently the number of officers, varying with the particular requirements of each defended locality. As a rule, in the mobile companies one officer was allowed to two lights, but in the fixed companies one officer to four lights was the more general scale.

The plant supplied for these companies consisted of 60 cms. projectors, operated by means of a large handwheel at the side and supplied with electricity from various types of semi-mobile generating sets mounted on lorries or trailers. In some of the districts the plant was completely non-mobile, the engines

---

[1] The companies of London Electrical Engineers in the London defences were at the same time reduced to six in number and numbered 1 to 6.

I

being permanently installed in engine sheds. Almost every type of engine and generating set in existence was pressed into service, those most generally met with being Gardners, Keighleys, Astors, Boulton & Pauls and Dormans. Not all of these types were found to be entirely suitable for the purpose for which they were required and technical difficulties of a minor nature were of frequent occurrence. A certain amount of trouble, for example, was experienced with the lubrication system of the Keighley engines and Lieutenant Robinson and Staff-Serjeant Scullard, of No. 9 (Tyne) Mobile Company, together produced a design for converting the oil feed to one of pressure type, which was quickly adopted by the manufacturers. There were also many cases during the winter of water freezing in the circulating systems of the Boulton & Paul engines, and many replacements of bent or broken impellor blades were found to be required in the rotary circulating pumps with which these sets were equipped.

The first companies in the new system to be completed were those in the North. On the 29th July No. 4 (A.A.) Company, Tyne Electrical Engineers, at Newcastle, was divided into two, the units so created becoming Nos. 34 and 35 (Tyne) A.A. Companies R.E. (T.F.),[1] which were commanded by Major C. M. Campbell and Captain K. S. M. Scott respectively. No. 33 Company continued for a time to serve the aerodrome barrage from Alnwick to Hartlepool, while Nos. 34 and 35 Companies manned the gun lights on Tyneside, the former to the south of the river and the latter to the north; No. 35 Company also detached seven lights, under Lieutenant R. G. Ellis, for the defence of the great munition factory at Gretna.

During June and July drafts of Tyne Electrical Engineers' personnel had been proceeding from Haslar to Leeds where they manned lights in the Bradford, Halifax, Huddersfield and York areas in co-operation with personnel of the Glamorgan Fortress R.E. (T.F.) under Captain Gordon and Lieutenant Pertwee of that unit. Among the vulnerable points in these districts were the large shell-filling factory at Barnbow and various explosive and acid works at Halifax and Huddersfield. At first, six lights[2] were manned by the Tyne personnel under Lieutenant R. H. Rooksby and Lieutenant E. Harrison, but the Glamorgan lights were gradually taken over, and on August 17th this group was formed into No. 37 (Tyne) A.A. Company R.E. (T.F.), with headquarters in the post office at Leeds, where Captain J. Young arrived from Haslar to take command. The establishment at this date was 3 officers and about 100 other ranks, and the Company now operated twelve 60 cms. searchlights, of which five were round Leeds, two each at Halifax and Bradford, and one each at Huddersfield, Wakefield and York. One of the subalterns supervised the five lights at Leeds and one at York, and the other the remaining six. The senior N.C.Os. with this Company were C.Q.M.S. Blair, Staff-Serjeant Simpson and Serjeant Unsworth. At the period of the formation of No. 37 Company, Zeppelin

---

[1] The establishment of No. 35 Company was 3 officers and 103 other ranks.
[2] At Southowram, Boothtown, Lindwell, Dalton, Rowley Lane (subsequently moved to Great Horton), and Eccleshill.

DORMAN 8 kW. GENERATING SET.

A typical example of the equipment with which all Home Defence A.A. Searchlight Units were provided. This photograph represents one of the mobile sets carried in a motor-lorry.

activity was considerable over this area, raids taking place on the nights of the 10th, 28th and 31st of July and on the 2nd and 8th August.[1] Captain Rooksby was relieved by Lieutenant A. Howell on the 1st October 1916. In November Captain Young was involved in an unfortunate motor cycle accident during an air raid warning, coming into collision with a tramcar and having his right hand mutilated. He was removed to hospital and was subsequently relieved in the command of this Company by Captain T. T. Tucker. All the plant of this Company was of a fixed nature, mounted on permanent concrete foundations.

No. 38 (Tyne) A.A. Company R.E. (T.F.) was formed in the Hull district from detachments drafted from Haslar, Captain K. A. Mountain being posted from the headquarters at the Tyne to assume command. The authorised establishment of this Company was 4 officers and 122 other ranks, all R.E., manning fourteen 60 cms. searchlights operated from various petrol generating sets installed on four-wheeled trailers of a semi-mobile nature. The headquarters of No. 38 Company was established in Hull. During the period that the Company was stationed in this locality there were three airship raids on that city but no airships were brought down by the defences. These raids were fairly innocuous, though, on one occasion, the A.A.D. headquarters was very nearly hit by a bomb. On this occasion one of the detachments of the Company was also severely bombed, but the light was maintained and continued to illuminate the airship, an action which earned for the Company the congratulations of the A.A. Defence Commander. This Company also put down and manned seven lights at the Howden air station and Selby airship sheds, but these were subsequently handed over to another company and the activities of No. 38 were thereafter restricted to the vicinity of Hull and the coast, two anti-aircraft lights being manned at Spurn Point. The formation of this Company at Hull in July liberated the detachment of No. 22 (Tyne) Aeroplane Squadron Searchlight Section which had been in that district since May and which was now enabled to rejoin its main body at Hornchurch in Essex.

Simultaneously with the formation of Nos. 37 and 38 Companies, No. 40 (Tyne) A.A. Company R.E. (T.F.) was formed at Sheffield where it gradually relieved certain personnel of the East Riding (Fortress) R.E. (T.F.). The Company consisted at first of four searchlight stations in Sheffield itself[2] and two in Lincoln[3] and was under the command of Captain J. R. Abbott with Lieutenant R. B. T. Pinkney and Second Lieutenant H. Hutchinson. After a few months the two stations at Lincoln were handed over to another company and the number of lights at Sheffield raised to six. The establishment of No. 40 Company was ultimately increased to 4 officers[4] and 96 other ranks

---

[1] These were still the days before the air defence intelligence system had been properly developed and each raid was followed by a series of rumours of an astonishing nature. Thus, on one occasion, a report was received to the effect that a householder, living in a remote rural district, had been knocked up one night to find a German officer at his door who asked him the way to Leeds. The man, in his astonishment, gave full directions to the officer, who then walked down the garden path, stepped into a small car and was immediately drawn up into the sky.

[2] Two of these were at Shire Green and Intake.

[3] At Burton Road and Washingborough.

[4] Lieutenant R. E. Jardine then joined the Company.

and the number of lights to eleven. At the outset, the searchlights were sited too near the centre of the town and they were very shortly moved, most of the stations subsequently erected being located on the outskirts of Sheffield where they were better sited to engage enemy aircraft before the latter could reach their objective. Sheffield was only once raided, two Zeppelins coming over the town on the night of the 25th/26th September, 1916. So effective were the lighting restrictions in this locality that the airships cruised about the vicinity for about an hour, evidently uncertain of their whereabouts. Eventually, a train entered one of the main railway stations at Sheffield and this appeared to give the enemy some idea as to his location, for bombs were immediately dropped. The city escaped lightly, a few works being damaged and some slum dwellings partially destroyed, while a few people were wounded. As luck would have it, it was not possible to make much use of the searchlights on this occasion, as the factories put whale oil on to their furnaces, giving off a dense smoke which hung like a pall over the town and which the searchlights could not penetrate. At Lincoln, on this same night, bombs fell within twenty yards of the projector at Washingborough. One gun opened fire but without apparent result, though a part of a propellor was later found in the vicinity.

The transfer of the personnel of No. 2 (Tyne) Company from the London defences to the Midlands proceeded throughout June and into July. The first detachment to be relieved was sent, under Lieutenant Sutthery, to Nottingham, and the manning of lights in this area, which came under the A.A. Defence Commander, Birmingham, was commenced. This detachment grew in numbers by the addition of men drafted from Haslar and scraped together from the sections in London, and, towards the end of June, complete searchlight detachments of No. 2 Company were replaced by personnel of the London Electrical Engineers and sent to the Birmingham A.A. defence area. By the middle of July the transfer was completed and the headquarters of the Company moved to Coventry where, under the new title of No. 42 (Tyne) A.A. Company R.E. (T.F.), it manned thirteen lights in the Coventry, Rugby, Banbury, Didcot and Nottingham areas.[1] This district was particularly vulnerable on account of the large steel works and munition factories, which included H.M. Filling Factory at Chilcomb, said to have been the largest in England.

In spite of the difficulties which No. 2 Company had experienced in taking over the equipment at the London stations they were enabled, largely due to the energy and efficiency of C.Q.M.S. Gill, to evacuate the London District without a single deficiency to account for, though it is recounted that, having obtained the signature of the relieving officer to the last of his vouchers, this C.Q.M.S. was observed waiting at Wembley Hill railway station a good two hours before the train for Coventry was scheduled to start !

---

[1] The lights manned by No. 42 Company were as follows :—

Coventry	...	...	...	... 3 lights	Banbury	...	...	...	2 lights.
(Wyken, Pinley and Radford).					Didcot	...	...	...	2 lights.
Rugby	...	...	...	... 2 lights.	Nottingham	...	...	...	4 lights.

On arrival, the headquarters (of No. 42 Company) was established at the light station at Wyken and the Company then came for operations under the A.A. Defence Commander, Birmingham. This officer had on his staff Captain B. S. Millard, Hants. Fortress R.E. (T.F.) (subsequently London Electrical Engineers) on whose charge all the equipment in the Birmingham A.A.D. Command was at first held, thus relieving the individual companies of much irksome responsibility. By September, however, it became apparent that this method of stores accounting was of far greater magnitude than had originally been anticipated and all stores had then to be taken on charge by the various companies. This involved a great deal of work as no inventories of stores on each station were available, but, in due course, the whole matter was put upon a sound footing and the company ledgers balanced. This Company was fortunate in having, at first, its generating sets of the same type, all being fitted with Coventry Simplex engines. After the Company had been a week at Coventry, Major Arnison arrived and took over the command, Captain D. A. Williamson returning to Haslar.

Up to this date no orders or instructions as to the tactical operation of the lights had been issued by the authorities, and each company commander was left entirely to his own devices in this respect. Uniformity was entirely lacking and this did not lead to the most efficient employment of the personnel and material at the disposal of the A.A.D. Commanders.

At the end of July two further aeroplane squadrons searchlight units were raised for duty with the aeroplane barrage, and the existing aeroplane squadron searchlight sections were re-named companies. Of those now created, No. 24 was a London company while No. 25 (Tyne) Aeroplane Squadron Searchlight Company R.E. (T.F.) was provided by the Tyne Electrical Engineers. This Company was formed at Haslar and proceeded to the Old Brewery, Great Smith Street, Westminster, to mobilise. The establishment of these companies, to man six lights, was now standardized at 4 officers and 70 other ranks, including 16 A.S.C. drivers. The officers of No. 25, on formation, were :—
Captain D. A. Williamson, Lieutenants E. F. Rendell and A. S. Burdis and Second-Lieutenant D. S. Anderson. The senior N.C.O. in the Company was Serjeant T. Rumney, Serjeant Galloway performed the duties of C.Q.M.S., and the two mechanists on the establishment were Staff-Serjeants C. Armitage[1] and Coxon. Three of the lights were commanded by serjeants and three by corporals.

No. 25 Company occupied some weeks in the process of completing mobilisation and during this period the men were billeted in London, the officers residing at " Sutties' Hotel " off Russell Square. These aeroplane squadron searchlight companies were mobile units and the equipment of No. 25 consisted of six 3-ton Daimler lorries with Coventry Simplex generators, 24″ searchlights

---

[1] Subsequently commissioned in the Tyne Electrical Engineers.

with portable resistance frames and switchboards, all of which were drawn from Woolwich Arsenal. This plant was all new when issued and the personnel of the Company was occupied in fitting it up in the lorries before it could be taken into use. The transport of the Company consisted of three 30-cwt. P.E. lorries, one motor car and five motor cycles,[1] all of which came from the vehicle depot at Grove Park. All these vehicles, with the exception of the car for the O.C., were old and in a bad state of dilapidation, the 30-cwts. in particular proving a hindrance rather than a help, while the motor cycles were never to be relied upon. The car, a Sunbeam, though part worn, was in very good condition and was the envy of other company commanders, who, possibly not endowed with such powers of persuasion as the O.C. of No. 25, had to content themselves with rickety old Ford " Lizzies " for their personal conveyance.

By September, the Company was ready for orders to move, all the plant having been fitted up and tried out on Wimbledon Common and the R.E. personnel trained to their work. The A.S.C. personnel, however, caused some anxiety as, with the exception of two belonging to a class known as " six-bob-a-day-men," they were not highly trained drivers and lacked experience.

After being in readiness for about a fortnight, orders were received late one afternoon for the Company to move next day with all speed to Selby in Yorkshire. The journey occupied two days and was not effected without incident. Staff-Serjeant Coxon unfortunately met with an accident, breaking his leg ; all the 30-cwts. lorries broke down on the first day and the Company finally entered Selby market place late on the second day with every serviceable Daimler lorry towing another vehicle ! The headquarters of the Company was established at the "Londesborough Arms" at Selby, but for two days all efforts to obtain further orders from the York headquarters failed. The delay, however, gave the Company time to lick its wounds and on the third day in Selby, having obtained the loan of three nondescript American trucks from an A.S.C. depot in York, and in the absence of any positive orders whatsoever, searchlight stations were established at York, Riccal, Escrick, Hemingborough, Howden and Barlow. These dispositions were confirmed from headquarters at York late that very afternoon, the object of the move of the Company to Selby having apparently been to take a temporary place in the defence of the airship stations at Howden and Barlow. Difficulties of both technical and tactical natures, such as those which have been described, were of common occurrence in the early days and called for the exercise of much initiative on the part of the officers concerned, and it is creditable to observe that these difficulties were invariably overcome in a highly satisfactory manner.

---

[1] These were still early days in the mechanisation of the British Army and the motor cycle as an officer's " mount " was still a novelty to some. An officer of this Company, having occasion to make a purchase when riding through London, wheeled his machine into a side street before proceeding with his business. On walking into the main street once more, he was confronted by an irate A.P.M. who set about explaining to him how impossible it would be to win the war if subalterns did not carry walking sticks ! Even the sight of the motor cycle failed to convince this myrmidon of military law that a walking stick was other than an absolute necessity and the incident subsequently developed into the only case of " indiscipline " which was brought to the notice of the O.C. of this Company.

CONVOY OF DAIMLER 3-TON LORRIES.

Vehicles of this type formed the bulk of the mechanical transport during the Great War and were used to equip the mobile searchlight units in the home defences. This photograph depicts No. 25 (Tyne) Aeroplane Squadron Searchlight Company on the way from London to Selby, September 1916.

Shortly after the arrival of No. 25 Company at Selby Lieutenant Rendell returned to Haslar to take command of an overseas section and was replaced by Second-Lieutenant H. V. Owen. At the same time Serjeant Galloway left to join an overseas section and the quartermaster's duties in the Company were then taken over by C.Q.M.S. Grix who arrived from Haslar.

In July also, an additional Mobile Company, originally styled No. 7 (Tyne) A.A. Brigade Searchlight Company and subsequently No. 10 (Tyne) Mobile Searchlight Company R.E. (T.F.), was formed at Millbank hospital under the command of Captain M. C. James. The establishment of the mobile companies (or brigade searchlight companies as they were sometimes known) was then reduced to 4 officers and 108 other ranks, and the lights from twelve to nine. At the same time Nos. 11 and 12 A.A. Brigade Searchlight Companies were raised by the London Electrical Engineers.

No. 9 (Mobile) Searchlight Company, which had hitherto been scattered all over the Country, was now concentrated and attached to No. 3 Mobile Anti-Aircraft Brigade R.G.A., and, for a time, the headquarters of the two Tyne Mobile Companies was in the officers' quarters of the R.A.M.C. barracks at Millbank, London.

At the beginning of September the enemy resumed his raids on London with the maximum number of airships available. On the 2nd September sixteen left Germany, but the attack dwindled on the way and only two vessels approached nearer than St. Albans. Of these, the S.L.11, a wooden ship, was attacked in the air by Lieutenant W. Leefe Robinson, of No. 39 Squadron R.F.C., Hornchurch, and fell flaming headlong to earth near Cuffley. The airship was actually being held in the beams of No. 22 (Tyne) Aeroplane Squadron Searchlight Company at the time she was attacked and was held by them until brought down. This marked the first occasion on which an airship was shot down by machine gun fire from an aeroplane, and it amply demonstrated that aeroplanes could only function successfully when the airships were well illuminated. For their share in this night's work the detachments of No. 22 Company received the congratulations of the Commander-in-Chief, Home Forces.

The German command was not, however, to be deterred by the loss of the S.L.11, and, on September 23rd, another raid on a grand scale was launched. This time eleven ships set out and succeeded in bombing Nottingham[1] as well as London, but the L.32, in endeavouring to avoid the main defences of London, was attacked by Second-Lieutenant F. Sowrey, also of No. 39 Squadron, and brought down in flames at the village of Billericay in Essex. Lights of No. 9 and of No. 22 Companies were in operation in connection with this raid.[2]

---

[1] During this raid a bomb fell to the west of the town and exploded at a point where the telephone wires to three of the searchlight stations were concentrated. The detachments carried on independently and, not being fully trained, they continued to search the sky for the Zeppelin for about two hours after it had left the area.

[2] Part of the framework of this airship is now exhibited in the drill hall at the Unit's Headquarters.

The third of these now famous raids took place on the night of the 1st October when another attack was made on London, this time with seven ships only. It was on this occasion that Kapitanleutnant Heinrich Mathy, the most redoubtable of the Zeppelin commanders, perished with his ship, the L.31, at Potters Bar. " At 11.30 he is over Ware. . . . He sets his engines going full speed and starts off due south to the attack. Ten minutes later, over Cheshunt, the lights find him. . . . So many lights that the ship looks from a distance to be floating on a pyramid of rays. . . . Mathy tries to escape from the searchlights by making a sudden turn to the right. . . . The lights still hold him."[1] . . . Second-Lieutenant W. Tempest, once more from No. 39 Squadron, R.F.C., attacked the L.31 twice, in spite of the severe anti-aircraft shell-fire being directed at her, and the ship, bursting into flames, sank slowly to the ground. Again, the searchlights of No. 9 and No. 22 Companies had the distinction of playing conspicuous parts in this action. " Although the projectors were still nearly all of the small 60 cms. pattern, the detachments were by now well enough trained to pick up and hold the airship of the period, in suitable weather. In addition to lighting up the target for the guns, search-light beams acted as valuable pointers to our pilots in the air. The defences, so far as London was concerned, had definitely overcome the airship menace, and from now on, no German airship intentionally approached London itself."[2]

In September, sufficient personnel and plant being available in the Birmingham A.A.D.C., an extra ' gun light ' company was formed, No. 42 Company giving birth to No. 41 (Tyne) A.A. Company R.E. (T.F.), which manned 12 searchlights in the Birmingham district, while No. 42 operated 17 in the Coventry, Derby and Nottingham areas.

Major Arnison remained in command of No. 42 Company, while Captain Sutthery was posted in command of No. 41 and established his headquarters in hutments erected on the post office grounds, Bardsley Green, Birmingham. In October the establishments of these companies were revised, as a result of which Major Arnison returned to Haslar. Captain Sutthery rejoined No. 42 Company as O.C. and Captain Millard took over the command of No. 41. The establishments of these two companies were then as follows :—

No. 41 (Tyne) A.A. Company R.E. (T.F.) ...3 officers ...103 other ranks.
No. 42 (Tyne) A.A. Company R.E. (T.F.) ...7 officers ...149 other ranks.

No. 10 (Tyne) Mobile Searchlight Company, which was equipped similarly to No. 9, was at first considerably split up. Following the air raid on Portsmouth on the night of the 25th September for example, two detachments, under Lieutenant Winter, (who had joined the Company from No. 22 at Hornchurch), were despatched to Fratton goods station to serve a 6" gun mounted on a railway truck. On arrival the G.O.C. Portsmouth ordered these lights to be transported to the Isle of Wight, but as this was contrary to the intention of

---

[1] " Air Defence " Major General Ashmore, p. 26.
[2] " Air Defence " Major General Ashmore, p. 27-28.

G.H.Q. Home Forces, the detachments were recalled to London. Two of their sets of equipment were subsequently handed over to the School of Electric Lighting at Stokes Bay on its being taken over in October by G.H.Q. Home Forces.[1] At the end of September Lieutenant Winter was detached with his section of three lights to join a mobile battery on the Essex and Suffolk coast stationed at Leiston. Subsequently, two of the lights of this section were moved for a few weeks to Sutton-on-sea and Skegness on the Lincolnshire coast and the section later withdrew to Yarmouth and Lowestoft. In November Captain James received an appointment abroad and Lieutenant Winter took over the command of the Company. In the middle of December the headquarters of the Company moved to Harwich and the searchlights took up positions on the coast line between Harwich and Yarmouth.

In the early winter of 1916, No. 3 Mobile Brigade, to which belonged No. 9 (Tyne) Mobile Searchlight Company, was allotted to that portion of the coast extending from Scarborough to Boston and the headquarters of this Company was moved to Hull. The population of this city, still the victim of many raids, was concerned for its own security, and strict orders were therefore given that no lorries should proceed through the streets of Hull itself lest it should be supposed that any portion of the defence was being withdrawn. The O.C. (Major Firmin) experienced great difficulty in visiting his sections in Yorkshire as this involved transporting his car by ferry across the Humber. It was therefore arranged for the company headquarters to be removed to Selby where they were accommodated over a chemist's shop, the workshop being set up in a fully-equipped garage which was taken over for the purpose. As an example of the volume of administrative work involved in the running of an 'independent' unit in time of war, it may be observed that no less than twenty-five weekly 'returns' had to be rendered to the headquarters of the Humber Garrison, many of them containing only the information " Nil " !

In October further aeroplane squadron searchlight companies were formed, and of these Nos. 27 and 29 were formed at Millbank Hospital from personnel of the Tyne Electrical Engineers and were commanded by Captain H. O. Rogerson and Captain W. H. James respectively. These Companies after the usual period of fitting-out in London left, on the 9th December, for their allotted stations at Selby and York. No. 29 Aeroplane Squadron Searchlight Company then established its stations between Catterick and York, (working in conjunction with No. 76 Squadron R.F.C. at Ripon), and No. 27 Company to the south of York in the area of Leeds and Selby.

Supplies of petrol were now becoming shorter and while with No. 29 Company Captain James devised a gas-bag for the utilization of commercial gas for the running of the searchlight engines. The gas-bag was mounted on a trailer and

---

[1] By this date the anti-aircraft searchlights in the Portsmouth defences had been taken over by personnel of the Hampshire Fortress R.E. (T.F.) and the lights there were grouped together to form No. 48 (Hants.) A.A. Company R.E. (T.F.).

was towed to the local gas-works for filling. This device was adopted to a considerable extent in the Leeds A.A. defence command and the name of Captain James was brought to the notice of the Secretary of State in connection with this invention.[1]

At the same time as these additional companies were formed for the barrage line, No. 22 Company moved from Hornchurch to Gainsborough in Lincolnshire where it was attached for operations to No. 33 Squadron R.F.C. No. 25 Company also moved at this date from Selby to Gainsborough and operated with the same squadron.[2] The headquarters of both companies were established at Tennyson Cottage, Gainsborough, and they came under Headquarters, Lincoln, for administration and under A.A.D.C. Sheffield for operations, though the latter left them severely alone and they worked at first entirely with the R.F.C. All ranks were in billets. At this stage the need for telephonic communication between headquarters and the searchlights in the barrage line became an urgent consideration, as the air defence intelligence system was beginning to be developed and there were many occasions during raids when the lights were in a position to communicate valuable information.

On the 23rd October Lieutenant H. S. Ripley, then serving with No. 42 (Tyne) A.A. Company at Coventry, was posted, with the temporary rank of captain, to the 6th Brigade R.F.C. as searchlight officer to assist in the organisation of the aeroplane barrage. This was now being further extended and was to consist of a double line of lights following a chain of aerodromes, which ultimately stretched from Edinburgh to Dover. All the country along this line had to be reconnoitred and the most suitable sites obtained for the lights. Captain Ripley also officiated as liaison officer between the R.F.C. and G.H.Q. Home Forces in arranging the many details of personnel.[3] This officer also carried out a considerable amount of experimental work with various types of projectors and was sent by the R.F.C. to Paris to superintend the construction of special 150 cms. lights built to British specification by Messrs. Saulter and Harlé. These lights were tested out in the Paris A.A. defences by arrangement with French G.Q.G.

The three lights of No. 33 (Tyne) A.A. Company, which were disposed in the neighbourhood of Hartlepool under Lieutenant C. B. Williamson, were in action on the night of the 27th November, 1916, when the Zeppelin L.34 was brought down in flames. The L.34 came in over the Black Hall Rocks at about half-past eleven and was at once picked up by the Hutton Henry light. She then turned in her course and bombed the light severely in an attempt to extinguish it. The searchlight was, nevertheless, maintained in action and this enabled Second-Lieutenant I. V. Pyott, of No. 36 Squadron,

---

[1] In some cases gas was laid on to the engines direct from the mains as there was no gas-bag available. At one small village where this was done the engine's demand for gas was so large that the cottages were left in semi-darkness to the extreme annoyance of the villagers.

[2] At this time the lights of No. 25 Company were at Blyton, Blyborough, Heapham, Fillingham, Saxilby and Upham.

[3] In the early stages of the development of the barrage line, in order to economise in personnel, certain of the lights actually situated on the aerodromes were operated by R.F.C. personnel, but this system did not long prevail.

R.F.C., to attack the airship with machine gun fire with the result that she fell, a mass of blazing wreckage, into the sea off West Hartlepool shortly before midnight. Her raid was not, however, entirely fruitless, as just before she caught fire she dropped her remaining bombs on West Hartlepool, killing four and wounding thirty-four persons and damaging some forty houses.

With the extension of the barrage line northwards from Alnwick, certain alterations were made in the organisation of the searchlight defences. Three new companies were formed of local personnel for the defence of Dundee, Edinburgh and Glasgow, and in December the personnel of No. 33 (Tyne) A.A. Company R.E. in Northumberland and Durham was absorbed into Nos. 34 and 35 Companies at Newcastle, and the company number was taken over by No. 33 (Renfrew) A.A. Company, which was formed at Glasgow by the Renfrewshire Fortress R.E. (T.F.). Nos. 34 and 35 Companies now operated both the gun and aeroplane lights, while the group of lights which had been detached to Gretna were, at the same date, increased from seven to eight[1] and constituted a separate company under Captain J. A. Lawther receiving the title No. 50 (Tyne) A.A. Company R.E. (T.F.).

No. 34 Company, still under the command of Major C. M. Campbell, now manned ten lights, mostly fixed, of which eight[2] were to the north of the River Tyne and two[3] to the south. No. 35 Company (Captain K. S. M. Scott) manned ten lights, all to the south of the river[4], and there were also in this area the three lights attached to No. 4 Mobile Battery with headquarters at Marsden Hall.[5] The C.Q.M.S. of No. 34 Company was C.Q.M.S. Preston and of No. 35 C.Q.M.S. Best. The mechanists were Staff-Serjeants J. and W. Walker. The aeroplane lights of No. 35 Company joined up in the south with those of No. 36 (N. Riding) A.A. Company whose headquarters were at Middlesbrough.

In December also the aeroplane squadron searchlight companies ceased to be employed solely in co-operation with aircraft and took over the operation of certain gun-lights in addition to the aeroplane lights. With the exception of four London units[6] these were then all re-named A.A. Companies R.E. the changes in title of the Tyne Companies being as follows :—

No. 22 (Tyne) Aer. Sqdn. S/L Company—No. 58 A.A. Company R.E.
No. 25 (Tyne) Aer. Sqdn. S/L Company—No. 57 A.A. Company R.E.
No. 27 (Tyne) Aer. Sqdn. S/L Company—No. 60 A.A. Company R.E.
No. 29 (Tyne) Aer. Sqdn. S/L Company—No. 56 A.A. Company R.E.

[1] At W. Scales, Howend, Batenbush, Springfield, Wether Hill, Bowness, Todholes and Glasson. All fixed.
[2] At Blyth, Seghill, Whitley, Wallsend, N. Gosforth, Benwell, West Denton and Heddon. One of the few war pictures dealing with anti-aircraft defence, Mr. (later Sir John) Lavery's painting " Anti-Aircraft, Tyneside," a copy of which hangs in the officers' mess at Headquarters of the Tyne Electrical Engineers at Tynemouth, was sketched at the Cullercoats A.A. gun during practice on the night of the 4th October, 1917.
[3] At South Shields and Cleadon.
[4] At Winlaton, Dunston Hill, Mount Pleasant, Wardley, Marley Hill, Birtley, Penshaw, Fulwell Quarries, Tunstall and Seaham.
[5] The three lights were at Newsham, Springwell and Seaham Grange.
[6] These were Nos. 20, 21, 24, and 26 Companies which combined to form two units styled No. 20/21 and No. 24/26 Aer. Sqdn. S/L Companies which continued until 1918.

The establishment of these companies now also underwent certain modifications. The number of lights manned by No. 56 A.A. Company under Captain W. H. James rose to twelve, scattered round York and extending northwards to Catterick, where they linked with those of No. 36 (N. Riding) A.A. Company. The headquarters of No. 60 A.A. Company under Captain Rogerson remained at Selby, taking over several gun-lights in that locality. To the south, No. 57 A.A. Company, with headquarters at Gainsborough, took over three gun-lights in the Scunthorpe area from the Hull Command.[1] At this time the Company was temporarily under the command of Lieutenant A. S. Burdis, Captain Williamson having been seconded to the R.F.C. in November.[2] No. 58 A.A. Company, still commanded by Captain Bowers, also increased its establishment by the addition of certain gun-lights and the headquarters were removed to Cranwell near Sleaford. The establishment of officers was however reduced to three, Lieutenant Murray transferring at this date to No. 38 Company at Hull. The most southerly of the barrage lights of No. 58 A.A. Company linked up with those of a London A.A. Company which had its headquarters at Peterborough.

The detachments for the new lights were supplied from Haslar but no corresponding increase in officers or headquarters personnel was authorised.

These companies soon ceased to retain their mobile characteristics and the lorries were handed in to the A.S.C., the engines and generators for the lights being thereafter housed in huts or on special four-wheeled trailers. The 30-cwts. vans were retained for administrative purposes and the mechanists were issued with sidecar motor-cycle outfits.

From July to November the A.A. searchlights at Nottingham and Derby were operated by No. 42 Company with headquarters at Coventry. There were five lights placed round Nottingham and three at Derby.[3] Owing to shortage of personnel these stations were at first manned at about half strength. There was no transport and the first officer in command of these lights was obliged to hire a taxi to take him round his stations. Lieutenant J. L. Batey took over these detachments shortly after their initial formation and was succeeded in November by Lieutenant D. E. Ross. By this time additional personnel had become available, and in December the Nottingham and Derby stations were brought up to strength and were formed into a separate company with the title No. 63 (Tyne) A.A. Company R.E. (T.F.). The first officer to command this Company was Captain J. F. S. Hunter who had become surplus on the absorption of No. 33 Company into Nos. 34 and 35 Companies in the Tyne Garrison. The establishment of officers was three and the headquarters was established in the G.P.O. at Nottingham, where the control was installed. The Company operated under the A.A.D.C. Birmingham, manning the eight lights above referred to, all fixed and all equipped with Dorman engines.

---

[1] At Brumley, Santon and Sawcliff.
[2] On February 1st, 1917, Captain C. F. Scott was appointed to command this Company which subsequently took over lights at Scampton, Stow, Burton Road, North Hykeham, Anborn, and Duddington, and established new stations at Scotterthorp, Thorne, Crowle and Canwick.
[3] These were :—Nottingham.—Bashford, Skeinton, Welford, Barton, Stapleford.
   Derby.—Chaddlestone, Allenton, Mickleover.

In the event of a raid R.E. officers used to visit one or other of their lights, generally by motor cycle. All officers engaged on the A.A. service were issued with the 'green pass,' which has already been mentioned, and this enabled them to use undimmed headlights when on the road. This was regarded in certain quarters as a doubtful privilege as, while it certainly rendered driving through the darkened streets and country lanes less dangerous, the civilian populace in some localities used to throw bricks at passing motor cycles with headlights, under the impression that the officer had been absent from his station and had to dash back to duty when the alarm was given !

By April 1917, the last month in which the army list attempts to shew the distribution of officers of the Tyne Electrical Engineers, the anti-aircraft search-light units at home which had been formed by them were as follows :—

No. 9 (TYNE) MOBILE SEARCHLIGHT COMPANY. (Hd.-Qrs. Selby).

Major N. H. Firmin.
Lieut. J. A. Power.
Lieut. J. Robinson.

No. 10 (TYNE) MOBILE SEARCHLIGHT COMPANY. (Hd.-Qrs. Harwich).

Lieut. R. P. Winter.[1]
Lieut. C. Russell.
Lieut. H. G. Buxton.
2nd-Lieut. T. S. Marshall.[2]

No. 34 ANTI-AIRCRAFT COMPANY R.E. (Hd.-Qrs. Newcastle-upon-Tyne).

Major C. M. Campbell.
Lieut. C. B. Williamson.
Lieut. O. W. E. Hedley.
Lieut. W. Dixon.
2nd-Lieut. H. S. Watson.

No. 35 ANTI-AIRCRAFT COMPANY R.E. (Hd.-Qrs. Newcastle-upon-Tyne).

Captain K. S. M. Scott.
Lieut. A. G. Dickson.
Lieut. W. Watson.
Lieut. J. R. T. Emerson.

No. 37 ANTI-AIRCRAFT COMPANY R.E. (Hd.-Qrs. Leeds).

Captain J. Young.
Captain T. T. Tucker.
Lieut. E. Harrison.
Lieut. A. Howell.

[1] Lieut. Winter handed over to Captain F. B. C. Sutthery in April 1917.
[2] 2nd Lieut. Marshall was killed in a motor cycle accident when on duty in Essex on the 1st August 1917.

No. 38 Anti-Aircraft Company R.E.   (Hd.-Qrs. Hull).
    Captain K. A. Mountain.
    Lieut. J. B. Murray.
    Lieut. C. A. Stephens.
    Lieut. F. S. Corby.

No. 40 Anti-Aircraft Company R.E.   (Hd.-Qrs. Sheffield).
    Captain J. R. Abbott.
    Lieut. R. B. T. Pinkney.
    Lieut. H. Hutchinson.
    Lieut. R. E. Jardine.

No. 41 Anti-Aircraft Company R.E.   (Hd.-Qrs. Birmingham).
    Captain B. S. Millard (Hants. (Fortress) R.E.).
    Lieut. H. J. Ingman.
    2nd Lieut. W. H. Winstanley.
    2nd Lieut. A. Woods.

No. 42 Anti-Aircraft Company R.E.   (Hd.-Qrs. Coventry).
    Captain F. B. C. Sutthery.
    Lieut. E. H. White.
    Lieut. J. L. Batey.

No. 50 Anti-Aircraft Company R.E.   (Hd.-Qrs. Gretna).
    Captain J. A. Lawther.
    Lieut. R. G. Ellis.

No. 56 Anti-Aircraft Company R.E.   (Hd.-Qrs. York).
    Captain W. H. James.
    Lieut. H. G. Campbell.
    2nd Lieut. F. W. Newman.

No. 57 Anti-Aircraft Company R.E.   (Hd.-Qrs. Lincoln).[1]
    Captain C. F. Scott.
    Lieut. A. S. Burdis.
    Lieut. D. S. Anderson.
    Lieut. H. V. Owen.

No. 58 Anti-Aircraft Company R.E.   (Hd.-Qrs. Cranwell).
    Captain F. H. Bowers.[2]
    Lieut. T. W. Crawford.
    Lieut. C. Graham.

No. 60 Anti-Aircraft Company R.E.   (Hd.-Qrs. Selby).
    Captain H. O. Rogerson.
    Lieut. F. L. Adendorff.
    2nd Lieut. C. B. Elliott.
    2nd Lieut. H. E. McDonald.

[1] Headquarters moved to Lincoln in March, and back again to Gainsborough in July.
[2] In June 1917, Captain Bowers was killed in a motor cycle accident and was replaced as O.C. of this Company by Captain K. A. Mountain.

No. **63** Anti-Aircraft Company R.E.   (Hd.-Qrs. Nottingham).
  Captain J. F. S. Hunter.
  Lieut. D. E. Ross.
  Lieut. H. Joseph.

At this time, there were no less than **17,000** anti-aircraft personnel in the British Isles, excluding the personnel employed with the naval aircraft round the coasts. By July there were in active existence twelve home defence squadrons of the Royal Flying Corps, operating from thirty aerodromes in the barrage line, and a total of forty-two anti-aircraft companies R.E.[1] scattered about the Country. Chains of observing stations[2] had been formed extending along the whole of the east coast, an efficient air defence intelligence system was in the making, and the organisation and tactical employment of the various components of the defence had by now been developed in accordance with a uniform and effective scheme of co-operation.

The searchlights of the defence were now classified according to their particular employment as follows :—

  (*a*) Aeroplane lights.
  (*b*) Mobile lights.
  (*c*) Advanced lights.
  (*d*) Linking lights.
  (*e*) Fighting lights.

---

[1] These were :—

No. 1 A.A. Company
No. 2 A.A. Company
No. 3 A.A. Company   London Electrical Engineers.—London.
No. 4 A.A. Company
No. 5 A.A. Company
No. 6 A.A. Company
No. 9 Mobile Searchlight Company—Tyne Electrical Engineers.—Selby.
No. 10 Mobile Searchlight Company—Tyne Electrical Engineers.—Harwich.
No. 11 A.A. Brigade Searchlight Company—London Electrical Engineers.
No. 12 A.A. Brigade Searchlight Company—London Electrical Engineers.
No. 20/21 Aeroplane Squadron Searchlight Company—London Electrical Engineers.
No. 24/26 Aeroplane Squadron Searchlight Company—London Electrical Engineers.
No. 31 A.A. Company—City of Dundee Fortress R.E. (T.F.).—Dundee.
No. 32 A.A. Company—City of Edinburgh Fortress R.E. (T.F.).—Edinburgh.
No. 33 A.A. Company—Renfrewshire Fortress R.E. (T.F.).—Glasgow.
No. 34 A.A. Company—Tyne Electrical Engineers.—Newcastle.
No. 35 A.A. Company—Tyne Electrical Engineers.—Newcastle.
No. 36 A.A. Company—North Riding Fortress R.E. (T.F.).—Middlesbrough.
No. 37 A.A. Company—Tyne Electrical Engineers.—Leeds.
No. 38 A.A. Company—Tyne Electrical Engineers.—Hull.
No. 39 A.A. Company—East Riding Fortress R.E. (T.F.).—Killingholme.
No. 40 A.A. Company—Tyne Electrical Engineers.—Sheffield.
No. 41 A.A. Company—Tyne Electrical Engineers.—Birmingham.
No. 42 A.A. Company—Tyne Electrical Engineers.—Coventry.
No. 43 A.A. Company—Essex Fortress R.E. (T.F.).—Harwich.
No. 44 A.A. Company—Essex Fortress R.E. (T.F.).—Pulham.
No. 45 A.A. Company—Kent Fortress R.E. (T.F.).—Sheerness.
No. 46 A.A. Company—Kent Fortress R.E. (T.F.).—Chatham.
No. 47 A.A. Company—London Electrical Engineers.—Dover.
No. 48 A.A. Company—Hampshire Fortress R.E. (T.F.).—Portsmouth.
No. 49 A.A. Company—Lancashire Fortress R.E. (T.F.).—Manchester.
No. 50 A.A. Company—Tyne Electrical Engineers.—Gretna.
No. 51 A.A. Company—Lancashire Fortress R.E. (T.F.).—Liverpool.
No. 53 A.A. Company—London Electrical Engineers.—Newhaven.
No. 55 A.A. Company—North Riding Fortress R.E. (T.F.).—Middlesbrough.
No. 56 A.A. Company—Tyne Electrical Engineers.—York.
No. 57 A.A. Company—Tyne Electrical Engineers.—Lincoln.
No. 58 A.A. Company—Tyne Electrical Engineers.—Cranwell.
No. 59 A.A. Company—Kent Fortress R.E. (T.F.).—Kent.
No. 60 A.A. Company—Tyne Electrical Engineers.—Selby.
No. 62 A.A. Company—City of Edinburgh Fortress R.E. (T.F.).—N. Queensferry.
No. 63 A.A. Company—Tyne Electrical Engineers.—Nottingham.

[2] Manned by Observer Companies of the Royal Defence Corps.

The aeroplane barrage now consisted of a continuous double line of lights spaced at intervals of approximately 3½ miles, the distance between each line of lights being about 5 miles. Each squadron headquarters controlled its own section of the barrage and the two lines were separately controlled so that they were seldom in action with the same target at the same time. Each light in the barrage line was numbered according to its section, this number being that of the squadron under the control of which it operated ; each individual light was further marked by a distinguishing letter denoting its position.

For the purpose of administration each light was in communication with squadron headquarters by means of the public telephone and was registered as an " air-bandit " post ensuring priority of connection in the event of a raid. The telephonic address of each A.A.D.C. headquarters was also " air-bandit." Warnings were issued to each light from squadron headquarters by a series of colours, viz. :—

Green	...	Field Marshal's warning.
Red	...	Take air-raid action.
Yellow	...	Resume normal conditions.
White	...	All clear.

As long as red was in force each light carried out an independent search in its own patrol area, and if a target was picked up the fact was notified by telephone to squadron headquarters and by the firing of a rocket in the direction of the target as a signal to pilots in the air. One light at a time only was to be exposed on each target, and the target was not to be illuminated at the moment of attack by friendly patrols in order to avoid the danger of illuminating the latter.

The mobile lights were employed on detached defences or on lines of communication between other defences. Their purpose was to take the enemy by surprise and, for this reason, they were frequently moved and were ordered not to expose until the target was well within the range of the guns with which they were co-operating, and then only by order of the gun commander.

Advanced, linking and fighting lights formed part of fixed defences established round important munition areas or large towns. The advanced lights furnished the outer ring and their main object was to turn the raiders away from the defence. If they failed in this object they retained the target under observation, passing it on to the linking lights which connected the outer ring with the inner, or fighting, lights. The fighting lights were always associated with guns and were sited 300 to 400 yards away from the guns they served. These lights were controlled entirely by the gun commander who alone ordered their exposure : the advanced lights were exposed at the discretion of the N.C.O. in charge. From the very first it was held that too many lights on a target obscured it from the gunners' point of view and that two, or at most three, lights on one target were generally sufficient.

SEARCHLIGHT PROJECTOR ON SLEDGE.

As used by No. 9 (Tyne) Mobile Searchlight Company for moving the searchlight across the ground from the lorry to its site.

To face page 144]

The summer of 1917 was marked by the very successful use by the enemy of aeroplanes for raiding London, and these raids had such effect on the public mind that the Government ordered a reconstruction of the defences to meet this fresh danger. Accordingly, on the 31st July, the London Air Defence Area (L.A.D.A.) was officially established under Major-General E. B. Ashmore and combined all the branches of the air defences in the South of England under one organisation. The area covered by L.A.D.A. was almost identical with that of the Home Defence referred to earlier.

The mobile companies continued to be scattered over wide areas and their lights were constantly being moved from place to place in accordance with the prevailing requirements. No. 9 (Tyne) Mobile Company, for example, moved position every few days and was constantly in action, as Flamborough Head, in the vicinity of which it was operating, was a favourite landfall for Zeppelins from which they used to set their courses for raids after crossing the North Sea. To assist in the rapid erection of their stations, this Company devised a form of sledge on which the projector could be moved across the ground from the lorry to its site.

No. 10 (Tyne) Mobile Company also covered large areas with its lights. At one time, in August, when the Company was commanded by Captain Sutthery, three lights under Lieutenant Parnall were disposed so widely apart as Chilworth, Farnborough and Dersingham. This last light was termed the " King's searchlight " being situated on the village green at Dersingham, near Sandringham House.[1]

During the latter half of the year it was decided that the existing system for providing A.A. searchlight defence was extravagant in personnel and a comprehensive scheme of re-organisation was embarked upon. Under the new scheme the personnel required for the searchlights co-operating with guns was to be attached to the various companies of the R.G.A. for operations but to their respective depots for administration. In each A.A.D.C. an R.E. officer was attached as second-in-command and as technical adviser. The result of this scheme was the saving of a considerable number of officers and administrative personnel which had hitherto formed part of the many A.A. companies, but we have seen what an immense amount of additional administrative work was thereby thrown upon the Depot Headquarters of the Tyne Electrical Engineers at Haslar. It was in October that the first move was made to give effect to this scheme of re-organisation by disbanding the company headquarters of the mobile companies attached to artillery brigades and the attachment of the searchlight sections to individual batteries. The system was not yet applied in its entirety, however, for one officer, the C.Q.M.S. and the pay serjeant of each company remained at brigade headquarters and continued, for a while longer, to carry on the administration of the sections.

---

[1] The defence for Queen Alexandra at Sandringham House had been manned by the R.N.V.R. Mobile Brigade as early as August, 1916.

K

By January 1918 the great re-arrangement of the R.E. personnel was completed, the headquarters of the gun-light companies being disbanded, while a new series of A.A. companies R.E., numbered 1 to 12, had come into existence and absorbed the searchlight personnel serving the aeroplane barrage.

Nos. 9a and 10 (Tyne) Mobile Searchlight Companies were now finally broken up, the searchlight detachments becoming attached to the guns of the 1st and 2nd Mobile Brigades respectively, with headquarters at Epping and Sevenoaks.

The distribution of Tyne officers with these brigades subsequent to this re-organisation was as follows :—

No. 1 Mobile Brigade. (Hd.-Qrs. Epping).
 No. 8 Battery ... 2nd Lieut. W. Rintoul—Three searchlights.
 No. 9 Battery ... Lieut. J. A. Power ,, ,,
 No. 10 Battery ... Lieut. J. Robinson ,, ,,
No. 2 Mobile Brigade. (Hd.-Qrs. Sevenoaks).
 Headquarters ... Captain F. B. C. Sutthery.
 No. 7 Battery ... 2nd Lieut. E. J. Parnall—Three searchlights.
 No. 11 Battery ... 2nd Lieut. P. V. Horler ,, ,,
 No. 12 Battery ... 2nd Lieut. C. Armstrong ,, ,,

The R.E. personnel at each battery headquarters now consisted of 1 officer, 1 mechanist and 1 section serjeant. The detachment for each searchlight consisted of 1 corporal, 1 second-corporal and 3 sappers. It was while attached to these batteries that 2nd Lieutenants Parnall and Armstrong received special mention for erecting during an air raid a complete searchlight station at Burnham-on-Crouch in the, at that date, unequalled time of two hours.

On the Tyne Nos. 34 and 35 Companies were disbanded, the personnel employed with the aeroplane lights in the districts formerly served by these two companies going to form No. 1 A.A. Company R.E. with headquarters at Newcastle. This Company consisted of eighteen 24″ aeroplane barrage lights (including one spare mobile set)[1] with an establishment of 4 officers and 117 other ranks and occupied an area from Alnwick to Hartlepool, working in co-operation with No. 36 Squadron R.F.C. The officers of No. 1 A.A. Company R.E. on its formation were :—

No. 1 A.A. Company R.E. (Hd.-Qrs. Newcastle).
 Captain K. S. M. Scott.[2] (From No. 35 A.A. Company).
 Lieut. W. Watson. (From No. 35 A.A. Company).
 Lieut. O. W. E. Hedley. (From No. 34 A.A. Company).
 2nd Lieut. G. T. B. Scullard.

---

[1] As to the light stations finally manned by this Company see page 156.
[2] Subsequently appointed M.B.E. Brought to the notice of the Secretary of State for War for valuable services.

The gun companies, R.G.A., in the Tyne A.A.D.C. were Nos. 19 and 20 at Newcastle and No. 21 at Gretna and to each of these Tyne Electrical Engineers personnel was attached to work the gun lights. The officers so attached and the lights with which they were supplied were as follows :—

No. 19 A.A. COMPANY R.G.A. (Hd.-Qrs. Windsor Terrace, Newcastle).
Lieut. W. Dixon. (From No. 34 A.A. Company).
9 Fighting 24" single searchlights.[1]
1 Spare mobile 24" single searchlight.

No. 20 A.A. COMPANY R.G.A. (Hd.-Qrs. Windsor Terrace, Newcastle).
2nd Lieut. F. Thompson.
9 Fighting 24" single searchlights.[2]
1 Spare mobile 24" single searchlight.

No. 21 A.A. COMPANY R.G.A. (Hd.-Qrs. H.M. Factory, Gretna).
Lieut. R. G. Ellis. (From No. 50 A.A. Company).
6 Fighting 24" single searchlights.[3]
1 Linking 24" single searchlight.

In April, Major Campbell, formerly O.C. of No. 34 (Tyne) A.A. Company, who had been second-in-command of the Harwich A.A. defences since September 1917, was appointed second-in-command at the headquarters of the Tyne A.A.D.C.

In the York and Leeds areas the old 37th, 56th and 60th A.A. Companies were broken up and Nos. 2 and 3 A.A. Companies R.E. came into being, the former to man 12[4] and the latter 13[5] aeroplane lights at Ripon and York respectively.

The establishment of No. 2 A.A. Company was 3 officers and 84 other ranks and of No. 3 A.A. Company 3 officers and 90 other ranks. Each company had one spare set of equipment which was mobile. The officers appointed to these companies on their formation were :—

No. 2 A.A. COMPANY R.E. (Hd.-Qrs. Ripon).
Captain T. T. Tucker. (From No. 37 A.A. Company).
Lieut. J. R. T. Emerson. (From No. 35 A.A. Company).
2nd Lieut. A. Stocks (City of Edinburgh Fortress R.E. (T.F.)).

No. 3 A.A. COMPANY R.E. (Hd.-Qrs. York).
Captain W. H. James. (From No. 56 A.A. Company).
2nd Lieut. E. Rendell. (From No. 56 A.A. Company).
2nd Lieut. L. S. Winkworth.[6] (From No. 56 A.A. Company).
2nd Lieut. C. B. Elliott. (From No. 60 A.A. Company).

---

[1] At Seghill, North Gosforth, West Denton, Benwell, Wallsend, Whitley, Blyth, South Shields and Cleadon.
[2] At Wardley, Mount Pleasant, Dunston Hill, Winlaton, Fulwell Quarry, Birtley, Penshaw, Tunstall and Seaham.
[3] At Todholes, West Scales, Springfield, How End, Wether Hill and Glasson.
[4] At Middleton Tyas, Neasham, Catterick, Cowton, Scruton, Yafforth, Theakstone Grange, Otterington, Brickyard Farm, Kilvington, Sessay and Rainton.
[5] At Helperby, Helperby Moor, Alne, Great Ouseburn, New Parks, Moor Moncton, Wiggington, Heworth Without, Crockey Hill, Copmanthorpe, Escrick, North Duffield and Cawood.
[6] Temporarily, *vice* 2nd Lieut. Rendell, while the latter was on leave.

These two companies were now under headquarters Leeds A.A.D.C., which also included Nos. 24, 25 and 28 A.A. Companies R.G.A. to which Tyne personnel was allotted to work the gun lights in accordance with the following scale :—

No. 24 A.A. COMPANY R.G.A. (Hd.-Qrs. G.P.O. Leeds).
    Lieut. A. Howell. (From No. 37 A.A. Company).
    5 Fighting 24″ single searchlights.[1]
    2 Fighting 24″ pair searchlights.[2]
    1 Linking 24″ searchlight.[3]

No. 25 A.A. COMPANY R.G.A. (Hd.-Qrs. York Bgds., Commercial St., Halifax).
    2nd Lieut. J. A. Ogilvy. (From No. 37 A.A. Company).
    6 Fighting 24″ single searchlights.[4]
    1 Fighting 24″ pair searchlight.[5]
    1 Spare mobile 24″ single searchlight.

No. 28 A.A. COMPANY R.G.A. (Hd.-Qrs. Labour Exchange, Selby).
    Captain E. H. Gibbon. (From No. 60 A.A. Company).
    7 Fighting 24″ single searchlights.[6]
    1 Linking 24″ single searchlight.[7]
    1 Spare mobile 24″ single searchlight.

The headquarters of the Humber A.A.D.C. was established at Hull and embraced Nos. 26 and 27 A.A. Companies R.G.A. as well as No. 4 A.A. Company R.E. This latter unit, co-operating with No. 33 Squadron R.F.C. at Kirton Lindsey, was formed with headquarters at Gainsborough[8] with an establishment of 4 officers and 116 other ranks including 2 A.S.C., M.T. drivers. The officers posted to this Company, which manned eighteen lights,[9] were :—

No. 4 A.A. COMPANY R.E. (Hd.-Qrs. Gainsborough).
    Captain C. F. Scott. (From No. 57 A.A. Company).
    Lieut. C. A. Stephens. (From No. 38 A.A. Company).
    2nd Lieut. N. K. Thompson. (From No. 38 A.A. Company).
    Lieut. F. E. Burnett (North Riding Fortress R.E. (T.F.)).

No. 26 A.A. Company R.G.A. had no Tyne personnel attached, but the following lights attached to No. 27 A.A. Company at Hull were manned by the Tyne Electrical Engineers :—

No. 27 A.A. COMPANY R.G.A. (Hd.-Qrs. Anlaby Road, Hull).
    Lieut. J. B. Murray. (From No. 38 A.A. Company).
    4 Fighting 24″ single searchlights.[10]
    3 Fighting 24″ pair searchlights.[11]
    1 Spare mobile 24″ searchlight.

---

[1] At Arthur's Dale, Great Preston, Beeston, Workhouse and Rothwell Haigh.
[2] At Scott Hall and Brierlands.
[3] At Lower Soothill.
[4] At Eccleshill, Great Horton, Moor Top, Booth Town, Bradley Mills and Mirfield.
[5] At Lindwell.
[6] At Acomb, Cliffe, Hemmingborough, Common Side, Brindleys, Villa Farm, Willitoft.
[7] At Woodhouse.
[8] The Headquarters of No. 4 Company was moved in June 1918 to Kirton Lindsey.
[9] At Potter's Grange, Cowick, Goole Fields, Thorne, Boltgate, Crowle, Amcotts, Brumby, Twigmore Grange, Scotterthorpe, Kirton Lindsey, Blyborough, Blyton, Heapham, Stow, Fillingham, Saxilby and Scampton.
[10] At Kilnsea, Harpings, Marfleet and Sutton Fields.
[11] At Spurn, Hessle Priory and Paull Battery.

In the Nottingham A.A. Defence Command, No. 5 A.A. Company R.E. was formed chiefly from personnel of No. 58 (Tyne) A.A. Company and had its headquarters at Grantham.[1] This Company, with an establishment of 5 officers and 143 other ranks, was the largest of the A.A. companies to be formed and manned twenty 24″ aeroplane searchlights[2], with a spare mobile set, operating with the patrols of No. 38 Squadron R.F.C. at Buckminster. The officers and senior N.C.Os of this Company were :—

No. 5 A.A. COMPANY R.E.   (Hd.-Qrs. High Street, Grantham).
Captain K. A. Mountain.   (From No. 58 A.A. Company).
Lieut. C. Graham.   (From No. 58 A.A. Company).
Lieut. T. W. Crawford.   (From No. 58 A.A. Company).
Lieut. H. V. Owen.   (From No. 57 A.A. Company).
2nd Lieut. R. P. Wallis.[3]   (From No. 58 A.A. Company).
C.Q.M.S. Jarvis.
S/Serjt. Henderson.
S/Serjt. Townsley.

At the same time as this Company was formed, Captain J. F. S. Hunter[4] from No. 63 A.A. Company was appointed second-in-command of the headquarters of the A.A.D.C. in Nottingham. The Tyne Electrical Engineers' lights attached to R.G.A. companies in this Command were then as follows :—

No. 29 A.A. COMPANY R.G.A.   (Hd.-Qrs. Peel Street, Groom Hill, Sheffield).
2nd Lieut. E. Birrell.   (From No. 40 A.A. Company).
6 Fighting 24″ single searchlights.[5]
3 Fighting 24″ pair searchlights.[6]
1 Spare mobile 24″ searchlight.

No. 30 A.A. COMPANY R.G.A.   (Hd.-Qrs. Burton Road, Lincoln).
Lieut. D. S. Anderson.   (From No. 57 A.A. Company).
6 Fighting 24″ single searchlights.[7]
1 Spare mobile 24″ searchlight.

No. 33 A.A. COMPANY R.G.A.   (Hd.-Qrs. Cavendish House, The Park, Nottingham).
Lieut. H. Joseph.   (From No. 63 A.A. Company).
10 Fighting 24″ single searchlights.[8]
1 Spare mobile 24″ searchlight.

---

[1] Moved in September 1918 to Buckminster.
[2] At Aubourn, Eastmere House, Wellingore, Scopwick, Leadenham, Carlton Scroop, Swarby, Sapperton, Loudenthorpe, Irnham, High Dyke Farm, Creeton, Buckminster, Streeton, Manthorpe, Barholm, Great Casterton, Helpstone, Stamford and Wansford.
[3] Relieved in June 1918 by a Lieutenant Fielding.   (Not T.E.E.).
[4] Subsequently appointed M.B.E.
[5] At Intake, Parkwood, Handsworth, Kimberworth, Ryecroft and Langwith.
[6] At Shire Green, Greystones and Bent Lathes.
[7] At Rauceby, Canwick, Burton Road, Paston Hall, Garton End and Moor Farm.
[8] At Clifton, Bramcote, Wilford, Skeinton, Aspley, Willsthorpe, Chaddesden, Allenton, Workhouse and Brauncewell (Cranwell).

In the Birmingham A.A.D.C., Captain Millard, formerly O.C. of No. 41 (Tyne) A.A. Company, was appointed to the headquarters as searchlight specialist officer. There was only one R.G.A. Company in this A.A. Defence Command, viz. :—

No. 12 A.A. COMPANY R.G.A. (Hd.-Qrs. Wolseley Works, Addesley Park, Birmingham).
Lieut. H. J. Ingman. (From No. 41 A.A. Company).
6 Fighting 24″ single searchlights.[1]
1 Spare mobile 24″ searchlight.

This completes the disposal of Tyne Electrical Engineers in the re-organisation of the air defences in January 1918. The internal organisation of the A.A. Companies R.E. was very similar to that of the old A.A. companies. Company headquarters consisted of a captain, C.Q.M.S., pay serjeant and one mechanist, while the scale of personnel allowed for the operation of the aeroplane lights was six to each station under a serjeant.

Each of the R.G.A. companies was allotted, besides the officer, one serjeant and one mechanist at headquarters, the searchlight detachment at each gun station consisting of six men for a single light and nine for a pair light, under a serjeant in the case of advanced or linking lights and under a corporal in the case of fighting lights. The R.E. officers attached to the R.G.A. companies officiated, in addition to their searchlight duties, as division officers R.E. in the various A.A. Defence Commands and were responsible for R.E. services affecting the A.A. defences including, subsequently, the installation and maintenance of the whole of the Patterson-Walsh electrical height-finding instruments used by the A.A. artillery in those localities.

On the 10th January 1918 Captain Ripley vacated the post of searchlight officer with the 6th Brigade R.F.C.[2] on transferring to the staff of the Director of Fortifications and Works at the War Office, (see Note " B " to Chapter III). Captain A. B. Williamson was now attached to the A.A. branch of G.H.Q., Home Forces, to work under Captain Forster, with effect from the 15th January. The duties of these latter officers were gradually extended to include all matters affecting searchlights and their accessories. Their branch finally dealt also with the distribution of searchlights in the various defences and with the establishments of the R.E. units manning them ; telephones required for air defence purposes (except G.H.Q. and internal communications) ; sound locators (as these instruments came into use) ; and with the establishment of listening posts. In June Captain Forster was promoted brevet major and appointed to take charge of the A.A. branch of G.H.Q., Home Forces, Captain Williamson replacing him in the post of G.S.O.3.[3]

---

[1] At Wolseley, Racecourse, South Yardley, Wyken House, Radford and Bodicote.
[2] The Air Force Bill was passed in November 1917 but it was only on the 1st April 1918 that the personnel of the R.F.C. was transferred to the R.A.F.
[3] The names of both these officers were brought to the notice of the Secretary of State for valuable services rendered in connection with the War, and Major Forster, in addition to his brevet promotion to that rank, was appointed O.B.E.

By March difficulty was experienced in obtaining from the manufacturers sufficient generating sets to equip all the searchlight stations. A proposal was then submitted to endeavour to obtain, for A.A. stations at home, supplies of electricity from local sources. Accordingly, on the 14th of that month, Captain Abbott was attached to G.H.Q., Home Forces, with the title of specialist electrical officer, to give effect to this proposal. The area over which this officer worked covered the whole of the London defences, the A.A. stations on the south coast from Dover to Southampton and those on the east coast as far north as Yarmouth. The work consisted of visiting each station to find out whether there was any generating or distributing plant installed in the vicinity. Where such existed, the local supply engineers were approached and the necessary details as to cost of supply to the searchlights arranged. In the state of development of electrical distribution at that time less than half the stations could be supplied in this way. Where no local supply company existed, an endeavour was made to obtain a supply of energy from any private source such as factories or large country houses generating their own electricity. In these cases it was necessary to ascertain the capacity of each individual plant, the hours during which it was operated and whether supply could be made available at any time.

Early in May 1918 the Portsmouth defences were included in the L.A.D.A. and the remaining anti-aircraft defence commands then coalesced to form what became known as the Northern Air Defences (N.A.D.), under Brigadier-General P. Maud with headquarters at Leeds, embracing all the ground troops in the areas of Forth, Tyne, Tees, Humber, Leeds, Birmingham, Manchester, Nottingham, Aberdeen and Cromarty.

There were now nine mobile batteries, all in the L.A.D.A.,[1] and the searchlight personnel with six of these was furnished by the Tyne Electrical Engineers. Apart from these, T.E.E. personnel on anti-aircraft service at home was now almost entirely concentrated in the northern air defences. The A.A. companies R.E. in N.A.D. were five in number and all of these were Tyne companies. The R.G.A. companies of the N.A.D. numbered twenty-three out of a total in the British Isles of fifty-five, and eleven of them were provided with searchlight personnel by the Tyne Electrical Engineers, who also provided details which were attached in the capacity of mechanists or for other duties to many other companies not specifically referred to. At the same time, as a result of the then prevailing shortage of man-power and of the many changes in the organisation of the air defences throughout the years of the War,[2] it cannot definitely be

---

[1] These were :—1st Mobile Brigade.—Nos. 1, 2, 3 and 8 Batteries.
2nd Mobile Brigade.—Nos. 7, 11 and 12 Batteries.
Dover A.A.D.C.—Nos. 9 and 10 Batteries.
[2] A table indicating the continuity of the various units through these confusing changes is given at the end of this chapter.

asserted that the whole of the R.E. personnel of the units officered by the Tyne Electrical Engineers[1] were at this stage found from the same source. The demand for man-power was so great at this period that very nearly all the high medical category personnel in the home defences had been withdrawn for service overseas and had been replaced by low category personnel drawn from all sources, while a number of women M.T. drivers, telephonists and clerks were included in the establishments of the A.A. organisations, thus releasing further men for abroad. The officers, grown weary of repeated fruitless applications to be released for foreign service, alone remained, as no suitable qualified men of low medical category could be found to replace them. The A.A. companies R.E. (as opposed to the R.E. detachments with the gun companies R.G.A.) were among the last to lose their A.1 men and some of them retained their original mechanists to the end. In August, as we have seen in an earlier chapter, the Headquarters of the London and Tyne Electrical Engineers were constituted parents of all A.A. and coast defence units in the United Kingdom and this resulted in large numbers of men in other units being attached to them. In this way the various companies and detachments may be said finally to have lost their unit identity by the end of the summer.

During this year, 1918, enemy aerial activity was confined almost entirely to aeroplane raids on London and the lights in the L.A.D.A. were frequently in action. In the raid on the night of the 29th January, the projector of one of the searchlights attached, under Lieutenant Parnall, to No. 11 Mobile Battery of the 2nd Mobile Brigade was struck by fragments of a bomb. Many of the fixed searchlights at this period were protected from bomb splinters by breastworks of sandbags or turf. In February Second-Lieutenant J. A. Metcalf was posted to No. 9 A.A. Company R.E. at Harrietsham, Kent, and had a section of five lights,[2] with headquarters at Eastwell Park. In May Captain J. F. S. Hunter was posted from Nottingham to take command of this Company which operated under the Dover A.A.D.C. Second-Lieutenant Metcalf was then in charge of eight lights in the outer barrage line with headquarters at Birchington.[3] This Company was in action on the night of the 19th May when the largest

---

[1] For the sake of clarity these may be recapitulated :—
   No. 7 Mobile Battery R.G.A.—2nd Mobile Brigade, Sevenoaks.
   No. 8 Mobile Battery R.G.A.—1st Mobile Brigade, Epping.
   No. 9 Mobile Battery R.G.A.—Dover.
   No. 10 Mobile Battery R.G.A.—Dover.
   No. 11 Mobile Battery R.G.A.—2nd Mobile Brigade, Sevenoaks.
   No. 12 Mobile Battery R.G.A.—2nd Mobile Brigade, Sevenoaks.
   No. 12 A.A. Company R.G.A.—Birmingham A.A.D.C.—Birmingham.
   No. 19 A.A. Company R.G.A.—Tyne A.A.D.C.—Newcastle.
   No. 20 A.A. Company R.G.A.—Tyne A.A.D.C.—Newcastle.
   No. 21 A.A. Company R.G.A.—Tyne A.A.D.C.—Gretna.
   No. 24 A.A. Company R.G.A.—Leeds A.A.D.C.—Leeds.
   No. 25 A.A. Company R.G.A.—Leeds A.A.D.C.—Halifax.
   No. 27 A.A. Company R.G.A.—Humber A.A.D.C.—Hull.
   No. 28 A.A. Company R.G.A.—Leeds A.A.D.C.—Selby.
   No. 29 A.A. Company R.G.A.—Nottingham A.A.D.C.—Sheffield.
   No. 30 A.A. Company R.G.A.—Nottingham A.A.D.C.—Lincoln.
   No. 33 A.A. Company R.G.A.—Nottingham A.A.D.C.—Nottingham.
   No. 1 A.A. Company R.E.—Tyne A.A.D.C.—Newcastle.
   No. 2 A.A. Company R.E.—Leeds A.A.D.C.—Ripon.
   No. 3 A.A. Company R.E.—Leeds A.A.D.C.—York.
   No. 4 A.A. Company R.E.—Humber A.A.D.C.—Gainsborough.
   No. 5 A.A. Company R.E.—Nottingham A.A.D.C.—Grantham.
[2] At Throwley, Shotton, Kennington, Cheeseman's Green and Eastwell Park.
[3] At Birchington, Reculvers, Wingham, Upstreet, Nonnington, Littlebourne, Bishopsbourne and Monkton.

raid of the War was made on London. On this occasion some forty enemy aeroplanes of the ' Giant ' and ' Gotha ' types set out from Germany each laden with no less than half-a-ton of bombs. Eight of these machines were shot down over this Country, one of them falling in flames at a place named Harty's Farm in the Isle of Sheppey, a moment after passing out of range of the Reculvers light. In all, twenty-four enemy machines were illuminated over the area of No. 9 Company on this night.

In the northern air defences enemy activity during 1918 was almost negligible. There were two Zeppelin raids in March, on Hull and West Hartle-pool on the 12th and 13th of that month respectively, and one larger raid over the midland counties on the 12th/13th April. This latter was the last airship raid on Great Britain if we except an inroad of Zeppelins over Norfolk on the night of the 5th August during which no bombs were dropped on land. Of the nine aeroplane raids on the British Isles during this year none penetrated north of Norfolk, so that the personnel of the northern air defences endured an uneventful and monotonous existence during this period. The ascendancy of the defence over the attack was now finally established and no enemy bomb was dropped on England after the night of the 20th May.

Certain developments in the equipment and organisation of the home defences were, nevertheless, proceeded with and, during this year, sound locators were first provided, the searchlight detachments being increased in establishment by two men each for their manipulation. A start was also made with the replacement of the 60 cms. projectors by those of 120 cms. diameter or larger. In place of the handwheels with which the smaller projectors had been directed these large searchlights were manipulated by means of a skew gearing attached to a " long-arm " projecting from the trunnions of the projector and carried round on a circular track or " raceway," the controlling number in the detach-ment being provided with a telescope which was now necessary to locate the smaller, faster and higher targets which the aeroplanes presented.

By July the number of A.A. companies R.E. had again risen to seventeen. Of the additional five companies No. 13 A.A. Company R.E. was formed at Cromarty and Nos. 14 to 17 in the L.A.D.A. Various Tyne Electrical Engineers' personnel found their way into some of these latter companies, among them being Captain Gibbon who was posted from No. 28 A.A. Company R.G.A. at Selby to command No. 16 A.A. Company R.E. at Sutton Farm, Hornchurch, Essex. This Company had an establishment of 4 officers and 90 other ranks and manned ten aeroplane lights[1] with a spare equipment. Second-Lieutenant H. Poland also had a section of five lights in No. 17 A.A. Company R.E.[2] at Biggin Hill near Westerham, Kent. At this date one of the sections[3] of No. 15 A.A. Company in the East London A.A.D.C., which had its headquarters

---

[1] At Tilbury, Orsett, Thorndon, Hutton, Stifford, Laindon, Dunton, Belhus, Ockenden and Brentwood.
[2] The lights of this Company were at Warlingham, Addington, Weike Farm, Crockenhill, Dartford, Outfall, Tatsfield, Downe, Bromley, Fawkham, South-fleet, Sevenoaks, Halstead, Chislehurst, Hartley and Greenhithe.
[3] With lights at Woodford Green, Lambourne End, Noakhill, Romford and Chadwell Heath.

at Upminster, Essex, was commanded by Second-Lieutenant Metcalf who had been transferred from No. 9 Company. All the lights of No. 15 A.A. Company R.E. were by now 120 cms. except that at Lambourne End which was provided with a 150 cms. projector with a gold reflector, which, it was hoped, would prove to have better powers of penetration on misty nights.

This Company, although nominally acting with No. 44 Squadron, R.A.F. at Hainault, worked in conjunction with the gun-lights in the command and was in direct touch with Headquarters at the Horse Guards. A most comprehensive and effective system of plotting and control was in action in L.A.D.A. at this time and, although this system was never called upon to stand the test of actual enemy raid, the defence of London at the end of 1918 was probably as nearly perfect as was at that time possible. While with this Company, Second-Lieutenant Metcalf developed a system of height determination which he had originally devised when with No. 9 Company in the Dover A.A.D.C. His system consisted of a small scale reproduction, on a specially constructed instrument, of the angles of elevation of two searchlights when exposed on the same target. This process was demonstrated before Major-General Ashmore, who declaring that the existing height-finders were "no good," took the whole apparatus away with him to his headquarters. It may be that this invention by an officer of the Tyne Electrical Engineers contributed materially to the more efficient working of the so-called "fixed azimuth" system of aircraft track-plotting which was subsequently practised by the air defences. Second-Lieutenant Metcalf was also honoured by being selected to give a demonstration of the existing method of plotting before their Majesties the King and Queen, H.R.H. Prince Albert, Field-Marshal Sir William Robertson and other distinguished personages on the occasion of a Royal tour of inspection of the East London Command. This officer also introduced several technical improvements into the plant at his searchlight stations, including bracket seats to carry the lamp operators on the large projectors and a system of illuminated bearing circles conveniently situated from the point of view of the telephonist. These innovations met with the approval of the G.O.C. with the result that an order was issued that "All officers of the London commands should avail themselves of any opportunity to inspect the Romford light."

Constant interchanges of officers took place, and in July the officers of the Tyne Electrical Engineers with mobile batteries and A.A. companies R.G.A. or commanding A.A. companies R.E. were as follows :—

LONDON AIR DEFENCE AREA.

1ST MOBILE BRIGADE	No. 2 A.A. Battery.—2nd Lieut. P. V. Horler.
(Nos. 1, 2, 3, 9 and	No. 9 A.A. Battery.—2nd Lieut. F. Braithwaite.
10 Batteries)	No. 10 A.A. Battery.—2nd Lieut. C. Armitage.
2ND MOBILE BRIGADE	No. 11 A.A. Battery.—2nd Lieut. E. J. Parnall.
(Nos. 7, 11 and 12	No. 12 A.A. Battery.—2nd Lieut. C. Armstrong.
Batteries)	

Harwich A.A.D.C. ...	No. 8 A.A. Battery.—2nd Lieut. W. Rintoul.
Dover A.A.D.C. ...	No. 9 A.A. Company R.E.—Captain J. F. S. Hunter.
East London A.A.D.C.	No. 16 A.A. Company R.E.—Captain E. H. Gibbon.

NORTHERN AIR DEFENCES.

Tyne A.A.D.C.... ...	Second-in-command—Major C. M. Campbell.[1]
	No. 19 A.A. Company R.G.A.—Lieut. W. Dixon.
	No. 20 A.A. Company R.G.A.—Captain J. Lawther.
	No. 21 A.A. Company R.G.A.—Lieut. R. G. Ellis.
	No. 1 A.A. Company R.E.—Captain K. S. M. Scott.
Tees A.A.D.C.	No. 2 A.A. Company R.E.—Captain T. T. Tucker.
Leeds A.A.D.C. ...	No. 24 A.A. Company R.G.A.—Lieut. A. Howell.
	No. 25 A.A. Company R.G.A.—Lieut. A. S. Burdis.
	No. 29 A.A. Company R.G.A.—2nd Lieut. E. Birrell.
	No. 3 A.A. Company R.E.—Captain R. B. T. Pinkney.
Humber A.A.D.C. ...	No. 4 A.A. Company R.E.—Captain C. F. Scott.[2]
Nottingham A.A.D.C....	No. 30 A.A. Company R.G.A.—Lieut. J. Joseph.
	No. 33 A.A. Company R.G.A.—Lieut. D. S. Anderson.

Captain W. H. James, formerly commanding No. 3 A.A. Company at York, was at this date in command of a special anti-aircraft searchlight unit R.E. which was attached to a " secret " formation assembled on the east coast, with headquarters at Colchester, with the object of effecting a landing on the Belgian coast behind the German lines. A suitable opportunity did not, however, present itself for putting this project into effect and this A.A. searchlight unit, consisting of three mobile sets, was employed, up to the time of the armistice, in the normal air defences of that locality under the A.A.D.C. Harwich.

The A.A. companies in the northern air defences were re-arranged to some extent during the latter half of the year 1918, but these re-arrangements were for the most part purely of an administrative nature. In the extreme north, Nos. 15[3] and 16 A.A. Companies R.G.A. (at Cromarty and Broughty Ferry respectively) were brought under the Edinburgh A.A.D.C., which now took the place of the former Forth A.A.D.C. In the same way, the Tyne and Humber A.A.D.Cs. were re-named the Newcastle and Hull A.A.D.Cs. respectively. The Tees A.A.D.C., now called the Middlesbrough A.A.D.C., was re-inforced by No. 2 A.A. Company R.E., at Ripon, formerly in the Leeds A.A.D.C., and this

---

[1] Appointed 16th February 1919 A.A.D.C. Middlesbrough.
[2] Captain Scott was shortly after this date posted to Malta. (See Note " B " to chapter III) and this Company was then taken over by Captain J. L. Winter of the Edinburgh Fortress R.E. (T.F.).
[3] A number of Tyne Electrical Engineers was attached to this Company.

latter Command now took over No. 29 A.A. Company R.G.A. at Sheffield. The units in the Northern air defences all remained in their former stations up to the end of the War.

The number of searchlights, and consequently the establishments for R.E. personnel attached to the R.G.A. companies, was scarcely affected by these changes, the only amendments being the addition to Nos. 21 (Gretna) and 26 (Grimsby) Companies R.G.A. of one linking light each.[1] The establishments of the five A.A. companies R.E. in the Northern air defences also remained unaltered with the exception of that of No. 1 at Newcastle which was reduced to the numbers sufficient to operate eleven searchlights in place of the seventeen formerly manned.[2]

The re-equipment of the home air defences was not completed by the time the armistice intervened in November. The scheme then accounted for no fewer than 629 anti-aircraft searchlights in the United Kingdom. Of these, 264 were in the Northern air defences, 357 in the London air defence area and 8 training lights at the school at Stokes Bay.

Immediately subsequent to the armistice when the question of demobilization began to be considered, schemes were advanced for the permanent retention of a proportion of the air defences of the Country. A conference held at the War Office on the 25th February 1919 minuted that " the arguments in favour of the upkeep and development of home defence against aircraft attack are as strong as those on which the maintenance of the navy and our coast defence system have hitherto rested," but the inevitable reaction after the stress of the war years combined with the prevailing demands for rigid economy prevented this principle from being applied. The R.E. personnel attached to the gun companies R.G.A. was the first to go and was followed almost immediately by that of the R.E. companies. By June the strength of the air defence troops in Great Britain had fallen to approximately one-sixth of that at the time of the armistice and, by the end of the year, the remnants of the great and efficient organisation of which we have traced the growth were of but trifling value. By 1920 the last traces were gone.

It became a popular habit with some of the more thoughtless persons during and immediately after the War to cry down those officers and men of the various services who did not see active service overseas but who were obliged, through force of circumstances, to carry out their duties at home. As the air defences at home absorbed the greater proportion of the total strength of the Tyne Electrical Engineers, it may not be without interest briefly to investigate the national importance of the work carried out by these men who were often deprived of the opportunity of undertaking more spectacular service overseas.

---

[1] At Bowness in the Gretna company and at Saxby All Saints in the Grimsby company.

[2] The lights stations finally manned by this Company were :—Plessey, Widdrington, High Clifton, Hirst, Fencehouses, Easington, Sherburn, Coxhoe, Castle Eden, Sedgefield and Benknowle.

During the whole of the war period there were 105 air raids[1] on Great Britain in which bombs were actually dropped over land. As a result of these raids, 1,413 persons were killed and 3,408 injured. How much more serious these casualties, mostly to civilians, would have been but for the development of the air defences may be inferred from the following analysis of the available statistics. The Zeppelin effort was, as we have seen, practically fought to a standstill by the end of the year 1916. Up to that time 42 raids had accounted for 1,724 casualties; yet, during the remaining two years of the War, there were only 11 airship raids and the casualties from this source amounted to no more than 190. Anti-aircraft gunnery and fighter aircraft each claimed a share in the defeat of the airship[2] but one may indeed ask how either of these weapons could have achieved this result by night (when all the Zeppelin raids took place) but for the presence of the searchlights to seek out and indicate their target.

The rise and decline of the German aeroplane campaign against England may similarly be related briefly. At its zenith, during 1917, 27 attacks were launched which caused 2,208 casualties. In 1918, when the air defences had been improved to meet this new form of attack, there were only six raids with a casualty list of 612.

Another important aspect of the case was the question of munition supply. Captain Joseph Morris[3] reminds us that " on receipt of air raid warnings, and there were many false alarms, work was suspended, sometimes over vast industrial areas, traffic was disorganised and an adverse effect was produced both on the workers and on the population. As a broad result, some one-sixth of the total normal out-put of munitions was entirely lost and the quality of a less proportion was effected." The urgent requirements, amounting frequently to dire need, of our Armies overseas and of those of our Allies, demanded the maintenance of uninterrupted supply of munitions from the factories at home. In so far as the anti-aircraft defence organisation reduced to a minimum the losses of time and consequently of output due to air raids over Great Britain, its efficiency may be said to have borne a very close relationship to the ultimate success of the armies in the field.

While the national importance of ensuring the uninterrupted operation of the munition factories cannot be exaggerated, it is possible to trace, in the service of the Territorial Force personnel in the home air defences, the very fundamental principle of the volunteer movement, for, to quote Walter Richards[4] " it needs no feudal tenure or obligation, still less hire or reward, to make men fight for hearth and home, for wife and children. The sturdy defenders of coast and marches and borderland of centuries ago were the predecessors of an army whose very existence . . . . rendered their country absolutely impregnable."

---

[1] These and the following statistics are taken from " The German Air Raids on Great Britain 1914–1918 " by Captain J. Morris. Certain other records, not so detailed, place the total casualties considerably higher.

[2] In the early days, at any rate, there was much rivalry between these two modes of defence and one or two of the first Zeppelins to be brought down were claimed by both Artillery and R.F.C., the rival claims being contested even to the point of official enquiry.

[3] " The German Air Raids on Great Britain," p. 159.

[4] " Her Majesty's Army," p. 268.

TABLE INDICATING THE CONTINUITY OF ANTI-AIRCRAFT SEARCHLIGHT UNITS OF THE TYNE ELECTRICAL ENGINEERS R.E. 1916—1918.

(Not shewn in the correct sequence of their formation.)

JUNE, 1916.	OCTOBER, 1916.	APRIL, 1917.	JANUARY, 1918. A.A. COMPANIES R.E.	JANUARY, 1918. A.A. COMPANIES R.G.A.
No. 33 (Tyne) A.A. Co. ...	No. 33 (Tyne) A.A. Co. ...	No. 34 (Tyne) A.A. Co.	} No. 1 A.A. Co. R.E.	{ No. 20 A.A. Co. R.G.A.
(Special) Tyne S/L Co. ...	No. 34 (Tyne) A.A. Co.	No. 35 (Tyne) A.A. Co.		No. 19 A.A. Co. R.G.A.
	No. 35 (Tyne) A.A. Co. ...	No. 50 (Tyne) A.A. Co.	—	No. 21 A.A. Co. R.G.A.
	No. 37 (Tyne) A.A. Co. ...	No. 37 (Tyne) A.A. Co.	No. 2 A.A. Co. R.E.	{ No. 24 A.A. Co. R.G.A.
				No. 25 A.A. Co. R.G.A.
	No. 38 (Tyne) A.A. Co. ...	No. 38 (Tyne) A.A. Co.	} No. 4 A.A. Co. R.E.	{ No. 27 A.A. Co. R.G.A.
	No. 25 (Tyne) Aer. Sqd. S/L Co....	No. 57 (Tyne) A.A. Co.		No. 30 A.A. Co. R.G.A.
No. 22 (Tyne) Aer.Sqd. S/L Secn.	No. 22 (Tyne) Aer. Sqd. S/L Co....	No. 58 (Tyne) A.A. Co.	} No. 5 A.A. Co. R.E.	—
	No. 29 (Tyne) Aer. Sqd. S/L Co....	No. 56 (Tyne) A.A. Co.	} No. 3 A.A. Co. R.E.	No. 28 A.A. Co. R.G.A.
	No. 27 (Tyne) Aer. Sqd. S/L Co....	No. 60 (Tyne) A.A. Co.		No. 29 A.A. Co. R.G.A.
	No. 40 (Tyne) A.A. Co. ...	No. 40 (Tyne) A.A. Co.		No. 12 A.A. Co. R.G.A.
No. 2 (Tyne) S/L Co. ...	No. 41 (Tyne) A.A. Co. ...	No. 41 (Tyne) A.A. Co.		
	No. 42 (Tyne) A.A. Co. ...	No. 42 (Tyne) A.A. Co.		No. 33 A.A. Co. R.G.A.
		No. 63 (Tyne) A.A. Co.		
No. 9 (Tyne) Mob. S/L Co.	No. 9 (Tyne) Mob. S/L Co.	No. 9 (Tyne) Mob. S/L Co....		} Nos. 1 & 2 Mob. Bdes. R.G.A.
No. 10 (Tyne) Mob. S/L Co.	No. 10 (Tyne) Mob. S/L Co.	No. 10 (Tyne) Mob. S/L Co....		

CHAPTER VII.

—

1915—1920. THE GREAT WAR.

THE ANTI-AIRCRAFT SEARCHLIGHT SECTIONS OVERSEAS.[1]

A.A. Defences with B.E.F., 1915 — Searchlight sections formed by London and Tyne Electrical Engineers — Defences at G.H.Q. and on the L. of C. — — No. 8 A.A.S.S. in the Salient — Improved defences for Audruicq — Procedure on formation of an A.A.S.S. — Expansion to 45 sections — Non-mobile nature of plant — " False Areas " — Sections reduced from 3 to 2 lights — Organisation of fixed defence — Defences at Dunkerque — The Equipment section at Calais — Development of Audruicq and Zeneghem defences — A.A. defence for Isbergues Steelworks — Technical developments — Concentration of searchlights in the salient during the third battle of Ypres — P.E. lorries — The searchlights bombed — Earthwork protection developed — Searchlights concentrated in the Somme area after the battle of Cambrai — Increase to 75 sections — and from 2 to 3 lights per section — Qualifications of personnel — Section with the Italian Expeditionary Force — Appointment of Assistant Inspectors of Searchlights — Further development of defences at Calais, Audruicq, Zeneghem and St. Omer — Deputy I. of S. Northern L. of C. — Searchlights in the retreat of Third and Fifth Armies on the Somme, March, 1918 — Lewis guns — Retreat on the River Lys, April, 1918 — Army lines of Searchlights — Searchlights with the Independent Air Force — Further increases approved — The " Camel Line " — Inspector of A.A. Defences appointed — Searchlights in the final advance, August to November, 1918 — Concentration after the armistice — Searchlights on the Rhine — Demobilisation.

———

ALTHOUGH the British Expeditionary Force in France was provided in 1914 with a certain proportion of anti-aircraft artillery, it was 1915 before the first anti-aircraft searchlights were employed overseas. In March of that year the Field-Marshal Commanding-in-Chief in France asked for " three or four searchlights suitable for use with anti-aircraft guns against airship attack by night."[2] The immediate result of this appeal was the formation at Chatham of a unit of Regular Royal Engineers under the title of No. 1 Anti-Aircraft Searchlight Section, R.E. This section, consisting of 1 officer and 19 other ranks and manning three 60 cms. searchlights, embarked in April and established itself in fixed positions round G.H.Q. at St. Omer. A second section, which was formed at Plymouth partly of Regular personnel and partly of London Electrical Engineers,[3] arrived in France in July and was also employed at St. Omer.

From these small beginnings was developed a great organisation of anti-aircraft searchlight units which, by the end of the War, consisted of no less than 76 sections with a total strength amounting to some 2,000 all ranks. This personnel was furnished almost entirely by the Tyne Electrical Engineers and by the London Electrical Engineers with a proportion of Regulars and about

---

[1] The maps at the end of the book will be found useful when reading this chapter.
[2] " Work of the R.E. in the European War " Vol. " Miscellaneous " page 20.
[3] Under Lieutenant G. Fergus-Wood, London Electrical Engineers.

600 men of medical category " B " transferred from the Infantry.[1] The very important part which the Tyne Electrical Engineers played in the development of the anti-aircraft service overseas forms the special subject of this chapter, but, as nearly all the anti-aircraft searchlight sections (A.A.S.S.) served only under a single officer, it is not surprising that, during the interval of time since their disbandment at the conclusion of active operations, the detailed stories of many of them should have become irretrievably lost. It has unfortunately been found impossible therefore to complete in every detail the list of sections actually formed for service overseas. (See end of chapter.)

The first of these sections formed of personnel of the Tyne Electrical Engineers was No. 6 A.A.S.S.R.E. which, under the command of Lieutenant R. L. Hunter with Serjeant W. Thompson as section serjeant and Staff-Serjeant K. McKenzie as mechanist, and in company with Nos. 3, 4 and 5 A.A.S.S. of the London Electrical Engineers,[2] embarked for France on the 8th May 1916. By this date the anti-aircraft resources in France had been increased by the arrival of the 50th (Field Searchlight) Company R.E.[3] comprising three mobile sections, one being equipped with two 90 cms. searchlights and the other two with one 90 cms. and one 60 cms. projector each. Before the end of May, No. 7 A.A.S.S., formed of London Electrical Engineers,[4] had also arrived in France, and the searchlight units then serving with the B.E.F. were so disposed as to cover the various echelons of G.H.Q. and several important localities on the lines of communication, as follows :—

50th (F.S.) Company R.E.					
1 Section	...	...	...	...	... Boulogne.
1 Section	...	...	...	...	... Audruicq.
1 Section	...	...	...	...	... Advanced G.H.Q.
No. 1 A.A.S.S.R.E.	...	...	...	...	... G.H.Q. 1st Echelon.
No. 2 A.A.S.S.R.E.	...	...	...	...	... G.H.Q. 2nd Echelon.
No. 3 A.A.S.S.R.E.	...	...	...	...	... Calais
No. 4 A.A.S.S.R.E.	...	...	...	...	... Abbeville.
No. 5 A.A.S.S.R.E.	...	...	...	...	... Etaples.
No. 6 A.A.S.S.R.E.	...	...	...	...	... Rouen.
No. 7 A.A.S.S.R.E.	...	...	...	...	... Abancourt.

It is not within the province of this account to trace the strategical or tactical progress of the operations of the British Forces in France, nor is it possible to devote more than occasional attention to the activities of those A.A. searchlight sections which were not formed of Tyne Electrical Engineers[5]. It is necessary, however, to record that from their first arrival overseas the sections were constantly in action with enemy aircraft and that, from the very outset,

[1] " Work of the R.E. in the European War " Vol. " Miscellaneous," page 320.
[2] No. 3 A.A.S.S.—Captain R. G. Madge. No. 4 A.A.S.S.—Lieut. T. H. Gotch. No. 5 A.A.S.S.—Lieut. A. W. M. Mawby.
[3] Formed chiefly of personnel of the 4th Coy. R.E. under Captain W. C. H. Prichard, R.E.
[4] Under Lieut. H. G. G. Clarke.
[5] The official account of the anti-aircraft searchlight organisation overseas is given in the " Work of the R.E. in the European War " Vol. " Miscellaneous." Some account of the tactical development of air defences in France is given in Major-General E. B. Ashmore's " Air Defence," pages 97 to 105.

the searchlights themselves were continually and severely bombed by the enemy in his night raids. During the whole of the remaining period of the War the British searchlights personnel " shewed great determination and courage in keeping the searchlights directed on to the enemy aircraft whilst being heavily bombed, a result which there is evidence to shew the enemy was never able to achieve in his own anti-aircraft defences."[1] At this early period it is recorded that the detection of enemy aircraft in the beams was an unusual occurrence and this fact is accounted for by the unsuitable type of equipment provided and the practical impossibility of hearing aircraft in the forward areas owing to the incessant gun-fire. From the very first, nevertheless, the presence of the searchlights was found of great value in driving raiders up to considerable heights, at which the accuracy of their bombing was seriously impaired.

In June 1916 No. 8 A.A.S.S. was formed at Haslar under Lieutenant R. P. Findlay, with Serjeant G. Hill as section serjeant and Staff-Serjeant W. Tate as mechanist. After equipping in London, this section embarked for France in the same month and, immediately on arrival, was sent to co-operate with No. 88 Section Anti-Aircraft Artillery (A.A.A.) in the protection of the Second Army ammunition dump between Abeele and Poperinghe. To this section was then attached certain infantry machine-gun personnel, but, as time went on, the need for these guns and men in the front line became so urgent that they were withdrawn in June of the following year. Lieutenant Findlay was mentioned in dispatches for distinguished conduct while the section was in this locality.

In July three further A.A. searchlight sections were formed for France. These were Nos. 9, 10 and 11 A.A.S.S.R.E., of which No. 9 was composed of personnel of the London Electrical Engineers[2] and the remaining two were formed at Haslar by the Tyne Electrical Engineers. The officers, section serjeants and mechanists of the latter at their formation was as follows :—

No. 10 A.A.S.S.	...	...	...	Lieut. A. B. Williamson.	
				Staff-Serjeant G. Robson.	
No. 11 A.A.S.S.	...	...	...	Lieut. D. Myles.	
				Serjeant A. Donkin.[3]	
				Staff-Serjeant T. Potts.	

These sections proceeded to the Headquarters of the London Electrical Engineers in Regency Street, London, to complete mobilisation. At this date, it was decided to reduce the establishment of the A.A.S.S. to man two lights only, as, owing to the shortage of men and equipment, it was found impossible to provide more than two lights for the defence of each locality ; moreover, the triangle formed by three lights installed round a vulnerable point presented an admirable aiming-mark for the enemy bombers. The establishment now approved[4] consisted of 1 officer, 1 serjeant, 1 staff-serjeant, 1 corporal, 2 lance-

---

[1] " Work of the R.E. in the European War " Vol. " Miscellaneous," page 316.
[2] Under Lieut. G. Owles.
[3] This N.C.O. was subsequently awarded the M.S.M.
[4] W.E. No. 309 " Fixed A.A. Searchlight Section (2 lights) " dated 19th July, 1916.

L

corporals and 13 sappers.[1] To each section was attached one A.S.C. driver for the box-van which constituted its first-line transport. The equipment, comprising two 60 cms. projectors (identical in type to those supplied to the home defence units. See Chapter VI.) with portable resistance and switchboards and Dorman two-cylinder petrol generating sets was drawn from Woolwich Arsenal, and arms from the Tower of London. These sections were issued with the means of providing their own inter-communication, four field telephones, together with the necessary cable, being included in the table of stores. This group of sections sailed for Havre from Southampton on the 17th August. On arrival in France they separated, No. 10 Section proceeding to St. Omer, where it relieved No. 2 A.A.S.S. in the defences of G.H.Q.[2] After a very short period at St. Omer, No. 10 Section moved to Dannes, which was the site of No. 19 Ordnance Depot, a very large ammunition dump situated on the railway halfway between Etaples and Hesdigneul.

No. 11 Section was posted to Hesdin, though for a period of about one week, during the visit of H.M. The Queen to the hospitals, this section moved by road to take up a temporary position in the defence of Rouen where it co-operated with No. 6 A.A.S.S., which was also a Tyne section.

At this period there was no night-flying fighter aircraft with the B.E.F. and the A.A. searchlights co-operated solely with the " Archies," as the A.A. artillery was colloquially known.

Two further A.A.S.S. were formed in August 1916, No. 12 A.A.S.S. being a London unit,[3] while No. 13 was formed of Tyne Electrical Engineers under Lieutenant E. F. Rendell.[4] The section serjeant of this section was Serjeant J. Laycock and the mechanist at its formation Staff-Serjeant T. Burns, but the latter transferred to the inland water transport branch of the R.E. very shortly after arrival in France and was replaced by Staff-Serjeant J. Brand. The establishment and equipment of this section were similar to those already formed with the exception that the engines were Coventry-Simplex and not Dormans. After the usual period of mobilisation at the Regency Street Headquarters, these sections embarked for France and were both posted immediately on landing to Audruicq, where a very large ammunition depot was established as well as the headquarters of the British railway organisation in France and other important depots and establishments. This locality had been very seriously raided, and for all practical purposes destroyed, on the night of the 22nd/23rd July, and, at the time of the arrival of Nos. 12 and 13 A.A.S. Sections, preparations were in progress for the more effective defence of this place against attacks from the air.

---

[1] The distribution of rank and file by trades was as follows :—Engine drivers 4, electricians 6, fitters and turners 2, joiner 1, miscellaneous trades 2, batman 1 ; total 16.

[2] During the battle of the Somme, which raged from July to November 1916, hostile bombing of horse-lines and battery positions in forward areas began to develop and No. 2 A.A.S.S. was one of two sections ordered into the area between Montauban, Trones Wood and the Somme, to assist in dealing with this fresh danger. The other was No. 5, which was relieved at Etaples by No. 9.

[3] Under Lieut. G. B. Adeney.

[4] From No. 25 (Tyne) Aer. Sqdn. S/L Coy.

There were now three A.A.S. sections[1] in this defence, No. 13 being stationed to the west of the town with its two lights adjacently sited. After a short time (early in 1917) the three sections were re-arranged in two triangles and one of the lights of No. 13 was moved to a point approximately eight kilometres north of Audruicq on the main road to Gravelines. The lights were connected by R.E. Signals with the A.A.A. headquarters north of Audruicq but there was as yet no detailed organisation for the control of the defences which, however, were subsequently very highly developed and will be referred to again later in this chapter.

At the end of August the Regular personnel with Nos. 1 and 2 A.A.S.S. was entirely replaced by personnel of the London Electrical Engineers.[2] Henceforth, the Regular R.E. personnel employed overseas with A.A. searchlights was concentrated in the 50th (F.S.) Company R.E., though a small number of Tyne Electrical Engineers who had served with the oxy-acetylene searchlights (See chapter IV.), and from other sources also, found their way into this Company.

Nos. 14 and 15 A.A.S.S. were formed together in September, the former of London Electrical Engineers[3] and the latter of Tyne Electrical Engineers under Lieutenant E. V. Baldwin, with Serjeant J. C. Vairy and Staff-Serjeant T. J. Cripps. The conditions under which the sections were formed and mobilised were now standardized, and the principle was adopted that they should be raised alternately by the Tyne and London Electrical Engineers. At this stage it will be found that the odd number sections were being furnished by the Tyne Electrical Engineers, though this was not always the case. The first intimation of the formation of a new section normally took the following form :—

121/8405 (M)

9th September 1916.

Sir,

I am directed to inform you that two anti-aircraft searchlight sections, to be designated Nos. 14 and 15 Anti-Aircraft Searchlight Sections, will be prepared forthwith at the Headquarters, London Electrical Engineers, 46 Regency St., Westminster, S.W. for service in France.

2. These sections will mobilize at war establishment for a " Fixed Anti-Aircraft Searchlight Section (2 lights) No. 309 dated S.D.2 19th July 1916."

3. The Royal Engineer personnel for these Sections will be provided from the London and Tyne Electrical Engineers respectively, and will be regarded as supernumerary to establishment.

---

[1] The third was a section of the 50th Coy. R.E.
[2] No. 1 Section was then commanded by Lieut. T. Rich, London Electrical Engineers.
[3] Under Lieut. J. D. Ross.

4. I am to request that you will arrange for the move of the necessary personnel for No. 15 Section to London, in communication with the General Officer Commanding-in-Chief, Northern Command, to whom a copy of this letter is being sent.

5. Instructions will follow regarding :—
   (a) Equipment.
   (b) The supply of the motor car with box body and driver for each section.

6. Instructions should be issued for indents to be submitted to the War Office, through the Assistant Director of Medical Services concerned, for any field medical equipment that may be required to complete the sections in accordance with the revised scale laid down in Appendix 51, Regulations for Army Medical Services, issued with Army Orders dated 1st December, 1915.

> I am,
> Sir,
> Your obedient Servant,
> (*Signed*) J. C. CHAMBERS, Capt., D.A.A.G. for
> Assistant Adjutant General,
> Mobilisation.

The General Officer Commanding,
London District.

The personnel for the section assembled at Haslar and proceeded to London, technical equipment was drawn from Woolwich Arsenal, the box-van and driver joined from the Mechanical Transport Reserve Depot at Grove Park, London, S.E. and, in due course, on the 20th November, the Director of Movements at the War Office issued orders for embarkation on the 27th of that month. The box-van proceeded by road to Southampton, heavy stores were loaded at Nine Elms station and the personnel entrained at Waterloo. The whole of Nos. 14 and 15 Sections arrived at Havre on the 27th November and by the 5th December all stores were unloaded and the sections accommodated in the details camp at Etaples. This procedure, with minor modifications, may be regarded as typical of the early stages in the histories of all the sections formed about this period.

On the 8th January 1917 No. 15 A.A.S.S. took up positions in the defences of Dannes. Here the lights were under the control of the A.A.A. and here the section remained until July, but, during this period, enemy activity over this part of the lines of communications was negligible. On the arrival of No. 15 Section at Dannes, No. 10 (under Lieutenant A. B. Williamson) was sent to the Second Army and stationed near Hazebrouck, in the vicinity of which it remained until the retreat of March 1918.[1] In this low-lying district

---

[1] In January 1918 Captain A. B. Williamson went to G.H.Q. Home Forces for duty with the A.A. branch (see chapter VI) and No. 10 Section was then taken over by Lieut. R. E. Jardine, Tyne Electrical Engineers.

ground fogs proved a serious nuisance and, though raids over this part were frequent and the section acquired considerable experience in locating targets, it was frequently found advisable not to expose at all.

The next of the A.A.S. sections to be formed by the Tyne Electrical Engineers was No. 17 A.A.S.S., which assembled at Haslar and proceeded to London on the 13th October 1916 under the command of Captain R. H. Rooksby.[1] The section serjeant was Serjeant N. Galloway[2] and the mechanist staff-serjeant J. W. Bonham. This section was accommodated, while in London, at Millbank Hospital. After some preparatory training, including some signals work for the post office (presumably in connection with the fixed air defences of London) at Denman Street, New Cross and Woolwich, and some difficulties in the provision of its plant, the section left Waterloo on the 20th December for Southampton, but did not sail until the 21st on account of the presence of mines in the Channel. No. 17 A.A.S.S. arrived at Havre and disembarked on the 22nd and marched to No. 2 Rest Camp. On Christmas Day the section entrained and proceeded via Rouen to Abbeville, where it remained for a month in inactivity, sharing a camp with No. 14 A.A.S.S. (London Electrical Engineers). The motor car had proceeded by road, accompanied by an interpreter, and the heavy searchlight plant (including the generating sets in immense packing-cases) followed by rail on the 29th December. Two stations were erected, on the 25th January 1917, near the recently formed ammunition depot at Saigneville, and designated S 1 and S 2, but these sites were not approved and it was the 3rd February, nearly four months after its first formation, before this section was ready for action in co-operation with No. 48 Section A.A.A. The group of sections in this locality was at this time under Captain R. G. Madge, London Electrical Engineers. The first Zeppelin alarm took place on the 16th March but enemy activity at this period was slight in spite of the immense accumulations of ammunition in the locality. On the 30th April Captain Rooksby handed over the command of this section to Lieutenant H. Hutchinson[3] and went to take over No. 8 A.A.S.S. at Abeele from Lieutenant Findlay.

No. 19 A.A.S. Section, which was the next of the Tyne sections to be formed, moved to Millbank from Haslar in November 1916 under Lieutenant W. G. Edwards, but, owing to the urgent need of strengthening the London and East Coast A.A. defences, plant could not be spared for this section until March 1917 when it embarked at Southampton for Havre. The War Office had, by now, formally approved the formation of a total of 45 A.A. searchlight sections for France and had also authorised an entire re-equipment with 90 cms. projectors, as it was found that the 60 cms. lights were not powerful enough. The provision of these larger projectors involved not only the substitution of heavier and more cumbersome searchlight apparatus but also more powerful, and consequently heavier, generating plant. The engines originally supplied

---

[1] From No. 37 (Tyne) A.A. Company at Leeds.
[2] From No. 25 (Tyne) Aer. Sqdn. S/L Company at Selby.
[3] From No. 40 (Tyne) A.A. Company at Sheffield.

for the use of the overseas sections were mostly 2-cylinder 8 k.w. Dorman sets, but even these had been found unduly cumbersome, weighing four tons, and some sections had been able to exchange them for lighter Austin or Astor sets or for the rather more popular Coventry-Simplex. For the 90 cms. projectors, themselves double the weight of the 60's, 16 k.w. generators were necessary, bringing the total weight of plant for a single station up to nearly seven tons.[1] This constituted a serious bar to mobility and made the moving of a station a long and heavy undertaking. No. 19 was the first of the Tyne sections to be equipped with 90 cms. projectors, the engines being Keighley 16 k.w. sets. These engines were found to need much expert attention, which was not demanded by the 16 k.w. Dormans which subsequently became available. During the winter months the chief concern of all sections was to avoid damage due to freezing, and more than one section had cracked water-jackets to replace in a hurry through failing to keep the engines warmed up during nights of inactivity.

No. 19 Section proceeded, on arrival in France, to the village of Brouckerque on the line of approach to Zeneghem, where further large ammunition depots were established. As soon as the searchlights were ready for action this section was given the task of establishing a " false area," understood to have been the first of its kind to be erected. The " false area " consisted in the first place of rows of electric lights slung on hop-poles and laid out to reproduce roughly the shape of the main ammunition depot at Zeneghem. On the approach of hostile aircraft all lights at Zeneghem were extinguished, but the "false area" lights remained on until it was certain that they had been spotted by the attacking aeroplanes. The lights were then switched off section by section in imitation of the ammunition depot receiving a belated alarm. As time went on, this area was elaborated by the addition of red and green railway signal lamps, sections of dummy railway track and several oil saturated flares which were lighted when bombing occurred. The flares were controlled from a central switchboard and were operated by blowing a fuse stretched across a small vessel containing petrol. As a decoy this " false area " was fairly successful and it is understood that, on one occasion, the German press published a glowing account of a devastating raid on this dump which, in fact, cost the British but a few gallons of paraffin in the flares and the replacement of a small number of fuses. On the other hand, as the fire of upwards of twelve guns could be, and frequently was, concentrated over it, sitting in the centre of the " false area " was not exactly a pic-nic and the section was not particularly popular with the neighbouring Flemish farmers. Being on the outskirts of Dunkerque and on the direct line of approach to Zeneghem, Audruicq and Calais, this section experienced a good share of enemy aerial activity.

No. 20 A.A.S. Section was, for some unexplained reason, also raised by the Tyne Electrical Engineers, the Section officer being Lieutenant C. W. Erskine, the section serjeant Serjeant Anderson and the mechanist Staff-Serjeant Gray.

---

[1] One section was issued with a 24 k.w. Mercedes engine weighing 7½ tons !

The exact date of formation of this section is not known, but it was in France before April 1917, moving to St. Omer, whence it supplemented the Audruicq defences to the south-east, one of the lights being situated at the village of St. Momelin about three kilometres north of St. Omer and the other a few kilometres further north again. The arrival of this section released the section of the 50th (F.S.) Company which had been at this station since May 1916 and which was now moved elsewhere.

Before either No. 19 or 20 Sections had landed in France, Nos. 21 to 24 A.A.S. Sections had been formed in that country from detachments rendered surplus on the reduction in establishment from three to two lights of the eight sections which were overseas prior to July 1916. No. 22 A.A. Section was formed of one detachment from No. 6 A.A.S.S. and one from No. 8, and was thus a Tyne section, but was commanded by Lieutenant Bruch of the London Electrical Engineers. The section serjeant was Serjeant Cole. Nos. 21, 23 and 24 Sections were London Sections but no record can be traced of the officers who commanded them or of the exact date of their formation, though they were all in existence prior to February 1917. These units were mainly allotted to supplement the defences of the large ammunition depots at Audruicq, Zeneghem, Saigneville, Abancourt and Dieppe.[1]

Some account may be inserted at this stage of the organisation of the A.A. defences in each locality. The senior searchlight officer in each defence worked in conjunction with the area gun commander, the latter taking executive control in action. The headquarters of the defence was habitually situated at one of the gun positions, the searchlights and guns in fixed defences being all in telephonic communication with headquarters. The area gun officer had a telephone post and headset some few yards behind the gun platform and from this spot he gave his orders to the gun station when in action. The searchlight officer had a similar post in close proximity to that of the gunner and was in communication with his lights. The gunner officer gave his orders for the tactical manipulation of the lights to the searchlight officer who in turn gave them to his detachment N.C.Os. In this way, the organisation for A.A. defences followed very closely that of coast defences.

In January 1917 the office of Inspector of Searchlights, under the Engineer-in-Chief at G.H.Q., was constituted to assist in the formation and training of the units and to advise the various staffs on the disposition of the available personnel and material. The officer selected to fill this post was Major W. C. H. Prichard, D.S.O., R.E.

No. 25 (London) and 26 (Tyne) A.A.S. Sections reached France in March, the latter under Lieutenant W. Fox with Serjeant C. Driver and Staff-Serjeant A. Burgess.[2] These sections were equipped with 60 cms. searchlights as

---

[1] " Work of the R.E. in the European War " Vol. " Miscellaneous," page 314.
[2] Subsequently commissioned and posted to command No. 13 A.A.S.S.

arrangements for the supply of sufficient 90 cms. projectors had not yet borne fruit. No. 26 Section proceeded to Calais.

Nos. 27 and 28 A.A.S.S. were both Tyne sections and were formed at Haslar in March 1917, the former under Lieutenant D. E. Ross[1] with Serjeant T. Young[2] and Staff-Serjeant N. Ward and the latter under Lieutenant H. G. Campbell.[3] These sections embarked together with No. 29 (London) Section in April. No. 28 Section proceeded first to Rouen, while No. 27 was sent forthwith to Dieppe where it assisted in the defence of a very large ammunition dump. The section remained there until 1918 and during this period laid out " false areas " similar to those already described. The "false area" at Dieppe was also claimed to have been the first, but, from the evidence available, it would seem that that established at Zeneghem by No. 19 (Tyne) A.A.S.S. was earlier by at least a month. At Dieppe, however, it is reported that the " false area," which was lit from one of the searchlight generators, drew no bombs during the nine months in which No. 27 Section was in the station. This section was equipped with Dorman engines and 60 cms. projectors. Nos. 30 and 31 A.A.S. Sections were formed by the London Electrical Engineers.[4] Of these, No. 30 Section was posted to the Audruicq and Zeneghem defences.

In April No. 11 Section, under Lieutenant Myles, was transferred from Hesdin to the Second Army area and established on the coast east of Dunkerque where it operated in conjunction with the French air defences. Air raids took place almost nightly on this place, Dunkerque being on several occasions during the War made the object of a combined enemy attack by sea, long-range land guns and air-craft. Fortunately the long-range bombardments fell just short of the town, but it has been said that Dunkerque suffered more severely than any other town in France from air-raids, though the military damage was slight considering the continual activity of enemy aircraft. Officers who visited Dunkerque declared that from the civilian point of view conditions in that locality were worse than in any other town behind the line. For his services in this defence Lieutenant Myles was decorated by the French with the Croix-de-Guerre.

We have so far traced the formation of the first thirty-one of the A.A.S. sections, and from this point onwards, of the next twenty sections formed, it will be observed that the even-numbered sections were raised by the Tyne Electrical Engineers and the odd-number sections by the London Electrical Engineers. Nos. 32, 33, 34, 35, 36 and 37 Sections were formed together

---

[1] From No. 63 (Tyne) A.A. Company at Nottingham.

[2] This N.C.O. was subsequently awarded the M.S.M.

[3] From No. 56 (Tyne) A.A. Company at York.

[4] No. 30 A.A.S.S.—Lieut. F. G. Hort. No. 31 A.A.S.S.—Lieut. H. A. Voss. The following officers of the London Electrical Engineers comprised the section officers of Nos. 16, 18, 25 and 29 Sections but their distribution is not known :—Lieuts. H. F. Hawarth, A. T. Sturgess, F. C. Stephens and H. L. Bazalgette.

in the early days of April and mobilised at Regency Street. The officers, section serjeants and mechanists of the three Tyne sections were as follows :—[1]

No. 32 A.A.S.S.	...	...	...	Lieutenant A. G. Dickson.[2]
				Serjeant G. A. Gill.
				Staff-Serjeant H. Kidner.
No. 34 A.A.S.S.	...	...	...	Lieutenant R. P. Winter.[3]
				Serjeant E. Wemyss.
				Staff-Serjeant J. Brand.
No. 36 A.A.S.S.	...	...	...	Lieutenant C. B. Williamson.[4]
				Serjeant R. S. D. Scrafton.
				Staff-Serjeant F. Ferguson.

It will be remembered that it had been decided to equip the whole of the overseas sections with 90 cms. searchlights and, to save delay, the personnel for the new sections was sent to France without equipment, which was provided through the agency of an equipment depot established under the Inspector of Searchlights at Calais. The 90 cms. equipment to replace the 60 cms. plant already in France was also dealt with through the same depot. The task of operating this depot was allotted to No. 26 A.A.S.S. (under Lieutenant Fox) which henceforth became known as the Equipment Section, the O.C. assuming the title of Searchlight Equipment Officer. The equipment duties of the section, which were in addition to its normal duties, included the receipt and issue of all plant and equipment for searchlight units in France and involved very heavy work[5] for which additional sections in the Calais defences were later made available, while No. 26 Section itself was strengthened by a number of infantry for general purposes.

Nos. 32 to 37 Sections were the first to be sent overseas without equipment, embarking together in the " Huntscraft " transport at Southampton on the 11th May for Havre and reaching Calais three days later. There they spent some weeks assisting the Equipment Section to unload 90 cms. plant and in various other duties, including, in the case of No. 34 Section, that of building camps for Egyptian labour battalions. During this period these sections were interested spectators of the many air-raids which took place over Calais and in the vicinity. The first of these Tyne sections to move away was No. 32 (under Lieutenant Dickson) which was posted to Abbeville defences and employed on " false areas " in that locality for a short while. The " false areas " at Abbeville were apparently not a success and were very soon abandoned, and the section then returned to Calais where it took up its position in the defences and assisted No. 26 Section in its equipment duties.

---

[1] The O.C. of No. 33 Section was Lieut. Bauman, L.E.E. That of No. 35 is believed to have been Lieut. Simpson ; of No. 37, Lieut. Squires.
[2] From No. 35 A.A. Company at Newcastle.
[3] From No. 10 (Tyne) Mobile Searchlight Company at Harwich.
[4] From No. 34 (Tyne) A.A. Company at Newcastle.
[5] Asked what he was doing, the O.C. No. 26 A.A.S.S. would inform enquirers that his duties consisted of taking delivery of Dorman generating sets, unloading them, and then loading them on to lorries of great height. He did not often give such detail, merely making the now historic remark " humping b——— Dormans." Everyone knew the weight of these sets, so further explanations were unnecessary !

No. 34 Section (Lieutenant Winter) left Calais in June, taking over two stations in the Zeneghem defences, and No. 36 (Lieutenant C. B. Williamson) left on the 28th of that month for Abancourt, where it relieved No. 22 Section (Lieutenant Bruch). This last section which, it will be remembered, was composed of two detachments of Tyne personnel now moved away to Dunkerque. The British military establishments at Abancourt included a very large ammunition depot and numerous labour and rest camps.

The Zeneghem defences to which No. 34 Section was now posted was, at this date, combined under one command with that of Audruicq. There were in the combined defences two A.A. batteries, six A.A. searchlight sections and the Audruicq and Zeneghem "false areas"[1], the former being operated by No. 19 and the latter by No. 13 A.A.S.S., both Tyne sections. The other A.A. sections in the area were Nos. 20 and 34 Tyne Sections and Nos. 12 and 30 London Sections. The officer commanding "A & Z" defences, as this organisation now became known, was Major Roscorla, R.A., and Lieutenant Rendell was detached from No. 13 Section (though he retained command of it) to control all the searchlights at Audruicq. Zeneghem was actually fought as a separate command, the searchlights in that area being under Lieutenant Edwards of No. 19 A.A.S.S. All the guns and lights were now connected by telephone to the defence headquarters at Hennuin, a proportion of the installation being carried out by the R.E. signal units in the district, but the greater part by the sections themselves who used their own portable instruments and 'D 3' cable. For convenience in telephoning each station was named, the guns being "Rifle," "Bayonet," "Sword," etc., and the lights "Madge," "Ethel," "Peggy," "Phyllis," and so on. This scheme worked quite satisfactorily and saved constant reference to individual sections but led occasionally to somewhat disconcerting messages being given and received. It is regrettable that such of these messages as are extant are considered unsuitable for publication !

The lights were not, at this period, equipped with any sound locators or other aid to searching, except night-glasses, and the percentage of targets picked up was not high. Accordingly, the object aimed at was the denial of the danger area to raiders by means of barrages combined with the use of "false areas," and the actual destruction of enemy aircraft was a secondary, though welcome, result.

No. 20 Section (under Lieutenant Erskine) is reputed to have achieved the distinction of being the first in the "A & Z" defences actually to pick up an aeroplane in the beam, but was also the first to suffer a casualty from bombs, No. 1 of one of the searchlights being wounded and the telephonist, the only man of the detachment under cover, being killed.

---

[1] The Audruicq "decoy dump" was near the village of Mannequebuerre and that of Zeneghem near Brouckerque.

Probably the most severe raid on Audruicq and Zeneghem occurred early in September when enemy attacks started shortly after sundown and continued until 3.0 a.m. No material damage was done to the Audruicq depot although a considerable number of casualties occurred to the depot personnel. The worst casualties occurred in a battalion of Chinese Labour Corps which was being marched into the open country for safety. Unfortunately, several bombs fell directly on the battalion, causing between 200 and 300 casualties. The searchlights proved reasonably efficient and a number of aeroplanes were illuminated, an A.E.G. bomber being shot through both engines and landing, otherwise intact, several miles to the west of Zeneghem, the crew being taken prisoners and handed over to the R.F.C. Intelligence Section. In all, the guns fired approximately 3,000 rounds during this raid and there is little doubt that, while their individual shooting did not result in the destruction or disabling of more than one aeroplane,[1] the system of barraging on what was known as the " S.O.S." plan, under which every gun came into independent action, was effective in diverting enemy raiders from the vulnerable areas.

After this, with one exception, no bombs were dropped in the vulnerable areas at Audruicq and Zeneghem up to the end of the War. The single exception took place on a night towards the end of 1917 which was deemed unsuitable for flying, when a single light aeroplane came over the defence with its engine shut off, so that no warning of approach was given, and dropped several bombs in the Zeneghem area igniting a dump of empty shell boxes and causing a serious fire which must have been visible for many miles, but fortunately the weather grew worse and no general raid followed.

In the " A & Z " defences, and indeed in all the L. of C. stations, both light and gun positions were permanent and in consequence conditions were definitely ' cushy ' (to use the expression of the time), all ranks being accommodated in huts, light emplacements being well built and maintained and the engine-exhaust silencing arrangements as nearly perfect as possible. Raids were frequent but, the principal anxiety of all concerned was to be transferred to an Army area.

After the foregoing somewhat lengthy digression we return to the consideration of the further expansions which the anti-aircraft searchlight organisation underwent in the early summer of 1917. The next group of sections to be raised, Nos. 38 to 45, completed the establishment of A.A. searchlight units which had been considered, in the fall of 1916, as likely to be sufficient for the requirements of the B.E.F. in France. Following the correct sequence, the Tyne sections in this group were Nos. 38, 40, 42 and 44, and the London sections Nos. 39, 41, 43 and 45.[2]

---

[1] Many were the suggestions made for the improvement of air defence. One infantry colonel expressed the opinion that if a searchlight could be arranged to throw a beam which changed colour every thousand feet, the gunners would have some idea of the height of the targets, and might be able to bet on a burst in their vicinity !

[2] No. 39 A.A.S.S.—Lieut. G. A. Wadham. No. 41 A.A.S.S.—Lieut. K. L. Wood. No. 43 A.A.S.S.—Lieut. C. H. Reed.

The officers and N.C.Os of the Tyne Electrical Engineers with this group were the following :—

No. 38 A.A.S.S.	... ...	Lieutenant F. S. Corby.[1]
		Serjeant J. Ireland.
		Staff-Serjeant A. Inglis.
No. 40 A.A.S.S.	... ...	Lieutenant E. H. White.[2]
		Serjeant Smithson.
		Staff-Serjeant J. Ogle.
No. 42 A.A.S.S.	... ...	Lieutenant J. L. Batey.[3]
		Serjeant McKenzie.
No. 44 A.A.S.S.	... ...	Lieutenant R. O. Porter.
		Serjeant J. Gair.
		Staff-Serjeant A. Leithead.

No. 38 A.A.S.S. was formed from personnel at Clifford's Fort, North Shields, and joined the other three at Haslar Barracks on the 16th May. The four sections proceeded together to London on the 21st May and, after undergoing at Millbank Hospital the short anti-gas course which was now a necessity for all troops proceeding overseas, embarked at Southampton on the 9th June, in company with the four London sections, in the "South-Western Miller 16" and "Marguerite" transports. The sections occupied only a small part of the accommodation in these two ships which also carried the first contingent of American troops to France.[4] Although there were senior American officers on board, it appeared that they were all medical officers or officers of other non-combatant branches of the American Army, and so it came about that one of the sapper subalterns was appointed O.C. troops in each of these transports. The Tyne Electrical Engineers can thus claim to have commanded, for the brief period of crossing the Channel, the first contingent of the American Army to enter the War.[5]

The sections disembarked at Havre on the 10th June[6] and No. 42 entrained that night for Calais, being followed on the 12th by No. 40 Section and on the 13th by Nos. 38 and 44, these latter sections having passed the intervening period at the Docks Rest Camp at Havre. At Calais the sections remained some weeks in No. 6 Rest Camp occupied in the exacting duties of "humping Dormans" for Lieutenant Fox, and in undergoing instruction in the most recent methods of air defence. Work for the Equipment Section commenced at 4.0 a.m. and ceased only when the last engine was loaded.

---

[1] From No. 38 (Tyne) A.A. Company at Hull.

[2] From No. 42 (Tyne) A.A. Company at Coventry.

[3] From No. 42 (Tyne) A.A. Company at Coventry.

[4] America declared war against Germany in April, 1917, and the first of her combatant troops were landed in France on the 26th June.

[5] Of the two transports, the "South-Western Miller 16" arrived at Havre first. Lieut. E. H. White was O.C. troops on this ship and Lieut. F. S. Corby, adjutant.

[6] As the ship came alongside the quay the men saw for the first time German prisoners working as dock labourers. This sight seemed to rouse the anger of one of the sappers of No. 38 Section, Sapper E. Dunn, a hardy rivetter of Tyneside, who immediately began to take off his jacket as he looked fiercely at the prisoners of war. Serjeant Ireland was dispatched to enquire what the sapper proposed to do and received the reply "Aam gannin to get at them Fritz's." This fire-eating sapper was then told that he couldn't do that as etiquette would not allow it, to which Dunn, looking somewhat rebuffed but still threatening, made answer "Whees etiquette? Cos aa can seun settle him an aal!" This incident appeared in the press at the time as typifying the spirit of the northcountrymen.

After about a fortnight in this place, No. 42 Section (Lieutenant Batey) left for Boulogne where it relieved a section of the 50th (F.S.) Company R.E. The A.A. artillery in this defence was furnished partly by the British but included also a battery of French " 75s." Raids took place over this locality whenever there was a moon, the objectives being the town of Boulogne one night and the defences the next. During this period the enemy caused very little material damage of military importance though on one occasion a water-main in the town was burst by a bomb and several civilians who had taken shelter in a cellar were drowned. Bombs occasionally fell unpleasantly near the lights causing casualties to the detachments, which served as a reminder that the searchlight units were on active service even at Boulogne. On the 26th August Lieutenant Batey left No. 42 A.A.S.S.,[1] but the section remained at Boulogne for the ensuing twelve months.

No. 38 A.A.S.S. was the next of these sections to leave Calais, proceeding to Romescamp in the Abancourt area on the 7th July. The sections in the area at this date included No. 7 (Captain Clarke, L.E.E.), No. 36 (Lieutenant C. B. Williamson) and No. 38 (Lieutenant Corby), grouped under Captain Clarke as A.S.O., but the area was, at this period, left rather severely alone by the enemy.[2]

No. 40 Section (Lieutenant E. H. White) was posted on the 11th July from Calais to Dannes where, on the following day, it took over the stations and plant of No. 15 A.A.S.S. (Lieutenant Baldwin). These lights were barrage lights for the defence of the ordnance depot, but, during the period when No. 40 A.A.S.S. was in this locality (July 1917 to March 1918), they never exposed in search of enemy aircraft, although during the famous raid on Etaples at Whitsun, 1918, when the hospital of the Order of St. John at that place was severely bombed, the barrage succeeded in bringing down a Gotha. No. 28 A.A.S.S. (Lieutenant H. G. Campbell) moved into the Dannes-Camiers district from Rouen in the autumn of 1917 and remained until the spring of the following year.

No. 44 A.A.S.S. (Lieutenant Porter) remained in reserve at Calais until August 1917 when it was detailed to the First Army XVIII Corps and proceeded, on the 24th of that month, to the vicinity of Berguette, Pas-de-Calais, taking up position to the east of the Aire Canal for the defence of Isbergues steel works in conjunction with " B " Battery A.A.A. and No. 14 (London) A.A.S.S. These works, though barely 20 kilometres from the Line, were at that time working at full pressure on munitions. This locality and that of Bethune, immediately to the south, were being " constantly bombed."[3] The low-lying ground between the steel works and the Line was unsuitable for searchlight work but the A.A. artillery put up a very effective barrage and the actual damage done to the works was small until February 2nd 1918, when the Germans

---

[1] Lieut. Batey subsequently rejoined No. 42 Section at Boulogne in August 1918.
[2] Lieut. Corby was appointed A.S.O. Abancourt on December 11th, 1917.
[3] " Work of the R.E. in the European War " Vol. " Miscellaneous," page 316.

made determined efforts to destroy them, even bombing ten times in one night. The following extract from the diary of the officer commanding No. 44 Section describes the action of the defence at this time :—

" All through the winter of 1917, every fine night meant air-raids from sunset to sunrise. The first alarm would be received at 4.30 p.m. and the last at, say, 7.15 a.m. The general procedure was to sit, dressed up in overcoats in the most convenient hut, field-glasses, barrage tables, etc., all ready to hand. Front line A.A.A. sections would report the approach of hostile 'planes by telephone :—' 70 Section reports hostile 'planes crossed the Line flying west.' 70 Section would be at, say, Laventie, which would give us 10 minutes warning. The gunner officer would immediately give the famous formula ' warn the factory.' The factory would be in full blast, but in a few minutes the hooter was blown, the whole place in darkness and the workmen all in excellent dug-outs. One would now hear the sound of the enemy sailing to the attack ; weird and wonderful orders would be given to the searchlights, and the gunner would give his orders for the barrage :—' Stand-by factory barrage, fuzes 9 and 10.' The enemy being now about a mile away, ' Go on ' would be given. Then things happened. Roar of guns, bursting shells, flashes and thunder of bursting bombs, machine-guns in frenzy, and searchlights waggling about in a helpless manner ; feeble voices from the factory look-out attempting to control the fire :—' Raise your barrage'; ' Lengthen your barrage'; ' Sweep right ' ; etc. The enemy, having unloaded his cargo of bombs, would then take his departure and information would be collected as to the damage caused and communicated by telegram to G.H.Q. After a decent interval the ' All Clear ' would be given to the works which would immediately spring into life again. It must have annoyed the enemy pilots to see the works going full blaze as they were re-crossing the Line on their homeward journey. Half-an-hour later, the dose would be repeated ! "

The whole process was much complicated by the presence of our own night bombers who had an advanced aerodrome in the vicinity and had to be challenged by searchlight, when they would reply by firing Very lights of the correct colours for the particular night. This challenging was the most arduous and difficult work meaning that the whole detachment had to stand-by and the engine be kept running most of the night. Often the searchlights had to be exposed as " lighthouses " to direct our own machines back to their aerodromes on foggy nights.

Two A.A.S.S. were also in the Bethune area, one of these being No. 17 A.A.S.S., under Lieutenant Hutchinson, which had moved into this district from Abbeville in June. Enemy aircraft action in this area was now, as has been mentioned above, considerable, and this district was also subject to serious " back-area " shelling.

At this period " long-arm " control for the projectors began to be introduced. Since June the searchlights, including the 60 cms. plants not yet exchanged, were fitted locally by each section with an extended pipe-control which largely overcame the difficulty that had been experienced in the direction of the beams by the use of the hand wheel on the projector. The pipe-control consisted of an extension of the right-hand trunnion by means of a 15′ shaft, made of $1\frac{1}{2}''$ piping, with a hand wheel at the end of the pipe by means of which the projector could be elevated and depressed. The end of the pipe travelled round on a " raceway " made of wood, the rail being about 4″ wide and surmounted by a round iron bar, about $\frac{1}{2}''$ in diameter, secured by staples filed down to feather edge on top. This arrangement permitted a reasonably smooth and accurate control of the beam in all directions. The raceway described suffered the disadvantage that it would only stand one or two removals before it had to be renewed and, later, iron supports made of $2\frac{1}{2}''$ angle iron pickets were supplied and the raceway itself was then made of $2\frac{1}{2}''$ flat iron bars bent to a true circle and bolted to the up-rights. This pattern of raceway subsequently continued in the service for many years though it suffered from two disadvantages which ultimately led to its being discarded. These were the difficulty in accurate levelling and that, after driving a few times in hard ground, the tops of the pickets bent over and the bolt holes became distorted. In the early months of 1918, the controlling handwheels began to be replaced by skew-gearing incorporating a telescope, the movements of which were co-ordinated with those of the projector. The invention of this apparatus is attributed to Captain Clarke, London Electrical Engineers, who is said to have constructed the first model when with No. 7 section at Abancourt in June 1917. A large number of these gears were made at Les Attaques work-shops, near Calais, under the supervision of the Searchlight Equipment Officer, Lieutenant Fox, and distributed to all the sections in France. Opinions as to the value of this mechanical control (as it was officially termed) varied somewhat. Most officers seemed agreed that it enabled slower and more accurate searching to be carried out, but it was found that, unless the projector was perfectly balanced in all positions, ease in manipulation was impaired and, owing to the " whip " in the pipe " long-arm ", the projector had a tendency to " hunt " when being elevated or depressed rapidly. One of the early difficulties experienced by the A.A. searchlights was in following targets passing directly overhead as it was found that the enemy aircraft invariably passed out of the beam while the operator traversed his light through 180 degrees in order to be able to depress it again. This difficulty was overcome by lengthening the lamp cables so that the projector could be turned on its back through a vertical arc of 180 degrees. It was also found that the use of the telescope required considerable training and, in common with most innovations, the skew-gear mechanical control was far from being universally popular when first introduced.

The 90 cms. searchlights were, of course, very much more powerful than the 60 cms. equipments[1] and the difficulty in detecting a target in the beam, occasioned by the amount of stray light in the neighbourhood of the projector itself, rendered the "long-arm" control an absolute necessity. The difficulty was still further countered by the introduction of stray-light deflectors, or "anti-dazzle devices," consisting of concentric rings of sheet metal built into the face of the projector. With the use of these appliances it was found that the controlling number of the detachment could hold a target much longer without the aid of spotters. No provision was officially made until late in 1918 for "dousing" the beams, arcs being struck when a target was heard approaching. Blankets were used largely for "dousing" the beam until the instant of exposure. Various experiments were made by the different sections to provide co-ordination between the spotters using binoculars and the numbers controlling the projector. Some of these devices consisted of sights fitted to the binoculars, in other cases the binoculars were mounted on graduated stands so that the angles of elevation and bearing could be read off and transmitted to the projector, while, in one section, No. 44 (Lieutenant Porter), an adaption of the Cardew Pane system of signalling by illuminated arrows was developed.

During the months of July, August and September very heavy bombing by enemy aircraft was experienced in the XV Corps area on the coast behind Nieuport, and in the Fifth and Second Army areas during the third battle of Ypres.[2] To cope with this, fourteen A.A.S.S. were concentrated in this part of the Front and the use of searchlights in forward areas was then developed in the Ypres salient. No. 8 A.A.S.S. (under Captain Rooksby) was already in this district, being in position between Abeele and Poperinghe. Late in June, one of the guns of No. 88 Section A.A.A., with which this section was at that time co-operating, was moved forward to Dickebusch to guard observation balloons which had been having a bad time from enemy aircraft, and one of the lights was also moved forward to a position where it could continue to be of service to this gun by night. On the night of the 6th July the enemy dropped no less than 88 bombs on the town of Bailleul and this light was then moved to that place. On the 14th July No. 15 Section (Lieutenant Baldwin), having been relieved at Dannes by No. 40 (Lieutenant White), arrived at

---

[1] The Army, as a whole, was ever very ignorant of all matters connected with the searchlights. Intense rivalry is said to have existed between neighbouring area commandants, town majors and others similarly placed, who had their own pet lights which, they asserted, were more powerful than the others (due, of course, to the fact that they were nearer to them). The nurse is perhaps to be excused who was overheard to say that she had seen "a long white light come out of a funnel and wondered where it was kept and how it was wound back." The staff officer who demanded to see where the carbide is kept when inspecting a searchlight detachment of electrical engineers is almost as hardy an annual as the thirster after knowledge who invariably fails to comprehend why the sound-locator has three trumpets on one side and only one on the other. The prize must be given, however, to the American who asked to be told what signal the aeroplanes gave to assist the searchlights to pick them up.

[2] The third battle of Ypres began on the 31st July, 1917. "Up to the date of the battle of Messines (June 7th, 1917) the Belgians had held the ground near the coast, and the five British armies had lain over their hundred-mile front in the order from the north of Two, One, Three, Five and Four. Under the new arrangement (i.e. prior to the third battle of Ypres) it was the British Fourth Army (Rawlinson) which came next to the coast, with the Belgians on their immediate right, and an interpolated French army upon the right of them. Then came Gough's Fifth Army in the Ypres area, Plumer's Second Army extending to the south of Armentieres, Horne's First Army to the south end of the Vimy Ridge and Byng's Third Army covering the Cambrai front, with dismounted cavalry on their right up to the junction with the French near St. Quentin. Such was the general arrangement of the forces for the remainder of the year." (Conan Doyle "The British Campaigns in Europe," page 524.)

" MOBILE ! "

Anti-aircraft section unloading searchlight
equipment in France.

Poperinghe and took up position at International Corner and Eykoek, three and six kilometres respectively to the north of that place. Early in August the second light of No. 8 Section was moved to the top of Boeschepe Hill. Enemy attacks over this sector now became more frequent, one in particular being made in great force on the night of the 1st/2nd September, when Staff-Serjeant Cripps of No. 15 A.A.S.S. was killed by an aircraft bomb whilst the light at which he was stationed was in action. The lights of this section had, in all, some fifty bombs dropped on or near them during the course of operations, and the total number of enemy aircraft held by them during the period they were in France was 95.

On the 4th September the Boeschepe light was moved further south and No. 8 Section was then concentrated round Bailleul. At this date No. 17 A.A.S.S. (Lieutenant Hutchinson) arrived in the area from Bethune and was posted in the neighbourhood of Ypres. Enemy aircraft were very active and these three sections were continuously in action every night, bombing commencing soon after dusk and continuing until daybreak.

In the same month of September Lieutenant Winter, leaving his own section, No. 34, in the Zeneghem defences, took over temporarily the command of No. 24 A.A.S.S. at La Panne in the Nieuport sector of the Line. This was one of the composite sections which had been formed on the reduction of the original eight sections from three lights to two and consisted entirely of London Electrical Engineers. This section also was constantly in action at this time, the enemy making nightly raids on Dunkerque and on the Handley-Page aerodrome at Couderkerque.

It was at this period that the process of replacing the non-mobile generating sets by Dennis-Stevens or Tilling-Stevens petrol-electric lorries was commenced in the forward sections. These lorries were landed at Dunkerque and sent forward immediately. Thus, on the 6th August, No. 8 Section had received its lorries, and the change over of projectors, which took place in this section at the same time, was completed by the 15th of that month. It was the intention that all the sections should be equipped with these vehicles, but, owing to difficulties of supply, it was April 1918 before the change was completed even in the forward sections. This fact added materially to the hardships which these sections underwent in the retreat on the Somme, while several of the L. of C. sections which subsequently took part in the final advance of the British Line from August to November 1918 did so still under the burden of their non-mobile plant. The 16 kW. sets were very cumbersome and difficult to move, the base-plates alone weighing as much as two tons each, and the arrival of the petrol-electric lorries greatly increased the mobility and efficiency of the sections. As an example, one of the lights of No. 8 Section was moved on the 27th September 1917 from Bailleul to Canada Corner and, although orders were only received for the move at 2.30 p.m., the light was in action at its new site that night, a great achievement when compared with the non-mobile nature of the old plant. This move was followed by two especially lively nights in action and, on the 29th, this light was moved again, this time to Fletre, arriving in its new position in time for an exceptionally active night's work in which, to quote the entry made at the time in the diary of the officer commanding this section, " umpteen machines were in the beams and the gunners had a good night's shooting." Another officer, commenting on this same night's activities, records that eleven enemy aircraft were picked up by the lights of his section (No. 15) and held for periods of from three to nine minutes duration. The following two nights are noted as having been " even busier."

The congratulations of the Fifth Army A.A. Group Commander were conveyed to all A.A. sections in his area in the following terms :—

FIFTH ARMY A.A. GROUP.

" The Group Commander wishes to congratulate all ranks of the A.A. searchlight sections on their excellent work during the recent raids. The ' picking-up ' has been exceptionally quick and all targets have been held until they passed out of range.

" Good searchlight work is the foundation of all A.A. defence at night."

A.A. Group Headquarters,
1st October, 1917.

Early in October No. 24 A.A.S.S. (under Lieutenant Winter) moved from La Panne to the Ypres salient and established searchlight stations near Brielen and Vlamertinghe some three kilometres to the north-west and west respectively of Ypres.

On the night of the 14th October the A.A.A. brought down an enemy Voisin and during a raid on the 19th the lights were severely bombed, bombs falling within 30 yards of the projectors, but the lights were all maintained in action. This bombing of the actual lights now became very prevalent, a particularly severe attack being made, apparently on the lights only, by twelve enemy machines on the night of the 7th November. At this stage of operations it became absolutely essential to dig in both the projector and the raceway to avoid casualties from bomb splinters and from shell-fire. Engines and lorries were also protected as far as possible, in localities where they could not be stationed in sunken roads or behind other natural cover, by sandbag breastworks built up to about three feet in height round the vulnerable parts. Heavy rains at this period also contributed to render existence in this locality far from comfortable.

The third battle of Ypres ended on the 10th November and the lights were then more closely concentrated round Ypres. At this period, and for the ensuing months, interchanges of personnel were made between the sections on the

SEARCHLIGHT EMPLACEMENTS.

Above—A 120 cms. projector with camouflaged sandbag breastwork erected in low-lying and water-logged land.

Below—A 90 cms. projector in excavated and revetted emplacement in permanent defence on L. of C.

L.O.C. and those in the forward areas. These interchanges were welcomed by all concerned as affording much needed rest to the men of the advanced sections and valuable experience to those on the L.O.C.

On the 9th December No. 8 A.A.S.S. (Captain Rooksby) moved away from the Salient and, travelling south by road via St. Pol and Arras, reached Bapaume in the Third Army area where the lights were brought into action at Villers-au-Flos and Fremicourt. In this locality the lights were constantly in action, an enemy A.E.G. bomber being brought down while in the beams of No. 8 Section during a raid on the night of the 23rd December. The A.A. defences were further reinforced on the 25th December, when Nos. 3 and 5 A.A.S.S. arrived in the vicinity. No. 11 A.A.S.S. (Captain Myles) was also in the Somme area at this date, being stationed in the new Fifth Army area[1] in the neighbourhood of Peronne, but this section was relieved in the opening days of February 1918 by No. 17 A.A.S.S. (Lieutenant Hutchinson) and moved north again, taking up positions to the east of Hazebrouck. No. 17 Section was not yet equipped with lorries and moved to Peronne from Ypres by rail, the transporting of their equipment from the railhead to their positions about two miles east of the town being carried out by the A.A. artillery.

On the 11th February 1918 No. 15 A.A.S.S. (Lieutenant Baldwin) left Poperinghe and proceeded south to Bapaume and took up positions in the defences there. At this period the enemy carried out much aerial reconnaissance by day and on one occasion early in the morning, when an enemy aircraft bombed some horse lines in the vicinity of one of the projectors of No. 17 A.A.S.S., Sapper G. Bage with a Lewis gun succeeded in turning the machine away. The aeroplane subsequently crashed about a mile away and burst into flames, being completely destroyed and the pilot killed. For this act Sapper Bage was awarded the Military Medal. Activity by night in the Somme area was at this time not very severe, but this proved merely to be the lull before the storm which was destined to burst upon this part of the Line a month later. Before describing the strenuous activities of the A.A. searchlight sections during the retreats of March and April 1918, it is necessary to chronicle certain developments which had, in the meanwhile, taken place in the general organisation of the air defences overseas.

In August 1917 a G.H.Q. committee on anti-aircraft protection had been convened, which estimated that the existing resources were inadequate and that at least 30 additional searchlight sections would be required in France.

Owing to the prevailing shortage of man-power, it was found possible to provide only about 30 per cent. of the additional personnel from England, the remainder being found in France from category " B " personnel transferred from the infantry and trained in the existing A.A. searchlight sections.[2]

---

[1] The battle of Cambrai ended on November 30th 1917. After this date the Fifth Army moved from Ypres and took over a portion of front in the Somme district, to the south of the Third Army, on the extreme right of the British Line.
[2] " Work of the R.E. in the European War " Vol. " Miscellaneous," pp. 21 and 317.

The personnel from home was provided by drafts of category " A " men withdrawn from the coast and anti-aircraft defence units and consequently the A.A. searchlight sections formed subsequent to No. 45 had no real unit identity, and it is probably due to this that only scanty information as to their activities is available. Although the " Work of the R.E. in the European War " records 75 A.A. searchlight sections as having been in France on the 1st January 1918, there is evidence to show that few of them existed as separate sections and that some drafts, amounting almost to complete sections, had not left England at that date. For these and other reasons it is not possible accurately to trace the formation and movements of these later sections, but such associations of officers or other ranks of the Tyne Electrical Engineers with any of them as have been recorded will be referred to later in this chapter. If these later sections never equalled the earlier units, it was probably due to the inferior qualifications of the men. The first 45 sections were, as we have seen, composed entirely either of London or of Tyne Electrical Engineers among whom the standards of education, physique and technical ability were unusually high. In the Tyne sections a high proportion of the men were actually " Tynesiders " with strong territorial associations and the great majority of them came from what may be described as the artisan as opposed to the labourer class. In the ranks of these sections were to be found many who might perhaps more usefully have served their Country as officers but for some inferiority complex or the fear of accepting responsibility.[1] The " esprit de corps " which such personnel possessed was a characteristic of the sections of Tyne Electrical Engineers, the value of which it would be difficult to over-estimate.

The G.H.Q. committee also authorised the addition of a third light to each of the existing sections in France and a revised establishment for " An Anti-Aircraft Searchlight Section R.E. (3 lights) " was published in February 1918, increasing the total rank and file in each section to 20.[2] The personnel for these third lights was provided but in some cases the third light itself never arrived. The expansion to provide the additional 135 lights went on all through 1918 and drafts of American and Canadian sappers were attached to the various sections for training in addition to the infantry personnel referred to above. The training of the category " B " men was found comparatively easy and they made excellent all-round sappers in duties which did not require high technical knowledge. It was also remarked that, in spite of their lower medical category, the prevalence of sickness among them was no greater than among the category " A " personnel. The men undergoing training were first of all instructed in their duties, both at the engine and at the light, by the N.C.Os. in charge of the various detachments. In the case of the American and Canadian drafts the men ran the whole station during the day and, after a short time, picked men took the places of certain of the responsible numbers in action at night and, in this manner, the feeling of being merely onlookers was discouraged.

---

[1] One such man was a lecturer in mathematics at one of our large Universities who did excellent service as an N.C.O. and was always one of the calmest men in a crisis. It has been suggested that this was possibly because he was able to work out quicker than anyone else the chances against being hit by a bomb.

[2] W.E. No. 1032 dated 27th February 1918.

ITALIAN PATTERN OF SEARCHLIGHT.
Installed above the pontoon bridge on the River Piave.
October 1918.

It has already been related in an earlier chapter how, in November 1917, the headquarters of the Second Army, together with two British Corps, were transferred to Italy to help in checking the Austro-German advance after the disaster of Caporetto.[1] Very shortly afterwards, the need for A.A. searchlights with the Italian Expeditionary Force (I.E.F.) began to be felt, with the result that in January 1918 No. 34 A.A. Searchlight Section was selected for transfer to Italy. This section subsequently proved to be the only A.A. searchlight unit to serve overseas outside France. The O.C. of this section, Lieutenant Winter,[2] was ordered to join the I.E.F. independently and proceeded on the 26th January to Padua, reporting on arrival to the Engineer-in-Chief and being posted to the A.A. Group Headquarters near Montebelluna in the Piave sector of the Line.

Two Italian Army searchlight companies which had been placed at the disposal of the British Army came under Lieutenant Winter, but did not prove a great success, chiefly on account of the fact that the plant available was not suitable for A.A. work.

In February No. 34 Section left Zeneghem and went to Italy with its numbers augmented to the new establishment for three lights, but took no plant with it from France. On arrival, the personnel was split up among the Italian detachments for the purpose of instructing these latter in the most recent methods of air defence, but this experiment did not prove very satisfactory. Accordingly, at the end of March 1918, three sets of Galileo-Fiat searchlight equipment were purchased from the Italian Government and these were operated by the personnel of No. 34 Section. These projectors were of a highly mobile pattern, being furnished with a light barrow-type carriage on two wheels by means of which they could easily be moved from their lorries to the actual sites to be occupied. On arrival in position the projectors were lowered on to their bases and the wheels removed.

When at the beginning of April the British forces took over the Asiago section of the Line in the north, No. 34 Section was sent ahead to assist in preparing sites for A.A. guns and, on the 12th April, Lieutenant Winter was appointed searchlight officer to the A.A. Group Headquarters, with the acting rank of captain, Lieutenant Armistead from the R.E. Base Depot taking over executive command of No. 34 Section.

Between April and September this section manned three stations in the vicinity of Schio and Thiene and the lights were in action on frequent occasions, though later, after the withdrawal of the German bombing squadrons during April, night bombing practically ceased. During most of this period Captain Winter was only partially occupied with searchlight work, two other interesting duties having been allotted to him. The first of these was to assist our own Air Force in dropping spies over the enemy's Line. For this purpose two searchlights

---

[1] The Austrian blow fell in October on the Isonzo Front and broke the strongest portions of the Italian Line. The Italian Army suffered such losses of men and material as made it doubtful, at that time, whether a stand could be made on the line of the Piave river.

[2] Lieut. Winter, whom we left in command of No. 24 A.A.S.S. in the Second Army area at Ypres in October 1917, had handed over this section to Lieut. North (L.E.E.) and reverted to his own section, No. 34, at Zeneghem, in December.

placed on a given alignment were exposed on receipt of wireless messages from the aeroplane and served as directing lights. The second was in connection with an experiment by the Italian searchlights in front-line lighting in co-operation with heavy artillery.

In October No. 34 Section returned to the Piave sector and took up positions to cover the pontoon bridges[1] which were constructed in the area of the Italian Tenth Army north-east of Treviso.[2] Here it remained until the 2nd November when the armistice between Austria and Italy was concluded. The section then went into billets in the Vicenza area until demobilised early in the following year. For his services in connection with A.A. defence in Italy Captain Winter was awarded the M.C.[3] A special medal was also struck by the Italian Government and presented to the members of the allied Italian expeditionary forces, but this was not recognised by the British military authorities as a service medal and permission to wear it as such was withheld.

In order to preserve a connected narrative the story of No. 34 A.A. search-light Section in its Italian venture has been carried to the conclusion of hostilities, and it is now necessary to return to the beginning of the year 1918 in order to continue the record of activities of the sections in France from that date.

As a further result of the G.H.Q. committee on anti-aircraft protection which was convened in August 1917, an assistant inspector of searchlights, with the rank of major, had been authorised for each Army[4], and for the lines of communication with the rank of captain. The appointments first made to fill these posts were as follows :—

ASSISTANT INSPECTORS OF SEARCHLIGHTS.

First Army	...	...	...	Major R. G. Madge (L.E.E.).
Second Army	...	...	...	Major D. R. St. J. J. ffrench-Mullen, R.E.[5]
Third Army	...	...	...	Major G. Fergus-Wood (L.E.E.).
Fourth Army[6]	...	...	...	——
Fifth Army	...	...	...	Major A. W. M. Mawby (L.E.E.).
L. of C. ...	...	...	...	Captain H. G. G. Clarke (L.E.E.).

[1] The crossing of the Piave by the 7th British Division furnishes one of the classic examples of pontooning during the Great War. The pontoon bridge over the Piave had the greatest overall length of any bridge erected during the war. (See " The crossing of the Piave in 1918 " by Major W. A. Fitz G. Kerrich, D.S.O., M.C., R.E., published in the R.E. Journal, Vol. 41, page 569).

[2] The Italian Tenth Army was composed of the British XIV Corps, 7th and 23rd Divisions and the Italian XI Corps, the whole under the command of Lieutenant-General the Earl of Cavan, K.P., C.B., M.V.O.

[3] Battle honours are, of course, not awarded to units of Royal Engineers, but it is interesting to note that, from the list of battle honours to which R.E. units could lay claim were this not the case, No. 34 A.A.S.S. would appear to be entitled to the honours " Piave " and " Vittorio Veneto," though the claim of this section to these honours is actually shewn in the list as " doubtful (no diary)." No. 34 is the only A.A. section figuring in the list except Nos. 1 and 2 A.A.S.S., though it would seem that several others must have qualified for various honours.

[4] The anti-aircraft defences in each Army were under a lieutenant-colonel of artillery with the title of Anti-Aircraft Defence Commander.

[5] Formerly O.C. of the 50th (F.S.) Company R.E.

[6] The Fourth Army was withdrawn into reserve in May 1917 and, after a period on the coast, virtually ceased to exist until it was reformed to relieve the much battered Fifth Army after the second battle of the Somme in April 1918. It is therefore improbable that any A.I.S. was appointed to the Fourth Army prior to that date, when the post was taken over by Major Mawby, formerly A.I.S. of the Fifth Army. After the Fifth Army had refitted it took over an area between the First and Second Armies south of Ypres and Captain E. F. Rendell (T.E.E.) was then appointed A.I.S. with the rank of major.

In February 1918 the defences at Audruicq and Zeneghem were further developed and included with those of Calais and St. Omer to form the northern air defences, the searchlights in the combined defences coming under a deputy assistant inspector of searchlights, Northern Lines of Communication. The first officer to fill this post was Captain E. F. Rendell who was appointed acting-captain R.E. on taking up this post on the 8th February 1918.[1]

There were, at this date, a large number of A.A.S.S. in the combined areas, including French and Belgian units, and the following Tyne sections :—

No. 13 A.A.S.S. ...	... Captain Rendell	... Audruicq.
No. 19 A.A.S.S. ...	... Lieutenant Edwards	... Zeneghem.
No. 20 A.A.S.S. ...	... Lieutenant Erskine	... Audruicq
No. 26 A.A.S.S. ...	... Lieutenant Fox ...	... Calais.
No. 32 A.A.S.S. ...	... Lieutenant Dickson	... Calais.

Four or five listening posts had also been established a short distance outside the vulnerable area and were connected by telephone with the defence headquarters.

All lights, listening posts and guns in the defence were now brought under rigid control in the event of action and a " plan " system, similar to those employed in harbour defences at home, was introduced. Under this system various combinations of lights were used so that the searchlight pattern could be varied from time to time by a single command from headquarters. The scheme was operated from a control table constructed locally. This was about six feet square and the top of it was hinged, the outside being painted with a map of the whole of the defence. Each gun and light position was marked by a hole in the table top in which was inserted a small electric lamp, all the lamps being controlled through switches, manufactured from cartridge cases, mounted at the side of the table. This was probably the first table of its kind to be used in France, its purpose being to enable the defence commander to see at a glance which of his lights were in action, and similar installations were subsequently made in other defences. All telephones were parallelled in action and the searchlight officer obtained approximate bearings and elevations of aircraft from the listening posts and lights and, by means of home-made protractors on the map at headquarters, plotted the course of the enemy.[2] Barrages of gun-fire were then put up on the appropriate map squares in front of the approaching targets and were nearly always successful in compelling a change of course.

At the same time as the co-ordinated defence was introduced the main ammunition depots were brought under control and alarm syrens provided, so that, on receipt of an air-raid warning, the depots could immediately be put in darkness and the personnel protect themselves by resorting to bomb-proof shelters specially provided.

---

[1] Captain Rendell was succeeded by Lieut. Hort (L.E.E.) in June and this officer was in turn relieved by Captain W. G. Edwards (T.E.E.) on the 30th August 1918. No. 19 A.A.S.S. was then taken over by Lieut. Curtiss.

[2] To avoid confusion, it was found necessary to limit the number of beams to three per target.

Experience showed that the control of so extended a defence from a single point was rather unwieldy, and the Calais and St. Omer areas were operated in action as separate organisations, the officer controlling the searchlights at each place receiving the title of Area Searchlight Officer (A.S.O.). It was also found that the combined defence headquarter telephone system, concentrated at a 6-gun battery position, was unworkable owing to the tumult and noise during action, and the headquarters were therefore removed to a house about two and a half kilometres north of the village of Hennuin.

On the 17th July Captain Fox was appointed A.S.O. Calais. In recognition of the arduous work carried out by the Equipment Section (No. 26), Captain Fox was mentioned in despatches and subsequently awarded the O.B.E., while Serjeant Driver received the Meritorious Service Medal. As reward for the distinguished service of this section in the Calais defences, Captain Fox was also decorated with the French Croix-de-Guerre and Corporal Rudge received the Military Medal. The award of the Croix-de-Guerre was notified in the orders of the Marshal of France, Commanding-in-Chief the French Armies of the East, Maréchal Petain, in the following terms :—

" A réussi, grâce à la vigilance et à l'energie de son commandement, à obtenir de son personnel des résultats remarquables. Plusiens fois des avions qui piquaient en attaquant à la bombe et à la mitrailleuse ont été maintenus dans les faisceaux, en particulier le 15 Septembre à Fort Vert et les 4 et 5 Juillet à Marck.
" Au Grand Quartier Géneral, 6 Fevrier 1919."

During 1918 the enemy air raids were concentrated on Calais and Dunkerque with occasional diversions to bomb offending searchlights or gun positions. Dunkerque, as has already been mentioned, was not included in northern air defences as the air defences of that locality were now provided entirely by the French Army. Track charts were prepared weekly and co-ordinated at A.A. defence headquarters.

Once established, the Northern Air Defence Area underwent no drastic change up to the end of the War. In the meantime, however, it had been decided that all the lights in France should be 120 cms. projectors and during the course of the year 1918 many of the stations were equipped with these, the provision of " Yorks " arc-centralizers being commenced in the summer of that year. Sound locators also became available towards the latter end of that year, the first, a home-made one, being introduced into the defence by No. 20 Section (Lieutenant Erskine). Shortly before the armistice, the Deputy Inspector of Searchlights, Northern Lines of Communication, (Captain Edwards), provided assistance in the installation of the Brock-Curling system of aeroplane position-finding on two 4.7″ naval type A.A. guns near Audruicq. This system was only once used in action, on a single target which was driven out of range after the firing of only a few shots.

184

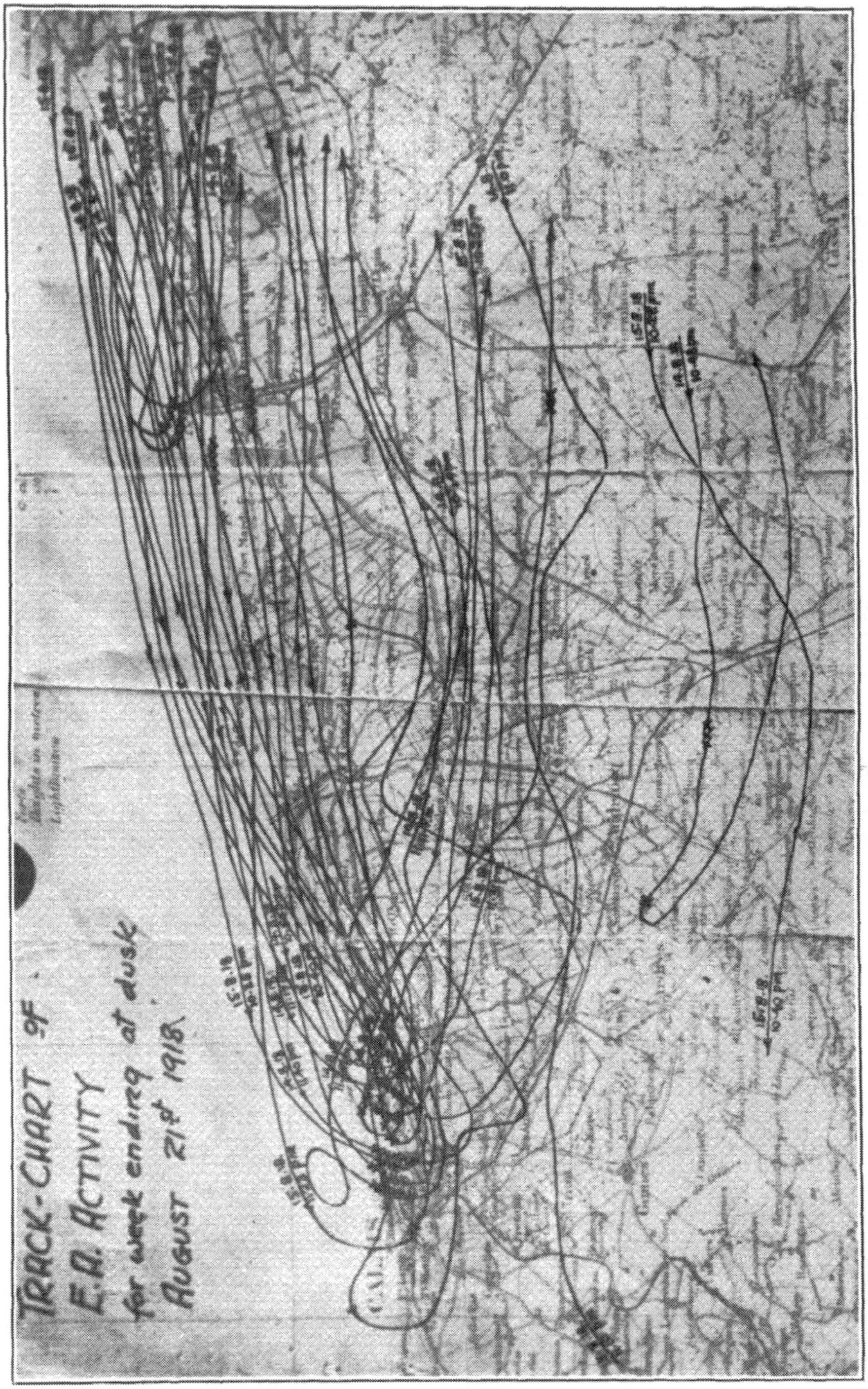

To face page 184]

It is now necessary to describe the operations of the A.A. searchlights in connection with the disastrous retreats during the second battle of the Somme and the battle of the Lys in March and April 1918. It is not possible here to trace in detail the progress of these battles, the earlier of which, the second battle of the Somme, opened on the 21st March 1918, after several days of intense bombardment by heavy artillery, in the Third and Fifth Army areas to the south. At this date the left of Sir Julian Byng's Third Army joined the extreme right of General Horne's First Army just north of the village of Fampoux, a few kilometres east of Arras, while the right touched the most northerly unit of General Gough's Fifth Army opposite Cambrai.[1] Following upon the successful battle of Cambrai in November 1917, several A.A. searchlight sections had been distributed to form a continuous lighted belt close up to the front line in the Third and Fifth Army areas, and the following Tyne sections were in this locality at the opening of the second battle of the Somme :—[2]

THIRD ARMY.

No. 8 A.A.S.S.	...	... Captain Rooksby	... Bapaume.
No. 10 A.A.S.S.	...	... Lieutenant Jardine	... Bapaume.[3]
No. 15 A.A.S.S.	...	... Lieutenant Baldwin	... Bapaume.

FIFTH ARMY.

| No. 17 A.A.S.S. | ... | ... Lieutenant Hutchinson | Peronne. |

The German attack began early in the morning of the 21st March[4] and so rapid was their advance that all the lights had hastily to be withdrawn. At this date, Captain Rooksby was officiating as assistant inspector of searchlights, Third Army, and upon him devolved the responsibility for the successful evacuation of the sections, while maintaining the most efficient co-operation possible with the artillery during the retreat. On the 22nd, No. 15 Section retired to a position at Miraumont and was in action that night when enemy aircraft activity was heavy and the whole area subject to active long-range artillery bombardment. This section and No. 8 were fortunate in being equipped with their P.E. lorries and the hasty retreat and constant dismantling and re-erection of plant did not tell so heavily on them as on the less fortunate No. 10 and No. 17 Sections who still had their cumbersome generating sets to move. The provision of transport by the A.S.C. for this service was also a matter of considerable difficulty owing to the inevitable disorganisation occasioned by the retreat ; No. 17 Section, at Peronne, was on the point of abandoning its plant when transport arrived and, in the face of the very greatest difficulties, the lights were removed and brought into action about three kilometres further to the rear. No. 10 Section, moving off about 4 a.m. on the 23rd, retired to Méaulte, a short distance from Albert, and the lights were temporarily in action there

---

[1] Conan Doyle " The British Campaigns in Europe," page 632.
[2] Ashmore, in " Air Defence," page 100, gives the total number of sections in the Third and Fifth Armies on the Somme as seventeen. Besides the four Tyne sections mentioned in the text, the 50th (F.S.) Company R.E. (three sections), and Nos. 2, 3, 9 and 45 Sections are known to have been in the vicinity.
[3] This section moved from the Second Army at Hazebrouck to Bapaume early in March.
[4] " The preliminary bombardment here as elsewhere broke out shortly after five in the morning and contained a large proportion of gas-shells which searched the rear lines . . . . as well as the front defences." (Conan Doyle " The British Campaigns in Europe," page 633).

that night. At this period, digging-in had to be abandoned and the lights were subject to severe bombing and to attack by machine-gun fire, the enemy firing down the beams at the projectors. The bullets could clearly be seen by those at the projector coming down the beam " like a flight of little white moths," but these sections were fortunate in escaping casualties from this cause. Méaulte was at that time the headquarters of the Third Army and while the section was here there was consistent bombing whether flying conditions were good or bad, " just a procession of machines arriving about 8 o'clock and continuing at fifteen to thirty minute intervals," to quote an officer of one of these sections. The usual approach was over the town, where a parachute flare would be released. The enemy, after proceeding about a mile further, then turned round and dropped his bombs on a well lighted target. Generally speaking, the A.A. artillery had little success and no machines were brought down, a fact which encouraged the enemy to attack at lower and lower altitudes to the increased discomfiture of the searchlight detachments. Some impression of the tense situation prevailing is conveyed by the following order, transmitted in manuscript, from the A.I.S. to No. 15 Section on the 24th March :—

> " If either of your positions becomes untenable, you will pack up at whichever position it is and proceed to ACHEUX, via ALBERT, reporting to me from there at TERRAMESNIL."

On the 25th March, on account of the continued rapid advance of the Germans and to the complete absence of transport, No. 10 Section was ordered to disable and abandon its plant. This was effected at midnight and the detachments withdrawn, but in the morning it was discovered that the enemy was not close at hand and the sets were subsequently recovered, though not by this section, which proceeded to Doullens and found itself, within a week, re-equipped with two mobile sets in P.E. lorries. Lieutenant Jardine of this section was mentioned in dispatches for his conduct during the retreat. By the 26th March the lights of No. 8 Section had been safely withdrawn as far as St. Pol and came again into action at Wanquettes and Avesnes-le-Comte. No. 15 Section, retiring by Beaumont-Hamel, Acheux and Doullens, reached St. Pol on the same date. Where all alike bore cheerfully the ardours of the retreat it is perhaps invidious to mention individuals, but two examples may be given as typical of the conduct of men of the Tyne Electrical Engineers during those strenuous days. Serjeant Vairy, when communication with his O.C. became no longer possible, moved his detachment from Bapaume to various pre-arranged positions in the rear, bringing his light unfailingly into action each night and only vacating these positions when they became untenable through shell-fire, and finally reported at Doullens with no casualties to personnel and his plant in perfect order. Sapper J. S. Hargreaves, a lorry driver who had been taught to drive only a week prior to the retreat, contrived by sheer determination and devotion to duty to bring his lorry to safety along heavily shelled roads and through congested traffic. Such men as these were never lacking among the sections when the need arose.

Meanwhile, No. 17 Section to the southward was constantly on the move and in action almost every night until the Line was ultimately consolidated at Villers-Bretonneux, the lights then being established outside Amiens and the projectors dug-in. These moves were rendered the more arduous by reason of exceptionally heavy shelling and bombing of roads by day and of continuous enemy activity by night, machines flying as low as 1,000 feet. Shortly after its arrival at Amiens this section too received its complement of P.E. lorries—just too late to be of real use.

Finally, by the 30th March, the German advance was brought to a stand-still and anti-aircraft defences were speedily developed for the important railway junctions of St. Pol and Doullens. The following was the message of the Army Commander circulated to all sections in the Third Army :—

"I should be glad if you would convey my heartiest congratulations to all ranks of the A.A. searchlight sections on their most successful operations during the last month.

(Sd.) J. BYNG, General,

31/3/18.   Third Army."

To St. Pol two sections were at first allotted, one of which was No. 15[1] and its O.C., Lieutenant Baldwin, was appointed A.S.O., being subsequently mentioned in despatches for his services in this capacity. At Doullens the sections, including Nos. 8 and 10, took up fourteen searchlight positions[2] and the defences were then developed on static lines, communications being constructed and controls dug-in. Track plotting was carried out as in the Audruicq and Zeneghem defences and, in order to simplify the determination of accurate angles of bearing and elevation of the projectors, No. 15 Section developed a system of illuminated scales, that for the angle of elevation being constructed from an "Archie" shell-case. On the 14th April the defences were reinforced by French A.A. artillery. Here, as in the Audruicq and Zeneghem defences, the various light positions were known by code names, those of No. 8 Section for example being distinguished by the prefix "EGG." On the 22nd May Captain Rooksby was appointed A.S.O. Doullens. At that date this locality was still very much under enemy shell-fire.

Lewis guns were now issued to each light and fixed to the projectors, thus providing the searchlight detachments at last with a means of replying to the fire of enemy aircraft. These were originally fitted near the barrel of the projector and aligned to move with it so that the fire could be directed along the beam at targets under observation. To allow for the movement of the target during the time of flight of the bullets from the Lewis gun, a special notched circle was provided in which moved the steel handle which replaced the butt of the gun. It was however generally thought better to mount the gun on a separate post where it could be manipulated with greater freedom and without interfering with the motions of the searchlight. It was also found

---

[1] The other was No. 3.
[2] "Work of the R.E. in the European War" Vol. "Miscellaneous," page 318, states that four sections were allotted to Doullens, which would give twelve lights only. The figure quoted in the text is from the diary of an officer present.

that, with the guns fixed to the projectors, the shock and vibration of discharge frequently caused the searchlight reflectors to crack. The ammunition with which these guns were supplied contained a proportion of tracer bullets, the trajectory of which could distinctly be seen at night thus enabling the gunners to correct their aim. In the early days the effective range of these weapons was apt to be grossly over-estimated and, from a distance, the trajectories of the tracer bullets could sometimes be discerned falling away apparently thousands of feet below the enemy machine.

Whereas, prior to the development of the defences at St. Pol and Doullens, enemy aircraft had obtained direct hits on railheads and depots, the experience gained by the searchlight detachments now enabled them to pick up most enemy machines coming within range, which as a result very rarely attacked below 6,000 feet and only inflicted slight damage of military importance. Heavy bombing continued over Amiens, 600-lbs. bombs being used in large numbers, but owing to the efficacy of the defences the majority of them fell outside the town.[1]

No sooner was the great German attack on the Somme brought to a standstill than the second blow fell in the First and Second Army areas to the north. The battle of the Lys opened on the 9th April and, during the ensuing twenty days, General Horne's First Army was driven back on a front of approximately twenty-one miles extending from La Bassée in the south as far as Hollebeke in the north. The resulting salient, some ten miles in average depth, enveloped the whole of the district lying between Bailleul and Bethune and seriously jeopardised the position of General Plumer's Second Army round Ypres.

The Tyne A.A. searchlight sections involved in this retreat were No. 11 (Captain Myles) and No. 44 (Lieutenant Porter). The former was stationed some few kilometres to the east of Hazebrouck in the second Army area and was subsequently obliged to abandon its searchlights owing to the impossibility of obtaining the necessary transport for their removal. This section was then withdrawn and re-equipped with mobile plant to occupy new positions on the eastern outskirts of St. Omer. Here it was very shortly joined by No. 28 (Lieutenant H. G. Campbell) which now moved up from Dannes and was also provided with three lorry sets.

No. 44 Section, situated with No. 14 (London) A.A.S.S. round the Isbergues steelworks in the First Army area, had also to withdraw on the 9th of April to the west bank of the Aire canal. This withdrawal was effected with considerable difficulty under enemy artillery fire, but in spite of the non-mobile nature of the equipment (some of which had to be dismantled and concealed until it could be fetched) this was carried out without loss of stores or personnel and the section took up new positions on the high ground to the west of Molinghem, overlooking the Lys valley. Immediately after the retreat the Isbergues steelworks were severely shelled and work had to cease.

The enemy's advance was finally stemmed on the 29th April and the battle of the Lys was at an end.

---

[1] Ashmore, " Air Defence," page 101.

With the object of intercepting enemy machines attempting to harass the L. of C., a continuous line of searchlights, two, three and, in places, four deep, was gradually established from St. Omer in the north (where it linked up with the Northern L. of C. defences) via St. Pol and Doullens to Amiens in the south. For this purpose, many more lights were required than had formerly been employed in the forward areas and these were provided by depleting the defences of stations on the L. of C. and by newly formed sections. The organisation of this lighted belt took some months to complete, but, before considering further the subsequent activities of the sections concerned, it is necessary to refer briefly to the circumstances under which the Independent Air Force came into existence and to recount the experiences of certain Tyne sections which were destined to be associated with that formation.

Up to the end of 1917 the shortage of material had precluded the use of aircraft on a sufficiently large scale for raiding enemy territory, but by the beginning of 1918 it was found possible to allot R.A.F. bombing squadrons specially for this purpose. In March the so-called " Southern Aerodromes " under construction in Lorraine numbered six, while the sites for five more had recently been selected.[1] At the end of May the R.A.F. establishments in the Southern Aerodromes were organised as an independent air force, under Sir Hugh Trenchard, with the object of carrying out reprisals on a large scale against the industrial centres of Rhineland. The provision of air defences for these aerodromes was early considered and the formation of twelve additional 3-light sections was authorised for this purpose, but none of these became available before the armistice. Meanwhile two A.A. batteries and two searchlight sections were moved from northern France, one of the latter being No. 27 (Tyne) A.A.S.S., under Lieutenant D. E. Ross. This section had been relieved at Dieppe, at the end of March, by one of the new sections, No. 52, under Captain W. A. Souter (Tyne E.E.). This latter section consisted mostly of London Electrical Engineers but contained a proportion of Tyne personnel, the section serjeant being Serjeant Taylor and the mechanist Staff-Serjeant R. W. Johnson. No. 27 Section was then stationed for a brief space at Abancourt, relieving No. 36 A.A.S.S. (Lieutenant C. B. Williamson) which moved to Abbeville. Early in May, No. 27 Section, together with another searchlight section and the two A.A. batteries mentioned above, the whole under the command of Major Michelli, R.E., entrained for Lorraine. On arrival at Vézelise (about 10 miles north of St. Firmin), No. 27 Section with one of the batteries was stationed at Ochet, near Toul, where there were an F.E. night-bombing squadron and a R.N.A.S. Handley-Page night-bombing squadron, as well as American, French and Italian squadrons. Here the section experienced a somewhat harassing time, one of the lights in particular being frequently put out of action by bombs, though only one man was killed. The aerodromes were constantly bombed and the British hangars had to be removed into the adjacent woods and carefully camouflaged, the aerodrome itself being used only for taking-off purposes.

---

[1] " Work of the R.E. in the European War," Vol. " Work under the D. of W.", pages 46–47.

In August two further searchlight sections arrived, one of these being No. 10 (Tyne) A.A.S.S., under Lieutenant Jardine, from Doullens, and Colonel Simon was transferred from the London defences to take command of the A.A. defences of the I.A.F. Owing to the long distance the lorries of No. 10 Section were placed on rail, the journey occupying over a week to accomplish. This section was established half-way between Nancy and Epinal where it remained until the armistice, but, although a number of enemy machines came over, the lights were ordered not to expose as it was thought that the enemy did not know the exact location of the aerodromes. In the meantime, No. 27 Section[1] moved to a Handley-Page night-bombing squadron which was subjected to spasmodic bombing by enemy aircraft. After the armistice, No. 27 Section moved to Calais and from there to Valenciennes where, in March 1919, it was demobilised. No. 10, after being stationed for short periods in various places, returned to England by train-ferry in the same month and was duly disbanded.

Having thus traced to its conclusion the association of the Tyne Electrical Engineers with the defence of the aerodromes of the Independent Air Force in Lorraine, we continue the narrative of the development behind the British Line in France and Flanders from April 1918 onwards. Before proceeding to describe in detail the experiences of the individual sections it will be advisable briefly to outline the general course of events up to the time of the armistice. The whole of the summer was occupied with the development of the lighted belt behind the Line. The sections thus employed were regarded as Corps troops and formed part of the various Armies forming the British Line. In the north, round Ypres, lay the Second Army, its area extending to the rear and enveloping part of the St. Omer, Audruicq and Zeneghem defences, while the Fifth Army, as soon as it had been reformed, took over that part of the Line to the south of Ypres, connecting the Second Army with the First. Southwards from La Bassée lay the First, Third and Fourth Armies, extending in the order named to a point some ten miles south-east of Amiens. Into this framework the sections were fitted as they became available. In June it was estimated that twenty additional sections would be required to complete the essential defences in France, bringing the total up to 285 lights organised in 95 sections of three lights each. Four sections of Canadian personnel were actually formed in France and were called the Canadian Searchlight Company.[2] Steps were taken to form the sixteen extra sections at home but none of these were ready before the armistice. By the 1st August 1918 there were, in fact, only 70 of the previously approved 75 sections in France[3] though the full number is shewn as having been serving in that theatre by the 1st November.[4] In

---

[1] A very highly placed Engineer officer came to inspect the stations of No. 27 Section in Lorraine. Commandeering one of the battery cars for the purpose, he proceeded with the O.C. of this section to the first position. On coming away there was no room to turn and the driver had to back out for a distance of about a quarter of a mile. This evidently displeased the colonel, who said, very fiercely, " What are you reversing so slowly for? Change up, man, I'm in a hurry."

[2] " Work of the R.E. in the European War " Vol. " Miscellaneous," page 21.
[3] Do. do. do. do. do. page 65.
[4] Do. do. do. do. do. page 69.
These were Nos. 1 to 33, 35 to 49, and 51 to 77, total 75. No. 34 was in Italy. There was no section with the number 50 owing to the fear of possible confusion with the 50th (F.S.) Company R.E. The sections of this company were apparently not included in the totals.

August, as a result of the introduction of the sound-locator and of P.E. lorries, a revised establishment for a " Mobile Anti-Aircraft Searchlight Section R.E. (3-lights, 120 cms. or 90 cms.) " was published, raising the total strength of other ranks to 27.[1]

Experience at home had shewn conclusively that the most effective method of dealing with air raiders was by the use of night-flying squadrons of aircraft working in co-operation with searchlights. By the summer of 1918 the state of the defences in England had rendered her vulnerable areas practically immune from attacks (see chapter VI.) and it was found possible to spare night-flying units for overseas. Thus, early in June a special night-flying squadron, No. 151, arrived in France and was stationed near Abbeville, where it co-operated with the lights of the First, Third and Fourth Armies. Results immediately improved and, thereafter, enemy aircraft very rarely passed through the belt unattacked and were frequently destroyed in the air. This success was the more remarkable from the point of view of the searchlights because the sound locators, now universally provided, were found, in forward areas, to be more of a hindrance than a help, occupying valuable space in transport, while, owing to the noise of bursting bombs and of shell-fire, men were not infrequently deafened in the attempt to use them. In the early stages of their development, sound locators were generally regarded as being, under the most favourable circumstances, of but little more value than experienced spotters and, under normal active service conditions, they proved to be quite useless.

Five night-flying squadrons were approved for France by the War Cabinet, but only one more, No. 152, reached France before the cessation of hostilities.[2] This method of defence was, nevertheless, gradually developed along the whole front, No. 152 Squadron being, in the middle of October, stationed in the neighbourhood of Lens. The machines used by these squadrons were single-seater Sopwith " Camel " aircraft, and, from this fact, the lighted belt became known as the " Camel line."

It had been felt for some time that a closer co-operation was required between the Air defences of the B.E.F., the French Army and Home Forces, as it was found that each was introducing developments in equipment and tactics without the knowledge of the others, so that the best use was not being made of all the knowledge and experience available. The French Army had exhibited a desire to co-operate in these matters and a close liaison already existed between their Anti-Aircraft Training Centre and our A.A. Searchlight and Sound Locator School at Stokes Bay, several senior officers of the French Army having visited the school to study the lines upon which the British defences were being developed. With the object, therefore, of co-ordinating the work at home and overseas, Lieut.-Colonel Monkhouse was appointed, in October 1918, Inspector of Anti-Aircraft Defences, B.E.F. and French Army.

[1] See table at end of chapter.
[2] Ashmore, " Air Defence," page 105.

Following upon this appointment, which was in addition to that he already held as Commandant of the school at Stokes Bay, Lieut.-Colonel Monkhouse visited a number of the A.A. stations in France, including French Army stations as far south as Nancy, but, as the armistice was declared the day he returned home from his first tour of inspection overseas, time did not permit of the reaping of the full benefit which should have accrued from the creation of this inspectorate.

Such, then, were the developments in the organisation of the British air defences in France between the early summer of 1918 and the end of the War. It remains to describe, as shortly as possible, the progress of the operations of the British Army in France and Flanders during this period, before being able satisfactorily to continue with the detailed narratives of the searchlight sections of the Tyne Electrical Engineers attached to the various Armies.

The final British advance commenced in the Fourth Army on the 8th August and was very shortly taken up along the whole tortuous hundred miles of the British Line. The " Camel-line " followed up as closely as circumstances permitted, the searchlights keeping their night-action stations roughly between 2,000 and 6,000 yards behind the front line and endeavouring to maintain continuous touch with the lights of neighbouring armies. During August, September and October, enemy bombing, particularly in the areas of the First, Third and Fourth Armies in the south, was maintained in gradually decreasing volume and severity and, by the end of October, had practically ceased. On the 11th November the Armistice was signed. At this date, the British Line ran approximately from Ghent in the north, through Mons, to a point some fifteen miles south-east of Maubeuge.

We are now in a position to take up once more the tale of the searchlight sections from April 1918, in which month, following the retreats on the Somme and on the Lys, the organisation of the continuous lighted belt behind the Line was commenced. By the time the British advance had commenced this belt, subsequently the " Camel-line," had absorbed the majority of the Tyne sections in France, but there remained on the L. of C. at this date No. 52 A.A.S.S. (Captain Souter) at Dieppe, and possibly others of the later sections of which unfortunately no record exists. Round Calais, Audruicq, Zeneghem and St. Omer, lay the following Tyne sections :—

Calais	...	...	No. 26 A.A.S.S.—Lieutenant Fox (A.S.O.).
			No. 59 A.A.S.S.—Lieutenant Robinson.
Audruicq	...		No. 13 A.A.S.S.—Lieutenant Burgess.[1]
			No. 20 A.A.S.S.—Lieutenant Erskine.
Zeneghem	...		No. 19 A.A.S.S.—Lieutenant Curtiss.[1]
			No. 38 A.A.S.S.—Lieutenant Corby.
St. Omer	...		No. 11 A.A.S.S.—Captain Myles.
			No. 28 A.A.S.S.—Lieutenant H. G. Campbell.
			No. 40 A.A.S.S.—Lieutenant E. H. White.

[1] Not an officer of the Tyne E.E.

Most of these sections had been in this region since their arrival in France but several only joined subsequent to the general re-organisation of April 1918. No. 38 Section arrived from Abancourt on the 17th April. The depletion of the defences of the L. of C. stations was carried out in many cases against the better judgment of the A.A. defence commanders, and the unwise nature of this move very shortly made itself apparent. For example, almost immediately after No. 38 had left Abancourt[1] the place was heavily bombed and the ammunition depot destroyed, although activity in that area had been negligible during the past twelve months. No. 59 A.A.S.S., one of the new sections, was in the Calais area from the 29th April. How No. 11 section found its way to St. Omer, following the retreat on the Lys, has already been described and the arrival of No. 28 has been duly chronicled. No. 40 Section, after various local moves in the Boulogne, Ecault, Condette and Hardelot district, exchanged plant with No. 51 Section at St. Omer on the 22nd August. During their thirteen months stay in the Boulogne area the lights of this section exposed in search of fifty-four enemy targets and illuminated twelve. At St. Omer, in two months there were ten exposures and three targets were illuminated. These figures are quoted as typical of the general standard which was at that time reached by L. of C. sections in France. "Our problem was to find 'Jerry'" writes an officer of one such section. "We knew from the dropping bombs that he was somewhere up in the sky, but which part of the sky he was occupying we did not know as we could only hear him occasionally owing to the noise of the 'Archie' barrage. Anyone who has operated a searchlight under such conditions knows what a big place the sky is and what little chance there is of finding the target."

The situation in this locality was complicated by the fact that parts of the St. Omer, Audruicq and Zeneghem defences were actually in the Second Army area and several of the sections belonged to this Army and advanced with it. No. 11 (Captain Myles), for example, followed the advance to positions south of Ypres and further east into Belgium, being established in stations on the road to Ghent when the armistice was signed. No. 20 Section (Lieutenant Erskine) moved forward and was the first to reach the line of the River Scheldt to the south of Ghent. No. 52 Section (Captain Souter) moved from Dieppe to Audruicq early in October 1918. No. 28 Section (Lieutenant H. G. Campbell) also moved forward to the neighbourhood of Ypres, where its Lewis gunners claimed an enemy bomber which was brought down in their beams. This section subsequently advanced to Courtrai, and thence further east, finding itself in positions a few kilometres north of Tournai by November 11th. No. 40 Section remained at St. Omer until the 27th October when it proceeded by rail to Menin but, being as yet unprovided with lorries, waited there for two days before being transported forward into two positions, one of which was two miles south of Courtrai and the other near Sweveghem, where it remained until after the cessation of hostilities. No. 46 A.A.S.S. (Lieutenant Ingman) was also in the Second Army area during the advance.

[1] No. 36 A.A.S.S. also left Abancourt at about the same date.

N

The Fifth Army, now under General Sir William Birdwood, flanked the Second on its right. To this Army Major E. F. Rendell was appointed, on the 30th August, Assistant Inspector of Searchlights, Major Mawby having remained in the south as A.I.S. Fourth Army. The headquarters of the A.A. defences, Fifth Army, were, in August, at Estrée Blanche, the defence commander being Lieut.-Colonel G. R. P. MacMahon. The defence consisted of five British searchlight sections and a Portuguese searchlight company which was extremely well equipped with an efficient mobile workshop. This company, consisting of 7 officers and 176 men, and manning six lights, was controlled by the British defence headquarters for all purposes except internal discipline. Portuguese personnel was attached to the British sections for instruction[1] and for his services in this connection Major Rendell was decorated with the Portuguese Military Order of Aviz, being granted the rank of Cavalier.

In the Fifth Army area was No. 44 Section (Lieutenant Porter) in positions overlooking the valley of the River Lys, about 6,000 yards from the Line. The aerial activity here was, in the weeks immediately following the battle of the Lys, considerable, and the new positions on the higher ground gave this section the opportunity of distinguishing itself, which was recognized by the award of the Military Medal to Serjeant Gair. While in these positions, the searchlight work improved immensely, and, on one occasion, a Gotha bomber was shot down in the beams of No. 44 Section.

The Fifth Army line remained until the end of August 1918 without alteration, but complete inter-communication was impossible and the introduction of organised controls could not be developed, each searchlight and gun acting independently in the event of alarm. As soon as the Fifth Army began to advance in conformity with the general advance of the Second Army in the north and the Third Army in the south, the searchlights and guns advanced, coming nightly into action in positions two or three miles behind the Line. Each evening, positions for the following night were selected from the map by the A.I.S. and instructions were issued at midnight to all the sections by dispatch riders, this system being continued until the German Line temporarily held at Tournai.

In October the Fifth Army searchlight line became part of the " Camel-line " and good results were then obtained. Early in this month Lille was evacuated by the Germans and intelligence reports indicated that the enemy would endeavour to destroy the town by concentrated raiding immediately after the evacuation. In consequence, as many searchlights as possible were advanced to the west of the town, together with guns, to form a night defensive organisation, but actually no attempt was made by the Germans to bomb the town. Some difficulty was experienced in getting the searchlights into position within 24 hours of the evacuation of Lille, as all the roadbridges and other crossings were destroyed by the enemy during his retreat.

---

[1] Lieut. H. G. Campbell was also sent from the Second Army to assist in the training of the Portuguese company for a period of six weeks in the spring of 1918.

GERMAN SEARCHLIGHT PLANT
captured at Ruysbruck.

To face page 194]

On October 19th No. 44, forming part of a group of four sections under Captain Porter, began a further advance and by the time of the armistice had reached Melles, some distance in front of Tournai. For this final work Captain Porter was mentioned in despatches. During the early stages of this advance, all the cross-roads were found to be cratered by enemy delay-action mines, and the section not infrequently found itself called upon to fill in the craters before proceeding with its more legitimate work. The advance was in this and other ways much delayed, one of the chief nuisances being occasioned by the height of the hoods of the P.E. lorries, which, having been specially heightened so as to contain a completely erected projector, were now found too high to pass under certain bridges on the route. All the sections succeeded, nevertheless, in getting forward to the satisfaction of the higher command, as is evidenced by the following letter from the A.A. defence commander to the A.I.S. dated the 25th October 1918 :—

"Will you please inform all the British and Portuguese searchlights in the command how much I appreciate their efforts in getting their lights forward so quickly during the last few days. I quite understand the difficulties with which they have had to contend, and am pleased to be able to tell you that the rapidity of their advance has been commented on by G.H.Q."

The Germans in their retirement were now abandoning large quantities of material and No. 44 Section was detailed to collect valuable electrical equipment for the Fifth Army. At Ruysbruck there fell into their hands in this way several sets of German searchlight equipment which were set aside for examination by the A.A. defence commander. This plant, consisting of converted horse-drawn equipment, was later handed over to Major Rendell, who consigned it to the O.C. Tyne Electrical Engineers at Haslar Barracks as a fitting memento of the activities of the A.A. searchlights in France.[1]

During the advance of the Fifth Army the searchlight organisation of the Second, Fifth and Third Armies endeavoured to co-operate so as to maintain an unbroken line over the whole front. After November 1st, however, it was found impossible to keep touch with the Third Army to the south where the transport was seriously held up by broken bridges and by inundations. At the time of the armistice the Fifth Army searchlight sections were practically out of the control of the defence headquarters, as the state of the roads frequently prevented the detachments reaching their pre-arranged locations. On several occasions in the last fortnight the whereabouts of sections was completely lost for periods of two or three days owing to the heavy traffic blocks, the searchlight sections being inextricably mixed-up with field batteries and other front-line units.

---

[1] On the departure of the Depot of the Tyne Electrical Engineers from Haslar in 1919, certain pieces of the equipment were deposited with the School of Electric Lighting, but the remainder was sent to North Shields where, for some years after the war, it was used for training purposes. Owing to the lack of suitable accommodation and the consequent action of the weather, this plant deteriorated, and, in 1923, the bulk of it was disposed of as scrap metal which fetched £18 towards regimental funds. The remainder, with the exception of a Siemens-Schuckert 8 k.w. limber generating set and a reflector and glass door from a 60 cms. projector (which now adorn the walls of the drill hall at Tynemouth), was disposed of in 1929.

The detachments, however, became so proficient in loading and unloading their lorries that it is reported that they could be ready for action within fifteen minutes of reaching a new position. For his strenuous work in charge of the searchlights of the Fifth Army during this final advance, Major Rendell was again mentioned in despatches and awarded the Military Cross.

Southwards of the Fifth Army lay the First with which were nine sections, Nos. 3, 5, 6, 9, 15, 18, 22, 29 and 32, under Major Madge, London Electrical Engineers. It is not known when No. 6 moved into this area from Rouen, or No. 22, which was last heard of at Dunkerque at the end of June 1917, but by August 1918 they were both somewhere in the First Army area and were no doubt contributing their share to the successes which the A.A. defences were achieving in that district.

No. 15 (Captain Baldwin), at St. Pol, was instrumental in bringing down several enemy machines, a Friedrichshafen being hit by gunfire while in the beams of this section on the 21st May, and a Gotha, returning from a L. of C. raid on the 1st June, being shot down in flames near Pernes after being picked up by its searchlights. This latter exploit called forth the following message from the Army Commander, General Lord Horne :—

" Congratulations to the St. Pol defences, both guns and searchlights, on your success last night. Keep on doing it."

In August, too, No. 32 Section (Lieutenant Dickson) arrived in the area from Calais and, with No. 15 and others, followed the successful advance of the First Army which commenced on the 26th August. The important success which was attained by the air defences during these operations is evidenced by the following message from Defence Headquarters, dated the 29th August 1918 :—

" G.H.Q. have written the following :—

' Hearty congratulations on the very excellent action reports for last week. The results obtained with the very scattered lights available are extraordinary and most creditable to all concerned.

' Three enemy aircraft were shot down in the beams last week and many other targets held for long periods. It is a great pleasure to submit such reports, which are the results of your own perseverance.'

Sd. :—R. G. Madge,

Major R.E.(T).,

A/Inspector of Searchlights, First Army."

At the end of August came No. 42 Section (Lieutenant Batey) from Boulogne. This section took over 120 cms. projectors and generating sets from another section at Doullens in the Third Army area and, three days later, was transported up to the neighbourhood of Monchy, a few miles south of Arras and 3,000—4,000 yards from the front line.

On the 27th September the advance of the First Army was resumed, forming the left flank of a huge movement in which the First, Third and Fourth Armies were all engaged. On the 30th, No. 15 Section, at Hendincourt near Arras, and No. 42, at Monchy, were in action with enemy aircraft for the last time. By the 20th October, they were in the region of Douai, the greatest difficulty being experienced in moving forward owing to the damage done by the enemy to roads, bridges and canals. Much of the country was under water and suitable positions for searchlights could not easily be found, while here more than anywhere the inherent disability of the petrol-electric system of transmission made itself apparent. The exceedingly poor starting torque which this system provides rendered it quite impossible to move a lorry without assistance once it got into heavy or waterlogged ground. Taking into account the exceptionally heavy demands made on the lorries, however, they were found otherwise to be very satisfactory and, although often overloaded with equipment and extra personnel, the steepest gradients were negotiated without any trouble.

The advance continued and the 26th October found the First Army search-lights on a line extending roughly from Somain to Marchiennes. At the former place No. 42 remained, but Nos. 15 and 32 advanced a few miles further, the former to Wallers on the 7th November and to Fresnes on the 10th, and the latter to St. Amand. In these positions, amid flat and desolate surroundings, the three sections welcomed the end of hostilities on the 11th November. "No one was in sight," says one who was there " but at eleven o'clock a cheer rang out simultaneously from all directions."

In the First Army area, southwards of the Third, four Tyne sections shared in the early successes of No. 151 Squadron, R.A.F. The original allotment of sections to this area was five, but this number was later much exceeded. The arrival of Nos. 8 and 10 Sections at Doullens has duly been chronicled. The establishment of searchlights was considerably increased during June and July 1918, among the earlier sections to arrive being No. 30 (London) Section, now under Lieutenant S. W. Newman, Tyne Electrical Engineers, and No. 76 Section, under Lieutenant A. D. Phillips, which formed in the area in June. Serious raids continued in this region, No. 8 A.A.S.S. having one of its lights bombed and a serjeant wounded on the night of the 29th June.

In August No. 10 Section (Lieutenant Jardine) was ordered away to join No. 27 in the protection of the southern aerodromes housing the Independent Air Force. The subsequent history of this section has already been recounted.

The attack in this part of the line commenced on the 21st August and, on September 4th, following the initial gains of the Third Army, all sections were on the move forward, No. 8 moving by road to the vicinity of Velu. On the following day one of the P.E. lorries of this section, while on the road, was struck by a shell, Sapper McPhee being killed and Sappers Payne and Turnbull wounded. No. 1 light of this section, under Corporal Swan, was in action that night when shells fell within twenty yards of the projector. For his conduct on this occasion Corporal Swan was subsequently awarded the Military Medal.

This period was signalised by intense enemy aerial activity and by a notable increase in the efficiency of the defences, four enemy aircraft being destroyed on the night of the 13th and one on the 15th September.

No. 55 A.A.S.S., a mobile two-light section, was also in this area at the time, being stationed near Bapaume, and on the 17th September was taken over by Lieutenant E. J. Parnall, Tyne Electrical Engineers, and moved up to the neigh-bourhood of Ypres on the 27th. At this period, and until the 2nd October, the lights moved forward daily,[1] No. 8 A.A.S.S. reaching Villers-Plonich, where one of the detachments was gassed during action on the night of the 3rd October. A week later, on the 10th, this section moved forward to Esnes, a distance of eight-and-a-half miles in one day, and was at that place when the armistice was signed. No. 76 Section, after following closely behind the line during the final stages of the advance, selected, on one occasion, a position some distance in advance of the line. The event passed off happily, however, and the armistice found this section not far from Cambrai, living in tents in about four inches of snow. Meanwhile No. 55 Section, after several intermediate stations, reached Le Cateau on the 24th October, where an emplacement, immediately after it had been excavated, was destroyed by a direct hit from a shell, but fortunately before the searchlight had been installed. Leaving here two days later, this section moved to Berlaimont where it received the famous message by despatch rider on the 11th November, " Cease hostilities at 11.0 a.m. and take defensive measures."

So rapid had been the advance that these sections had, at this time, to send between 30 and 40 miles for their rations, no mean undertaking for the anti-quated motor-vans which constituted the sole first-line transport of the sections, when the almost complete absence of proper roads in the heavily shelled forward areas is taken into account.

During the latter stages of the advance the forward areas were subjected to an intensive campaign of enemy propaganda, leaflets being dropped from aircraft over the lines. One of these depicted Paris being visited by a number of aeroplanes at night and the Angel of Peace being burnt at the stake by Clemen-ceau. The inscription read as follows :—"Qui est-ce qui attire les avions ? " (" Who is it that is attracting the aeroplanes ? "). " Parisiens ! Vous avez beau tirer vos rideaux, c'est Clemenceau qui montre le chemin aux Gothas en brulant la paix." ("People of Paris ! You may well draw your curtains ; it is Clemenceau who is shewing the way to the Gothas by burning Peace.") Another leaflet, with similar purpose, illustrated Paris being bombed from the air and the Angel

---

[1] In certain localities the A.A. searchlight sections earned the nick-name " Pickfords " from the fact that they were constantly " removing."

198

of Peace being held by Clemenceau. This effort is worded :—" Les conséquences du refus." ("The result of refusal"). " Pauvre Paris . . . mais pour accepter la paix il ne faudrait avoir les mains libres." (Poor Paris . . . but to accept peace I should have to have my hands free."). These particular examples of propaganda were aimed at the French troops, but similar leaflets, written in poor English, were also distributed from the air, urging the soldiers to refuse to continue the war and emphasising Germany's wish for peace while laying the blame for any further bloodshed and destruction on the Allies if they did not conclude an immediate and honourable peace with Germany. This propaganda might have had more serious results but for the exceptionally high morale of the troops at this particular stage, engendered by the very rapid advance of the past three months.

We pass now to the extreme right wing of the British line in order to continue the story of the sections with the Fourth Army from the period immediately following the second battle of the Somme up to the time of the Armistice. We left No. 17 Section (Lieutenant Hutchinson) in position before Amiens, the sole representatives at that time of the Tyne Electrical Engineers with this Army if we exclude a few other ranks in the 50th Company R.E. Early in April No. 36 Section (Lieutenant C. B. Williamson), newly equipped with 120 cms. searchlights but as yet without lorries, arrived at Abbeville from Abancourt in time to take part in what has been described as " probably the most intensive German bombing in our back areas," during which the town was practically destroyed and three miles of the seven-mile long ammunition dump at that place was blown up. After the first few raids, balloon aprons, consisting of captive balloons with steel ropes suspended between them, were sent up, but it was a remarkable fact that the enemy raids were confined to nights on which this precaution was not taken.

No. 60 A.A.S.S., one of the new sections, reached Abbeville late in June but its O.C., Lieutenant L. S. Winkworth, Tyne Electrical Engineers,[1] having been delayed through sickness at the base, did not join till some time later. This officer was subsequently A.S.O. Abbeville from the 11th September until the 1st November, when the section moved further south and Captain Winkworth (as he then was) became A.S.O. Amiens,[2] the section remaining there until after the armistice.

On the 13th July another exceptionally severe raid took place over Abbeville, but, by this date, the reorganisation of the A.A. defences was bearing fruit, many aeroplanes being brought down by the " Camels " of No. 151 Squadron.

As soon as the advance began on August 8th the sections were on the move forward, No. 36 Section having by this date received its lorries and been

---

[1] From No. 27 A.A. Company R.G.A. at Hull.
[2] This officer was subsequently A.S.O. and D.O.R.E. Dieppe from 27th December 1918 to the 13th February 1919.

re-equipped with 90 cms. projectors.   During the ensuing months these sections moved nearly every day to new positions and were in action every night. Everywhere is told the same story of many enemy aircraft destroyed by the Air Force machines assisted by the searchlights, and of well earned congratulations from the Army Commander.   The searchlight work had by now reached a standard of efficiency at which the detachments were well capable of holding a bomber in the beams without illuminating our own fighters.   The method then adopted was for a triangle of lights to hold the target, leaving its tail in darkness in order to give the attacking machine an opportunity of approaching it from the rear without itself becoming visible to the occupants of the bomber. The A.A. guns were not entirely displaced by the fighter aircraft.   As soon as an enemy machine came in range, the guns attacked it, ceasing fire only when it flew out of range or had been shot down, or upon a pre-determined signal-light from one of our own aircraft patrols.

In September, No. 151 Squadron moved forward to Vignacourt.   By this time No. 17 Section was once more occupying the very positions round Peronne from which it had retired in March, and No. 36 Section, moving by way of Amiens, had proceeded some ten kilometres to the south and reached positions in the vicinity of Athies.   From this point the advance followed a north-easterly direction, No. 36 Section being on the extreme right of the British Line, mixing at times with French troops.   The Sections continued to move daily and were in action frequently within 1,000 yards of the Line, being often detailed to positions which were still in the enemy's hands and which were only evacuated by the Germans a few hours before the searchlights arrived.

Throughout September the German raids continued, but so striking was the success of the night-flying operations that by the end of the month, though the enemy could still be heard passing to the south over the French area, their activity in this part of the Line had almost entirely ceased.[1]   Heavy shelling of roads was, however, maintained, and progress became increasingly difficult owing to the efficient demolitions of cross-roads and bridges carried out by the retreating enemy.   Until the armistice these sections worked night and day getting into and evacuating positions, the prevailing conditions necessitating frequent long detours and involving serious delays in negotiating demolished roads, while the presence of countless delay-action mines left by the enemy, in the area to be traversed, contributed to render the last few days of hostilities not the least strenuous.   The Fourth Army, the first to commence the advance in August, actually moved further forward than any other of the British Armies, traversing 75 miles before the armistice and, in the last week of the operations, advancing no less than 25 miles, bringing the Line on to the Franco-Belgian frontier east of Avesnes by the 11th November.   The searchlights

---

[1] " Between the 13th September and the end of the month the Squadron, working in excellent co-operation with the anti-aircraft guns and searchlights, destroyed fourteen bombers . . . Altogether, during its short stay in France, No. 151 Squadron accounted for twenty-six bombers in night combat."   Ashmore, " Air Defence," page 104.   During 1918, the A.A. guns brought down 176 enemy machines as compared with 95 in 1917, 50 in 1916 and 20 in 1915.

Searchlight plant at the Fifth Army Depot at Canteleu, Lille, March 1919.

Above—Projectors and generating sets.

(Note the heavy base sections referred to in the account of No. 27 A.A. Section).

Below—Early sound locators.

were by this time rather left behind and immediately after the armistice concentrated at Avesnes, some 10 miles behind the Line, proceeding thence to Namur, where the A.A. defences, Fourth Army, came under one roof for the first time in their existence.

The Third Army sections concentrated immediately after the armistice at the village of Amfroipret and moved to Le Quesnoy where, on the 3rd December, they were honoured by being inspected by H.M. The King. A week later they moved up to Chaleroi and reached Namur on the 13th December, being accommodated with the Fourth Army Group in the infantry barracks. On the 30th December all plant was unloaded and taken into store in a large factory situated opposite the barracks.

Valenciennes was made the rallying point of the A.A. defences in the First Army area, headquarters being established in a very large house which had just been finished when the war broke out and which had, by some strange chance, almost entirely escaped damage, one corner only having been hit by a shell. Near the house was found a screw factory admirably suited to accommodate the plant, while the workmen's houses stood ready to receive the troops. Before pulling-in to Valenciennes, No. 15 Section (Captain Baldwin) operated its searchlights in connection with the reconstruction of bridges and railway track between Valenciennes and Mons. On the 17th November this section was attached to No. 2 Canadian Overseas Railway Construction Company which was working solely at night and generally laid about one mile of track each night between the hours of 6.0 p.m. and 5 a.m. The lights were, when possible, run on the lorries themselves, which were placed on level-crossings or bridges about $1\frac{1}{2}$ miles apart, the cable being flaked out near the lorry. Where this arrangement was impossible, the lorry was left on the nearest road and the projector erected actually on the track. Concentrated beams were used with the lamps facing each other in pairs and just sufficiently elevated to keep the fringe of the beams off the track 2,000 yards away, thus providing a reflected and non-blinding light on the works. In this manner excellent illumination was afforded along the whole of the distance between each pair of lights. For piling and bracing in connection with bridge work open arcs without the projector-barrels were used, one on each side of the gap, thereby eliminating shadows. By the light of these arcs the working parties were able to work as fast by night as by day, with the result that the work was finished in considerably shorter time than would have been possible without the lights. By the 28th November the line had been repaired as far as Mons and the section then moved back to Valenciennes. To this place also returned No. 27 A.A.S.S. (Lieutenant Ross) from Lorraine.

The Fifth Army searchlights collected at Lille and those of the Second Army at Roubaix, and certain of the sections here, as elsewhere, were first employed on electric lighting work in various camps, hospitals and concentration areas. No. 40 Section (Lieutenant E. H. White), for example, was sent back to Hazebrouck on this duty, while No. 44 Section (Captain Porter) was similarly employed between Hals and Brussels, arriving in that district on the 2nd December. The following extract from the diary of the officer commanding this section illustrates the nature of the work which was now being undertaken :—

> " The village of Enghien supply was two small Petter engines with 200 volt generators in a barn behind an estaminet. Our mains were overhead through the village street and those to be provided with electric light were fed off the mains through the nearest window. The neighbouring village had no supply until one day I saw the local Belgian contractor poling along the road to join onto my limited supply. I must say I admired his cheek—and cheek he had—for, though frustrated in this attempt to supply the next village, he continually contracted to supply civilians in Enghien. I had at last to send out patrols every night to cut off with pliers any new consumers that had been connected that day by our worthy and energetic friend."

At Christmas at Lille the chief occupation of the Fifth Army searchlights was the supply of electric light to the army theatre, where Leslie Henson and his company were appearing in the review " Alladin."

Shortly after the armistice it was decided to send the Second and Fourth Armies to the Rhine, but within a week this plan was revised, the Second Army alone forming the British Army of Occupation in Germany and the Fourth Army forming a supporting force in Belgium behind the German frontier.

Several searchlight sections, including Nos. 11 (Captain Myles) and 28 (Lieutenant H. G. Campbell), moved up by road with the A.A. guns of the Second Army into Germany, No. 11 being stationed eventually at Bonn, on the Rhine, and No. 28 at Cologne. On the 17th January 1919 a second group of four searchlight sections from the Fourth Army also moved into Germany. These included Nos. 17 (Lieutenant Hutchinson) and 36 (Lieutenant C. B. Williamson) which travelled via Liége, Aix la Chapelle and Cologne to Bonn. Here the lights were used, in co-operation with the British Rhine river police, for controlling river transport and preventing smuggling. Anti-aircraft positions were also selected on the German side of the Rhine for occupation in the event of a renewal of hostilities, and these positions were actually manned for a few nights when the Army of Occupation moved forward to occupy the eastern circumferences of the bridge heads at the time of the ratification of the peace treaty at Versailles on the 28th June, 1919. The work of the searchlight sections with the Rhine Army was not arduous, consisting of but two hours training daily, with plenty of games and sport. The control of navigation was carried out from towns at the northern and southern limits of the British occupied territory. Two lights were installed at each end and were manned by each of the sections in that area in turn. Godesberg, situated some five miles south of Bonn, was the southernmost point and here a searchlight was

"THE WATCH ON THE RHINE."
Searchlight at Godesberg.

To face page 202]

installed. The lighting of the Rhine had the effect of illuminating the historic castles with which the famous gorge abounds and was a source of great attraction to the inhabitants who came from miles around to enjoy the sight. Occasionally the searchlight detachments were subjected to sniping from the opposite bank but fortunately this did not result in any casualties to plant or personnel. Among other duties, the searchlights were used in connection with the various torchlight tattoos which were produced by the Rhine Army.

The sections, in common with the rest of the Army of Occupation, were excellently accommodated in the best buildings available, the detachments on duty at Godesberg being billeted at the Rhine hotel, Driesen, situated on the water front, and the remainder at Villip some 2½ miles away from the river.

Leaving them to enjoy this comparative leisure after the strenuous existence of the past three years, we must return to France and Belgium in order to consider the course of events in the various army searchlight depots which were left behind. Soon after the 1st January 1919 the process of demobilisation set in and the searchlight sections slowly faded away as their members returned to civil life. There was, at this time, a great deal of unrest developing among those members of the army who were not, in their own opinions, being sufficiently rapidly demobolised. This unrest was no doubt largely encouraged by the enforced idleness of the great bulk of the troops in France following the strain of the years of war, and it is gratifying to record that this unrest was noticeably less among the members of the searchlight sections, largely by reason of the fact that for them interesting work was available almost up to the day of demobilisation. Those, who were not occupied in re-instating the electric light supplies in the various occupied localities, were engaged in packing-up and dispatching the searchlight plant to England. This work went on all through January, February and into March, the P.E. lorries being returned in convoys by train-ferry from Dieppe to Southampton and thence by road to the Ordnance depot at Hilsea. In March 1919 the remains of the Fifth Army searchlight depot was established in a disused cotton factory at Canteleu in the northern part of Lille. This place, the First Army depot at Valenciennes and the L. of C. depot at Audruicq were the last to go and by the end of March or early in April the searchlight personnel in France and Belgium had all been demobilised.

In order to maintain the searchlight sections on the Rhine, personnel desirous of, and due for, demobilisation was replaced by volunteers for service in the Armies of Occupation. Thus, during the early months of 1919, several changes took place among the officers on the Rhine, while several more sections were sent into Germany. Among the latter were Nos. 6 and 8 Tyne Sections formerly of the First and Third Armies.[1] Early in March Lieutenant Hutchinson handed over No. 17 A.A.S.S. at Bonn to its third commander, Lieutenant P. V. Horler, and Lieutenant E. H. White arrived from Lille to take over No. 9 (London) Section from Lieutenant Owles. In the same month Lieutenant

---

[1] The officers in command of these two sections at this date are not known. Captain Rooksby, formerly O.C. of No. 8, was demobilised on the 19th February.

C. B. Williamson was succeeded in command of No. 36 by Lieutenant P. F. Fawcett at Godesberg. This latter officer had come out from England in November 1918 and had been employed in the intervening four months at the First Army searchlight stores depot at Valenciennes. On the 4th April Lieutenant A. G. Dickson, formerly in command of No. 32 A.A.S.S., took over No. 11 from Captain Myles at Bonn. There were now six searchlight sections in the southern area of the British zone of occupation and Lieutenant Horler was appointed Group education officer to these, functioning under the supervision of the Rhine Army educational school. The work of this officer, which was in addition to that of a section officer, consisted of forming classes, arranging lectures, preparing syllabi and acting as instructor and lecturer, with an N.C.O. as his assistant. Other officers of the Tyne Electrical Engineers who were employed with searchlights on the Rhine were Lieutenants J. Welsh, Emerson, Ross and A. D. Phillips.

Towards the end of 1919 reductions in the Corps troops of the Rhine Army were commenced and certain reorganisations took place. The searchlight sections were now grouped into two A.A. searchlight companies in the northern and southern areas of the British zone of occupation respectively. Thus, on the 7th September, No. 11 A.A.S.S. was absorbed into No. 1 A.A. Searchlight Company at Cologne and Lieutenant Dickson then returned home whence he was sent to Mesopotamia.[1] At this date Lieutenant E. H. White was also transferred to Cologne and remained there as O.C. searchlights, Northern area, until he also proceeded to Mesopotamia on the 31st October.[1] In October 1919 five or six of the sections returned to England under Lieutenant Phillips and were disbanded. No. 17 A.A.S.S. (Lieutenant Horler) concluded its active military career at the end of the year, when, together with No. 6 Section, under the same officer, it commenced its homeward journey from Cologne by road. The route lay through Aix la Chapelle and Lille to Dunkerque, whence the sections proceeded by train-ferry to Richborough in Kent. From this port they travelled via Canterbury and Aldershot to Haslar Barracks, Gosport, and then the plant was finally handed into store at Hilsea and the men demobilised.

No. 36 Section (now under Lieutenant Fawcett) remained in operation at Godesberg until it was flooded-out in the rising of the Rhine at the end of 1919, when the section was withdrawn to Cologne and joined No. 8 A.A.S.S. (now under Lieutenant H. G. Campbell). These two sections, the last of the searchlight sections of the Tyne Electrical Engineers overseas, returned together to England in March 1920 and were demobilised at Haslar Barracks, but Lieutenant Campbell was first charged with the duty of delivering, at the Headquarters at Clifford's Fort, two P.E. lorries formerly the equipment of No. 8 A.A.S.S.

With their disbandment this record of the anti-aircraft searchlights in France, Belgium, Italy and Germany, must be brought to its conclusion, but no account, however imperfect through lack of historical detail, could be closed

---

[1] See Note " D " at the end of chapter VIII.

without reference to that fundamental characteristic of the whole organisation which conferred unusual degrees of responsibility upon the comparatively junior officers and the non-commissioned ranks to whose good fortune it fell to command the sections and detachments. During the war it was, it is true, not uncommon for lieutenants to command battalions, for junior N.C.Os. to hold isolated posts against determined enemy attacks and for companies to be brought out of battle by serjeants, but these responsibilities were generally the result of casualties sustained in action. With the anti-aircraft searchlight sections it was different. Here it was no sudden chance of battle which temporarily invested juniors with the cloak of authority until such time as more experienced men could be provided once more to control the destinies of their units. It has often been said that a non-commissioned officer of Royal Engineers habitually shoulders responsibilities equivalent to those of officers in other arms of the service and in no case has this been more marked than in the anti-aircraft searchlight sections, in which the exceptionally isolated nature of the work rendered each section officer and each detachment commander so often dependent on his own powers alone for the very existence and continued efficiency of his command.

It would be unseemly, therefore, not to pay due tribute to the high standard of discipline and of technical and administrative ability displayed by all ranks of these sections, properties which have ever been characteristic of personnel of the Tyne Electrical Engineers.

TABLE OF ANTI-AIRCRAFT SEARCHLIGHT SECTIONS.

FORMED FOR SERVICE OVERSEAS, 1915–1918.

(76)

Number of Section.	Parent Unit.	Officers.
No. 1 A.A.S.S. ...	R.E.	Lieutenant T. Rich (L.E.E.)
		Lieutenant Mason (13th Oct., 17)
No. 2 A.A.S.S. ...	R.E. & L.E.E.	Lieutenant G. Fergus-Wood (July, 15)
No. 3 A.A.S.S. ...	L.E.E.	Captain R. G. Madge (April, 16)
No. 4 A.A.S.S. ...	L.E.E.	Lieutenant T. H. Gotch (April, 16)
No. 5 A.A.S.S. ...	L.E.E.	Lieutenant A. W. M. Mawby (April, 16)
No. 6 A.A.S.S. ...	T.E.E.	Lieutenant R. Ll. Hunter (April, 16)
No. 7 A.A.S.S. ...	L.E.E.	Lieutenant H. G. G. Clarke (May, 16)
No. 8 A.A.S.S. ...	T.E.E.	Lieutenant R. P. Findlay (June, 16)
		Captain R. H. Rooksby (5th May, 17)
		Lieutenant H. G. Campbell (Dec., 19)
No. 9 A.A.S.S. ...	L.E.E.	Lieutenant G. Owles (July, 16)
No. 10 A.A.S.S. ...	T.E.E.	Lieutenant A. B. Williamson (July, 16)
		Lieutenant R. E. Jardine (Jan., 18)
No. 11 A.A.S.S. ...	T.E.E.	Lieutenant D. Myles (July, 16)
		Lieutenant A. G. Dickson (14th April, 19)
No. 12 A.A.S.S. ...	L.E.E.	Lieutenant G. B. Adeney (Aug., 16)
No. 13 A.A.S.S. ...	T.E.E.	Lieutenant E. F. Rendell (Aug., 16)
		Second-Lieutenant Burgess (?) (Aug., 18)
No. 14 A.A.S.S. ...	L.E.E.	Lieutenant J. D. Ross (9th Sept., 16)
No. 15 A.A.S.S. ...	T.E.E.	Lieutenant E. V. Baldwin (9th Sept., 16)
No. 16 A.A.S.S. ...	L.E.E.	(a)
No. 17 A.A.S.S. ...	T.E.E.	Captain R. H. Rooksby (13th Oct., 16)
		Lieutenant H. Hutchinson (30th April, 17)
		Lieutenant P. V. Horler (March, 19)
No. 18 A.A.S.S. ...	L.E.E.	(a)
No. 19 A.A.S.S. ...	T.E.E.	Lieutenant W. G. Edwards (Nov., 16)
		Lieutenant Curtiss (?) (30th Aug., 18)
No. 20 A.A.S.S. ...	T.E.E.	Lieutenant C. W. Erskine (, 16)
No. 21 A.A.S.S.(b)...	L.E.E.	———
No. 22 A.A.S.S.(b)...	T.E.E.	Lieutenant Bruch (L.E.E.)
No. 23 A.A.S.S.(b)...	L.E.E.	———
No. 24 A.A.S.S.(b)...	L.E.E.	Lieutenant R. P. Winter (T.E.E.) (e) (Sept. to Dec., 17)
		Lieutenant North (L.E.E.) (Dec., 17)
No. 25 A.A.S.S. ...	L.E.E.	(a)
No. 26 A.A.S.S. ...	T.E.E.	Lieutenant W. Fox (Feb., 17)
No. 27 A.A.S.S. ...	T.E.E.	Lieutenant D. E. Ross (March, 17)
No. 28 A.A.S.S. ...	T.E.E.	Lieutenant H. G. Campbell (March, 17)
No. 29 A.A.S.S. ...	L.E.E.	(a)

Number of Section.	Parent Unit.	Officers.
No. 30 A.A.S.S. ...	L.E.E.	Lieutenant F. G. Hort (March, 17)
		Lieutenant S. W. Newman (T.E.E.) (June, 18)
No. 31 A.A.S.S. ...	L.E.E.	Lieutenant H. A. Voss (March, 17)
No. 32 A.A.S.S. ...	T.E.E.	Lieutenant A. G. Dickson (April, 17)
No. 33 A.A.S.S. ...	L.E.E.	Lieutenant Bauman (April, 17)
No. 34 A.A.S.S. ...	T.E.E.	Lieutenant R. P. Winter (April, 17)
No. 35 A.A.S.S. ...	L.E.E.	Lieutenant Simpson (April, 17) (c)
No. 36 A.A.S.S. ...	T.E.E.	Lieutenant C. B. Williamson (April, 17)
		Lieutenant P. F. Fawcett (March, 19)
No. 37 A.A.S.S. ...	L.E.E.	Lieutenant Squires (April, 17) (c)
No. 38 A.A.S.S. ...	T.E.E.	Lieutenant F. S. Corby (16th May, 17)
No. 39 A.A.S.S. ...	L.E.E.	Lieutenant G. A. Wadham (May, 17)
No. 40 A.A.S.S. ...	T.E.E.	Lieutenant E. H. White (16th May, 17)
No. 41 A.A.S.S. ...	L.E.E.	Lieutenant K. L. Wood (May, 17)
No. 42 A.A.S.S. ...	T.E.E.	Lieutenant J. L. Batey (16th May, 17)
No. 43 A.A.S.S. ...	L.E.E.	Lieutenant C. H. Reed (May, 17)
No. 44 A.A.S.S. ...	T.E.E.	Lieutenant R. O. Porter (16th May, 17)
No. 45 A.A.S.S. ...	L.E.E.	——
No. 46 A.A.S.S. ...	T.E.E.	Lieutenant H. J. Ingman (c)
No. 47 A.A.S.S. ...	(d)	——
No. 48 A.A.S.S. ...	L.E.E.	Lieutenant Napier (c)
No. 49 A.A.S.S. ...	(d)	——
(No Section allotted to the number 50 owing to the existence of No. 50 (F.S.) Company R.E.)		
No. 51 A.A.S.S. ...	(d)	——
No. 52 A.A.S.S. ...	(d)	Captain W. A. Souter (T.E.E.) (March, 18)
No. 53 A.A.S.S. ...	(d)	——
No. 54 A.A.S.S. ...	(d)	Lieutenant Evans (L.E.E.) (c)
No. 55 A.A.S.S. ...	(d)	Lieutenant E. J. Parnall (T.E.E.) (e) (Sept., 18)
No. 56 A.A.S.S. ...	(d)	——
No. 57 A.A.S.S. ...	(d)	Lieutenant Compton (L.E.E.) (c)
No. 58 A.A.S.S. ...	(d)	——
No. 59 A.A.S.S. ...	(d)	Lieutenant J. Robinson (T.E.E.) (27th April, 18)
No. 60 A.A.S.S. ...	(d)	Lieutenant L. S. Winkworth (T.E.E.) (June, 18)
No. 61 A.A.S.S. ...	(d)	Lieutenant E. Rendell (T.E.E.) (c)
No. 62 A.A.S.S. ...	(d)	Lieutenant F. G. Hort (L.E.E.) (May to Aug., 18)
		Lieutenant Frost (L.E.E.) (Aug., 18) (c)
No. 63 A.A.S.S. ...	(d)	——
No. 64 A.A.S.S. ...	(d)	——
No. 65 A.A.S.S. ...	(d)	——
No. 66 A.A.S.S. ...	(d)	——
No. 67 A.A.S.S. ...	(d)	——
No. 68 A.A.S.S. ...	(d)	——
No. 69 A.A.S.S. ...	(d)	——
No. 70 A.A.S.S. ...	(d)	——

Number of Section.	Parent Unit.	Officers.
No. 71 A.A.S.S. ...	(*d*)	——
No. 72 A.A.S.S. ...	(*d*)	——
No. 73 A.A.S.S. ...	(*d*)	——
No. 74 A.A.S.S. ...	(*d*)	——
No. 75 A.A.S.S. ...	(*d*)	——
No. 76 A.A.S.S. ...	(*d*)	Lieutenant A. D. Phillips (T.E.E.) (June, 18)
No. 77 A.A.S.S. ...	(*d*)	——

NOTE.—In the foregoing table, the ranks given are those held by officers at the time of joining the sections. The dates of joining, where known, are given in brackets after the name of the officer. The date given after the name of the first officer in each section, is approximately the date of formation of the section. Officers are of the parent unit except where specifically shewn otherwise. Officers of London Electrical Engineers are those on formation of sections only. No attempt has been made to trace subsequent changes of command in these sections.

(*a*) The officers of these four Sections were Lieutenants H. F. Haworth, A. T. Sturgess, F. C. Stephens and H. L. Bazalgette, but their distribution is not known.

(*b*) These four sections were formed in France from surplus detachments in Nos. 1 to 8 A.A.S.S.

(*c*) Doubtful.

(*d*) Parentage uncertain, and probably mixed L.E.E., T.E.E., and medical category " B " infantry.

(*e*) Not the first officer to command the section.

In addition to those mentioned in the above list, the following officers of the Tyne Electrical Engineers are understood to have served with searchlights overseas :—

Lieutenants R. G. Ellis, H. Joseph, C. G. Huntley, A. S. Burdis, J. R. T. Emerson (France 12th April 18 to 30th Jan. 19 ; Rhine 30th Jan. 19 to 5th Nov. 19), and F. Braithwaite.

FIXED ANTI-AIRCRAFT SEARCHLIGHT SECTION (2 lights)
FRANCE
WAR ESTABLISHMENT.

Detail.	Officers.	S/Sjts. and Serjts.	Rank and File.	Total.	Remarks.
Subaltern	1	–	–	1	
Mechanist staff-serjeant ...	–	1	–	1 (a)	(a) May be mechanist or elec-
Serjeant	–	1	–	1	trician.
Corporal	–	–	1	1	
Sappers	–	–	14	14 (b)	(b) Includes 2 L/Corporals.
Batman	–	–	1	1	
Total	1	2	16	19	
Attached :— Driver, A.S.C. (M.T.) ...	–	–	1	1	
TOTAL SECTION (including attached)...	1	2	17	20	

TRANSPORT.

Detail.	Vehicles.	Drivers.	Remarks.
Motor-car with box body ...	1	1(c)	(c) Provided by A.S.C.
(e) Power lorries (petrol–electric)...	2	2(d)	(d) Provided from unit.
Total	3	3	

DISTRIBUTION OF RANK AND FILE BY TRADES :—

Engine drivers	4
Electricians	6
Fitters and turners	2
Joiner	1
Other trades	2
	15

Supersedes W.E. No. 351/49 as far as France is concerned.

(e) For sections equipped with power lorries in lieu of generating sets.

WAR OFFICE (S.D.2.)
    24th August, 1917.

O

ANTI-AIRCRAFT SEARCHLIGHT SECTION, R.E. (3 lights).

FRANCE.

WAR ESTABLISHMENT.

(i) PERSONNEL.

Detail.	Officers.	S/Sjts. and Serjts.	Rank and File.	Total.	Remarks.
Officer	1	–	–	1	
Staff-serjeant mechinist ...	–	1	–	1	
Serjeant	–	1	–	1	
Corporal	–	–	1	1	
2nd Corporals	–	–	2	2	
Sappers and Pioneers(a) ...	–	–	14	14	(a) Includes 2 L/Corporals.
Batman	–	–	1	1	
Total Section (excluding attached)...	1	2	18	21	
Attached :— Drivers A.S.C. (Mech. Transport) ...	–	–	2	2	
TOTAL SECTION (including attached)	1	2	20	23	

(ii) TRANSPORT.

Detail.	Vehicles.	Drivers.	Remarks.
30 cwt. Lorry	1	2 (b)	(b) Provided by A.S.C.

Issued with amendments to War Establishments Part VII A No. 1010.

Supersedes W.E. No. 351/48 so far as France is concerned.

WAR OFFICE (S.D.2.)

27th February, 1918.

MOBILE ANTI-AIRCRAFT SEARCHLIGHT SECTION, R.E.
(3 Lights, 120 cms. or 90 cms.)
FRANCE.
WAR ESTABLISHMENT.
(i) PERSONNEL.

Detail.	Officers.	S/Sjts. and Serjts.	Rank and File.	Total.	Remarks.
Officer	1	–	–	1	
Mechanist staff-serjeant(a)	–	1	–	1	(a) May be mechanist engine driver or electrician.
Serjeant	–	1	–	1	
Corporals	–	–	2	2	
2nd Corporal	–	–	1	1	
Sappers (b)...	–	–	20	20	(b) Includes 2 L/Corporals.
Batman	–	–	1	1	
Total Section (excluding attached)	1	2	24	27	
Attached :— Driver A.S.C. (Mech. Transport) ...	–	–	1	1	
TOTAL SECTION (including attached)	1	2	25	28	

(ii) TRANSPORT.

Detail.	Vehicles.	Drivers.	Remarks.
Motor cycle	1	–	
Motor car with box body ...	1	1(c)	(c) Provided by A.S.C.
Power lorries (petrol-electric) ...	3	3(d)	(d) Provided from rank and file of unit.
TOTAL	5	4	

DISTRIBUTION OF RANK AND FILE BY TRADES :—

Engine drivers	6
Electricians	9
Fitters and turners	3
Joiner	1
Other trades	4
	23

Supersedes W.E. Nos. 1032 and 685.

WAR OFFICE (S.D.2.)
   17th August, 1918.

CHAPTER VIII.

1919—1923.

THE TYNE ELECTRICAL (FORTRESS) R.E. (T.A.).

Progress of demobilisation — The " Victory " march — Reunion dinner originated — Reconstitution of the Territorial Force — Establishments for electric light and works companies — Signals duties removed from the Corps — The plea for an anti-aircraft unit — The question of medals for home defence troops — Recruiting — The " Territorial Army and Militia Act " — Conditions of service — The plea for a regimental headquarters — Composition of the reconstructed unit — The reserve of officers — Training — Drill hall at Whitley Bay — The coal strike — The Tyne E. & M. Companies, Defence force — Camps at Hartley and at Monkseaton — The Crest — Detachment at Catterick — Rifle shooting successes — Lieutenant-Colonel Robinson.

T HE evacuation of Haslar and the return of the Depot Headquarters to Clifford's Fort in April 1919 really conclude the War history of the Tyne Electrical Engineers.

On the 3rd June Lieut.-Colonel Robinson was demobilised and, after a short time, Captain Sharp also returned to his civil occupation, handing over the duties of adjutant to Captain D. A. Williamson with effect from the 7th June.

The Unit was now without any fixed establishment and the vast bulk of the personnel which had comprised the many detachments formed for the War had been demobilised, while the very companies and sections themselves were disappearing from the pages of the army list. Certain of the A.A. searchlight sections, however, still remained with the Army of Occupation on the Rhine, and personnel for these was provided largely from volunteers who surrendered their claims to early demobilisation.

Some of the A.A. companies R.E. were also in active existence as well as several of the R.G.A. companies to which personnel of the Tyne Electrical Engineers was attached.

About 65 officers, mostly demobilised, but including a few in special employment or seconded to other arms and services,[1] remained on the active list of the Unit, the remainder having by this time resigned their commissions or transferred to the reserve. The adjutant and the quartermaster, Captain A. Reed, were alone actively engaged in the final clearing up of matters at the Headquarters at Clifford's Fort.

---

[1] See note " D " at end of chapter IV.

COLONEL ERNEST ROBINSON, C.B.E., T.D., D.L., J.P.
Commanding Officer, 1918–1925.
Honorary Colonel, 1930.

To face page 212]

The work during this period of 1919 was uneventful and entirely of a routine character, in striking contrast to the state of ceaseless activity and strenuous effort which had prevailed during the past four and a half years.

The 16th Company R.E., which had been occupied as a works company in the Tyne garrison, had temporarily ceased to exist, and the various ranks of the Royal Engineers who were still employed under the D.Os. R.E. at Frenchman's Bay and at Hartley were posted to the Tyne Electrical Engineers for administration until the 16th Company was resuscitated.

The Unit was represented in the " Victory " march which was held in Newcastle-upon-Tyne on the 19th July. A band was collected by Bandmaster Richley and the detachment, commanded by Major O. M. Short, numbered several hundreds. The captured German searchlight equipment, to which reference has been made in an earlier chapter, had proceeded independently by road to Newcastle from North Shields and followed the detachment in tow of a steam traction-engine.

The last draft of Tyne Electrical Engineers' personnel left Clifford's Fort on the 1st November for No. 3 A.A. Company at Leeds. This Company, together with Nos. 1 and 13 at Newcastle and Edinburgh respectively, were among the last to be retained and by the end of the year they also had been disbanded, though several of the A.A. searchlight sections remained on the Rhine until April, 1920.

On the 21st November 1919 Captain Williamson handed over his duties to Captain Reed, who then became acting adjutant in addition to his duties as quartermaster.

Captain A. J. Sergeant was seconded to the R.E. and remained at Haslar until the 19th September 1922 for duty as adjutant and quartermaster of the reconstituted S. of E.L. and of the 4th and 22nd Companies R.E.

On the 17th April the first of the officers' re-union dinners, which have since continued as an annual function, took place at Tilley's restaurant, Newcastle-upon-Tyne, at which 64 of the serving and retired officers attended.

The re-birth of the Territorial Force dates from the 30th January 1920 and recruiting commenced on the 16th February of that year, but at first without any fixed establishment.

In November a provisional peace establishment for two companies, viz. :— one works company and one electric lights company, both destined for the Tyne Garrison, was included in the revision of peace establishments under the new title of Tyne Electrical (Fortress) Royal Engineers (T.F.).

The provisional establishment so granted was for a total of 15 officers and 169 other ranks, only about half of the pre-war establishment. These numbers were divided between the two companies in the following proportions :—

No. 1 Works Company, 11 officers and 145 other ranks.
No. 2 Electric Lights Company, 4 officers and 39 other ranks.

The officers allotted to the works company were 1 major, 3 captains, 3 lieutenants and 3 second-lieutenants ; and, in the electric lights company, 1 captain, 1 lieutenant and 2 second-lieutenants. In the group of electric lights companies comprising the Tyne, Tees and Humber, the senior officer might be a major. The establishment also provided for an adjutant to be attached to the electric lights company and a permanent staff instructor to be posted to each company.

There had been previously circulated, in February, 1920, a draft organisation for the allotment of Territorial units to coast defences, in which the war strength of the companies was to have been :—

No. 1 Works Company, 11 officers and 291 other ranks.
No. 2 Electric Lights Company, 4 officers and 80 other ranks.

In this draft provision was also made for a fortress signal section with a war establishment of 1 officer and 105 other ranks but, before the final promulgation of the peace establishments in November, the formation of the Royal Corps of Signals from the signal companies, Royal Engineers, had taken place, and this section was, accordingly, removed from the establishment of the Tyne Electrical Engineers. It followed that the responsibility for the construction, maintenance and operation of fortress telephone communications, which had formed part of the technical duties of the Unit since 1907, were handed over to this new corps.

The strength to which recruiting was now authorised was 60 per cent. of the war establishment.

It was pointed out at the time that the Tyne Electrical Engineers had established for themselves, during the War, a great reputation as specialists in anti-aircraft searchlight work ; that the change from the somewhat highly specialized duties of this service to those of works companies would prove far from popular ; and that some difficulty might be experienced in getting men to rejoin under these conditions.

The reply of the War Office was to the effect that the position was fully appreciated, but that it was unlikely that any Territorial Force anti-aircraft units would be formed, owing to the then prevailing economic conditions.

Another element, which subsequently was found to militate against successful recruiting into a unit destined primarily for home defence, was the fact that the services of troops who were necessarily retained in this country during the War, though they were ready and anxious to serve abroad, were not recognised by the issue of a medal. This grievance was all the more aggravated by the fact that the R.N.V.R. personnel, who had been engaged on identical work in the early stages of the defence of London, by virtue of being administered by the Admiralty, had received the General Service Medal. Lieut.-Colonel Robinson personally took a leading part in endeavouring to obtain adequate recognition of the extremely valuable war services of home defence units, but unfortunately without success.

Recruiting was, nevertheless, proceeded with from the Headquarters at Clifford's Fort, with the result that by April 1921 the actual strength of the electric light company represented 98 per cent., and that of the works company 60 per cent. of their respective peace establishments.

In the meantime two alterations of great importance had been introduced by the passing of the " Territorial Army and Militia Act " of 1921. The first was the change of name from Territorial Force to Territorial Army and the second dealt with conditions of service. At the time of the old Act of 1907 there were two distinct organisations, a regular army producing an expeditionary force and a home defence army consisting of the Territorial Force. Under the new Act, there was to be but one army, consisting of the professional portion, the Regular Army, and the second-line portion, the Territorial Army, through which all expansion was to take place in the event of war.[1] The conditions of service can be gathered from the following recruiting advertisement inserted in the local press.

Tyne Electrical Royal Engineers.

Recruits Wanted.

Tradesmen wanted to enlist in the above Unit for " Electric Light " and " Works " Companies, Royal Engineers, Territorial Army (Coast Defence). Full particulars and conditions of service can be obtained on application to the Adjutant, Clifford's Fort, North Shields, between the hours of 7.30 p.m. and 9.30 p.m. on Tuesdays and Thursdays. Past members (trained men) of the Unit are particularly desired and will be given priority in filling vacancies. Men must be between the ages of 18 and 38 (members of recognised Cadet Units over 17). A limited number of selected men for appointment for Warrant and N.C.Os. rank of and above the rank of Serjeant can be taken up to 45.

---

[1] Re-affirmed in W.O.L. Gen. No./5065 (S.D.2) dated 23 February. 1925.

Periods of Engagements.

      Soldiers who have served during the War for 6 months...   3 years.

      Other men, i.e., untrained men and cadets   ...   ...   4 years.

      Every man must be formally enlisted as a sapper and will have to be physically fit for general service.

      Annual training will consist of (*a*) drill, (*b*) musketry, (*c*) annual training in camp. (Not less than 8 and not more than 15 days).

      By fulfilling conditions, bounties can be earned as follows :—

          Trained men up to £5 each year.

          Recruits 1st year £4.

      Pay and allowances at the full current army rates whilst attending camp (including separation allowances as may be allowed for the Regular Army).

      Rations in kind wherever possible will be issued.

It will have been observed that the establishment of the reconstituted Unit did not provide for any headquarter personnel and that the Commanding Officer, though actually a substantive lieutenant-colonel since June 1st 1916, was acting as O.C. of the works company in the rank of major. It was not until July 1923 that an attempt was made to obtain authority for an establishment of personnel for regimental headquarters.

The case for the additional establishment rested largely on the precedent of pre-1914 days and also on the presumption that, on embodiment, the Unit would be subject to a considerable and rapid expansion as in the case of the Great War. In these circumstances it was considered that a depot establishment would be a need of the first importance. The application was not, however, approved by the War Office and efforts have since repeatedly been directed, but so far without success, towards the realisation of this object.

A great difficulty which presented itself at this time was the selection of officers to complete the very attenuated establishment of the new Tyne Electrical Engineers. It had, however, been agreed by the War Office that, in order to retain the services of those best qualified, officers could be appointed, irrespective of the ranks they held, on the understanding that, while retaining their rank, they would receive pay and allowances for the position they occupied.

As a result, several of the officers of the reconstituted Unit were senior to the ranks they occupied under the new establishment, a circumstance which must have contributed largely towards the successful manner in which the Unit quickly reached its former efficiency.

The officers ultimately appointed to the reconstructed Unit under W.O. Authority dated the 30th July 1921, were :—

1. Lieut.-Colonel E. Robinson. (In command).[1]
2. Major N. H. Firmin.[2]
3. Major C. M. Forster.[2]
5. Captain I. F. Fairbairn-Crawford.
5. Captain R. Sharp.
6. Captain H. Sherlock.[3]
7. Captain K. S. M. Scott.[3]
8. Captain E. H. E. Woodward.[3]
9. Lieutenant R. P. Winter.
10. Lieutenant C. B. Williamson.
11. Lieutenant H. G. Campbell.
12. Lieutenant J. L. Batey.
13. Lieutenant B. H. Leeson.
14. Lieutenant P. F. Fawcett.
15. Captain and Qr.-Mr. A. J. Sergeant.

The appointment of Captain A. Reed as adjutant was confirmed, with effect from the 6th January, in the *London Gazette* of the 2nd May 1921.

The medical officer attached to the Unit was Captain J. B. Williamson, R.A.M.C. (S.R.), Lieut.-Colonel Gibbon having left the Unit in February 1919 to take up another appointment.

The Rev. H. L. Lloyd continued with the Unit until the 28th May 1924 when he retired, on appointment to the living of Chevington, and was not replaced, as the establishment of the Tyne Electrical Engineers no longer provided for a chaplain.

Of the officers on the Active List who were not appointed to the reconstructed Unit, the great majority now transferred to the Territorial Army Reserve of Officers, Tyne Electrical Engineers ; a few to other units, while the remainder relinquished their commissions.

A problem similar to that of the officers presented itself with the other ranks, of whom over half were re-enlisted Tyne Electrical Engineers. Many of these men found themselves serving as sappers in a Unit in which they had previously held warrant or non-commissioned rank. Ten months elapsed after recruiting was opened before any promotions could be made, and it was nearly 18 months before J. Chambers and I. T. Womphrey were promoted company-serjeant-major and regimental-quartermaster-serjeant respectively of No. 1 Works Company, to fill the posts of the only warrant officers in the new establishment. The senior N.C.O. in No. 2 Electric Light Company was Lance-Serjeant A. W. Smith. The permanent staff instructors at this time were Serjeant H. B. Keily, R.E., and Serjeant R. Bwy, R.E.

---

[1] In the rank of Major.
[2] In the rank of Captain.
[3] In the rank of Lieutenant.

The official date of taking over the command of the reconstituted Unit by Lieut.-Colonel Robinson was the 12th April 1920, but drills were not commenced until the 14th September in that year.

It was, at this period, necessary for a man to attend during the training year (which terminated on the 31st October) at least twenty drills of one hour's duration each, of which twelve were to be technical and eight infantry training. In order, however, to qualify for the bounty referred to previously, it was necessary to complete fifty drills and fire the open range musketry course.

Drills at first took place for periods of two hours on each night in the week except Saturday and Sunday.

The musketry course included firing 55 rounds of ball ammunition, of which not more than 20 rounds were to be fired by an individual on any one day.

The technical training was at first carried out, under the R.E. mechanists of the 16th (Fortress) Company R.E.,[1] on the coast defence electric light plant installed at Tynemouth Castle and the Spanish Battery (the parties parading in Front Street), and the rest of the instruction at the Headquarters, Clifford's Fort. In November 1920 a mobile searchlight projector (90 cms.) and a petrol-electric lorry became available for training and were taken into use at the Fort. Later, when a second lorry was available, this was used on occasions to carry working parties to and from the Blyth defences.

Owing to the inconvenience of Clifford's Fort, due both to lack of space and by reason of its inaccessibility, authority was obtained for the use, in addition, of the drill hall at Rockcliffe Avenue, Whitley Bay. This hall had formerly been used by " G " Company of the Northern Cyclist Battalion which had been disbanded.

The hall, which was shared with a squadron of Northumberland Hussars (Yeomanry), contained a miniature rifle range and was used for musketry instruction, the first parade taking place there on Tuesday the 24th November 1920. The annual musketry course for trained men was resumed at Ponteland in December of that year.

The band was also re-formed under the direction of Bandmaster Richley, authority having been obtained to employ 16 sappers as musicians.[2]

In January 1921 it was decided to confine evening training to Tuesday and Thursday evenings, and on the 7th and 8th of May there took place at Ponteland, for the purpose of firing the musketry course, the first week-end camp to be held after the War.

---

[1] Commanded at this time by Major W. Barr, Co. Bn. R.E.
  The Coast Battalion organization was not revived after the War and the coast battalion companies became fortress companies. The remaining coast battalion officers completed their service, mostly with these companies, but there were no further Commissions into this establishment.
[2] W.O.L. 9/Gen. No./74 (A.G.I.) dated 9th June 1920.

Regular Army courses at the School of Military Engineering, Chatham, and at the School of Electric Lighting, Gosport, were also recommenced, at which members of the Unit attended.

Hardly, however, had the Unit begun to settle down to its new organisation and recommenced training, when the coal strike, declared on the 1st March, 1921, created a fresh distraction.

The Government immediately proclaimed a state of emergency in Great Britain and set about the formation of a national defence force. For this purpose the permanent staff of the Territorial Army was used as a framework, and both serving Territorials and others were enlisted on special attestations for a period of not more than three months. Particular attention was naturally paid to the selection of a reliable class of men.

In spite of the fact that the threatened strikes of the transport-workers and railwaymen in support of the miners were averted, it was considered advisable to provide the defence force with technical troops. Instructions were, accordingly, issued for the raising of six Electrical and Mechanical (E. & M.) Companies, Royal Engineers, Defence Force. Two of these, and a headquarters, were to be raised by the Tyne Electrical Engineers and four, and a headquarters, by the London Electrical Engineers. At a later date the raising of further E. & M. Companies was authorised, these being allotted to important manufacturing and industrial areas. One of them, the company for Newcastle, was added to the two already allotted to this Unit, but was never raised.

The establishment was as follows :—

    Headquarters : 3 officers and 8 other ranks.
    Each company : 5 officers and 113 other ranks.

Transport was to be provided locally under Command arrangements. Stores, with certain exceptions, were drawn up to 50 per cent. mobilisation scale.

No trade establishment was laid down, though the trades thought to be most in demand were stated to be electricians, engine drivers, fitters and turners.

Certain difficulties immediately presented themselves, chief among which were the lack of workshop accommodation and tools, and the absence of definite instructions as to the duties to be carried out by the companies.

Recruiting was, however, proceeded with on the assumption that the duties would be similar to those of the E. & M. companies which were formed during the Great War for service in France and that they might be required to run electric generating and water-supply plant for the maintenance of public services in the event of grave emergency.

Skilled men of all trades associated with electrical, mechanical and general engineering were selected, and training in the overhaul, repair and maintenance of all machinery available was set in hand.

The command of these companies fell to Lieut.-Colonel Robinson, while Captain R. Sharp and Captain Woodward commanded the individual companies. Major Burton took this opportunity of renewing his long association with the Tyne Electrical Engineers by taking over the duty of adjutant, and Major Forster acted as second-in-command. The other officers who served were Captain C. F. Scott, Lieutenants C. B. Williamson, Winter and Fawcett, and Captain and Quartermaster Reed.

By the 14th April all the officers were selected, 71 men were attested and 59 others registered, a large proportion of whom were on the point of being attested when orders were received to suspend recruiting. Only about 30 per cent. of these other ranks were serving in the Tyne Electrical Engineers at the time and these were discharged from the Territorial Army on enlistment into the Defence Force.

Recruiting was undoubtedly hindered by a fear of subsequent victimisation and of possible loss of employment due to delay in disembodiment.

The first duty of the companies was the erection of barbed wire entanglements for the protection of the wireless station at Cullercoats, which had been occupied by a detachment of the Royal Navy, but the wire was shortly after removed when the detachment was withdrawn.

Beyond this, the companies were never actually employed on service, though no doubt their presence assisted to shorten the duration of the strike which was happily declared at an end on the 1st July. Thereupon, the Tyne E. & M. Companies, after a brief existence of just over three months, were disbanded as quickly as possible. By the 5th July all the men had been discharged from the Defence Force and those who had left the Tyne Electrical Engineers for this service were forthwith re-enlisted into the Territorial Army.

On the dispersal of the Defence Force each officer received a copy of the following letter from the Prime Minister :—[1]

> " 10 DOWNING STREET,
> LONDON, S.W.1.
> 4th July, 1921.
>
> Sir,
>
> I have much pleasure in personally thanking you and in conveying to you the thanks of my colleagues, for the services you have rendered during the recent national emergency.
>
> The readiness which you and others have shown in coming forward to defend the public was an effectual guarantee for the maintenance of law and order, and, after the preservation of the people from threatened privation and misery, it will be remembered with gratitude by all sections of the community.

---

[1] The officers of the Defence Force were gazetted out on the 20th September 1921.

I trust no further demand will have to be made on your services. If, however, a fresh necessity should arise, I feel sure you will respond with no less public spirit and zeal than you have shown during the past crisis.

> I am, Sir,
> Your obedient Servant,
> (*Signed*) D. LLOYD GEORGE."

The opportunity now seemed suitable once more to attempt to improve the technical importance of the Unit's work and a strongly worded plea was forwarded to the War Office for the retention, permanently, on the peace establishment of the Territorial Army, of E. & M. companies R.E., and for the conversion of the ever unpopular Works company of the Tyne Electrical Engineers to that service. Once again, however, the existing need for economy rendered it impossible for the War Office to act upon this suggestion, though the proposals were " noted for consideration when conditions permit of the expansion of the Territorial Army."[1]

The Unit now returned to normal training and duty, and the annual training in camp, which had been interfered with by the coal strike and the formation of the E. & M. companies, took place at Hartley, in the vicinity of Roberts Battery, from the 20th August to the 3rd September.

The Unit, 128 strong, was joined at Hartley by the North Riding (Fortress) R.E. (T.A.), under Major H. Wintersladen T.D., and the Tees (Fortress) Signal Section, Royal Corps of Signals, under Lieutenant G. Mackenzie, the whole force being commanded by Lieut.-Colonel Robinson.

Training was carried out in camp with mobile plant, and by day and night on the coast defences at Blyth and Tynemouth. No. 1 Works Company also trained to some extent in the branches of field engineering appropriate to its service. The distribution of officers in camp, between the two companies, was as follows :—

No. 1 Works Company.
>   Captain R. Sharp O.B.E.
>   Captain H. Sherlock.
>   Captain E. H. E. Woodward M.C.
>   Lieutenant H. V. Owen (T.A.R.O.).

No. 2 Electric Light Company.
>   Captain K. S. M. Scott M.B.E.
>   Captain C. F. Scott (T.A.R.O.).
>   Lieutenant C. B. Williamson.
>   Lieutenant H. G. Campbell.

On the 30th August the units in camp were inspected on parade and on the works by Colonel G. D. Close C.B.,[2] Chief Engineer, Northern Command.

---

[1] W.O.L. 9/Engrs./8696 (A.G.7) d/12/12/21.
[2] Born 8/7/66; 1st Commission 29/4/85; Colonel 16/12/13; Retired (Brig. Gen.) 31/12/21.

The Tyne companies manned Blyth, the Spanish Battery and a mobile set in the camp, while the North Riding company manned Tynemouth Castle.

On his departure the inspecting officer caused the following order to be published :—

" It is notified for information, that Colonel G. D. Close C.B. (Chief Engineer) who inspected the Tyne Electrical Engineers R.E. (T.), North Riding (Fortress) R.E. (T.), and Tees (Fortress) Signal Section, Royal Corps of Signals, on 30th instant, expressed his entire satisfaction with the steadiness and general smartness of all ranks on parade, as well as with the progress made by all ranks with their technical duties."

The regimental rifle prize meeting was revived at Ponteland on Sunday the 18th September. This was the first occasion on which the handsome silver bowl presented by Major M. W. Buck for the officers' revolver championship was competed for, being won by Lieut.-Colonel Robinson.

In the same month was held the first meeting since the War of the Northumberland County Rifle Association, many members of the Unit taking part with considerable success. The Earl Grey Shield for quick-firing was won for the first time by a team representing the Tyne Electrical Engineers, and it was remarked at the time that they were the first unit other than infantry to hold this trophy. This successful opening season for the Tyne Electrical Engineers' marksmen culminated with the winning by them, on the 23rd October, of the Ladies' Challenge Trophy on the first revival of this event since the War.[1] The Unit had thus lost nothing of its former efficiency with the rifle and continued from this date to occupy a prominent position among rifle shooting units in the county.

In December authority was received for the wearing of Royal Engineer buttons in place of the " general service " buttons hitherto issued. At the same time, attention was directed to the fact that certain men were wearing Regular R.E. shoulder titles in place of the regulation $\frac{\text{T}}{\text{R.E.}}$ titles. An application was therefore submitted, and, early in 1922, authority was obtained for the taking into use again of the special $\frac{\text{T}}{\frac{\text{R.E.}}{\text{TYNE}}}$ titles which had been used before the War.

The annual prize giving and dance was held for the first time since the War on the 3rd February 1922 at the Empress ballroom, Whitley Bay, at which the prizes, together with numbers of war medals and decorations, were presented by Sir Francis D. Blake, Bart., C.B., D.L., M.P., Chairman of the Northumberland County Territorial Association.

---

[1] The team was as follows :—Lieut. Colonel Robinson, Major Firmin, Captain Woodward, R.Q.M.S. Womphrey, Serjeant Forsyth, Lance-Serjt. Young, Lance-Serjt. Ellis, and Sapper Dick.

On the 22nd March the appointment of His Grace The Duke of Northumberland C.B.E., M.V.O. as Honorary Colonel reached its normal termination, but, greatly to the satisfaction of all ranks, was extended by War Office authority for a further period.[1]

On the 5th May, Captain T. Henderson M.C., A.F.C., was appointed to the reconstructed Unit, surplus to establishment, but to remain seconded while serving with the Royal Air Force.[2]

Annual training in camp this year (1922) took place on the Beverley estate at Monkseaton, the advance party arriving on the 13th June to be followed by the main body on the 17th of that month. The strength of the Unit in camp was 13 officers and 149 other ranks.

Firm in the belief that the mission of the Unit was of a more technical nature than that offered by the duties of the works company, both companies were treated unofficially, for the duration of this camp, as electric light companies.

The training, though carried out by companies, was divided into two categories, the supervision of the technical work being placed in the hands of Major Forster, assisted by Captain Woodward, and the regimental instruction falling to Captains Sharp and K. S. M. Scott. The technical instruction so provided included lectures and night manning with both fixed and mobile coast defence and anti-aircraft searchlights, although the latter formed, as yet, no part of the Unit's official duties.

The inspection took place on the 25th and 26th June, and Colonel H. L. Pritchard, C.M.G., D.S.O.,[3] Chief Engineer, Northern Command, who carried out the inspection of both companies on parade and at their technical duties, expressed himself " particularly impressed with the steadiness of all ranks on parade and with the general proficiency of the Unit in its technical work." This opportunity was taken for the Chief Engineer to present various war medals and decorations to members of the Unit.[4] Several new silver cups and trophies were presented to the Unit about this time by various retired officers. Among these were the Toomer cup, allocated for infantry drill, and the Forster and Porter cups for which competitions in technical work were organized.

The rifle club continued to thrive and held, during the period of camp, miniature rifle competitions in the quarry, Hill Heads, Whitley Bay. It was for the highest aggregate in these competitions that the Hall cup was first awarded, and this cup, the gift of Captain W. Hall, continues to be competed for annually in the events comprising the miniature rifle championship of the Unit.

---

[1] Extended once more on 20/3/27 for a further period of five years.
[2] Captain Henderson relinquished his commission in the T.E.E. on appointment to the R.A.F., 31st July 1926.
[3] Born 16/11/71 ; 1st Commission 13/2/91 ; Colonel 31/12/21 ; Major General (Cmdt. S.M.E., G.O.C., T. & M. Area and Inspector of R.E.) 5/11/28 ; Col. Cmdt. 18/2/31.
[4] Among these was Sapper H. G. Anderson who had been awarded the Military Medal under the following circumstances. On November 10th and 11th, 1918, while serving with a Lewis gun section of the 6th (Queen's Own) Royal West Kent Regiment, he was holding a bridge across the river Burdon (on the Franco-Belgian border) against hostile machine-gun fire, and, though wounded, he gallantly continued in action with his Lewis gun which he eventually brought out of action with him.

At the Northumberland County Rifle Association meeting this year, members of the Tyne Electrical Engineers carried off over £70 in prize money, representing over a quarter of the total prize list.[1] Major Firmin, who was shooting exceedingly well, was selected to shoot in the Territorial Army Eight at Bisley.

In October authority was given for the senior of the two permanent staff instructors (at that time Serjeant H. B. Keily) to be appointed acting company-serjeant major.

The crest of the Tyne Electrical Engineers was registered at the College of Arms on the 31st October 1922. The Blazon, as approved by Portcullis, is as follows :—

> " Issuant out of a mural crown or, a dexter cubit arm grasping a winged arrow enflamed proper."

The crest is derived from the crest of the Master-General of Ordnance, which is similar to that formerly used in the submarine mining service and may still be seen in the fly of the blue ensign worn by R.E. vessels.

The old crest was carved on the stone above the main gateway of Clifford's Fort, and the Blazon is understood to be :—

> " Issuant out of a mural crown, a cubit arm grasping a thunderbolt."

A thunderbolt in armoury (according to Berry's definition) is represented by :—

> " A twisted bar, in pale, enflamed at each end, surmounting two jagged darts in saltier, between two wings expanded, with streams of fire issuing from the centre."

Comparing the crest formerly used by the Submarine Miners with that now approved for the Tyne Electrical Engineers, it will be seen that they are very similar, the main difference being that the latter bears an arrow instead of a thunderbolt. (See illustration.)

On the 2nd February 1923 the Honorary Colonel, His Grace The Duke of Northumberland, presented decorations, war medals and the prizes won during the past year to members of the Unit at the Empress ballroom, Whitley Bay, the function being attended by some 1,500 guests.

The Unit was represented by a Guard of Honour, under the command of Captain Woodward, on the occasion of the unveiling, by the Duke of Northumberland, of the Tynemouth War Memorial on Saturday, 28th April, the whole parade, including detachments of the R.N.V.R. and other units of the T.A., being commanded by Major Firmin.

On the 1st June, 1923, the Unit was completely rearmed with the Mk. III* short Lee-Enfield rifle, but a percentage of E.Y. bombing rifles and D.P. weapons was retained.

---

[1] Among the more successful shots were :—Lieut.-Colonel Robinson, Major Firmin, Capt. Sharp, Lieut. Leeson, Lieut. C. B. Williamson, R.Q.M.S. Womphrey, Serjt. Forsyth, L/Sjts. Dick and Young, and Sappers Purtle, Atkinson and Wilkinson.

The annual camp was held this year again at Monkseaton, the main body marching in on the 23rd June. The full complement of 16 officers attended this camp and the 186 other ranks present were divided between the companies in the following proportions :—

No. 1 Works Company 141
No. 2 Electric Lights Company 35

The plant available for technical training had by now been increased by the provision of a Crossley high-speed generating set, and the number of P.E. lorries was raised to four.

The possibility of again taking up A.A. searchlight work had never been lost sight of and, on this occasion, training was carried out in camp with mobile A.A. searchlights in addition to that with the fixed coast defence plant. On the 26th June a party left for Catterick under the command of Lieutenant Winter, with Lieuts. C. B. Williamson and Parnall and about 36 other ranks, to obtain night practice with aircraft. The equipment taken by this party consisted of three of the four P.E. lorries, two carrying 90 cms. A.A. mobile lights, and the third towing the captured German set, which was used for training purposes at that time. The detachment was accommodated with the 50th (Northumbrian) Divisional R.E., who were encamped at Catterick aerodrome.

The aircraft provided for co-operation were high-speed Bristol fighter machines from the Signal Co-operation Flight[1] R.A.F., a unit justly proud of its night-flying and claiming never to have been held in a searchlight beam. Very satisfactory results were, nevertheless, obtained by Lieut. Winter's detachment on the two nights when practice took place. On the third day the party returned to Monkseaton. This episode served to demonstrate that the Tyne Electrical Engineers were well capable of again undertaking this type of work and the results must materially have contributed towards the conversion of the Works Company to the anti-aircraft service, which was destined shortly to be effected.

On Monday, the 2nd July, the Chief Engineer, Northern Command, Colonel H. Biddulph C.B., C.M.G., D.S.O.,[2] inspected the Unit on night manning when, in addition to the coast defence searchlights at Blyth[3] and Tynemouth, anti-aircraft practice was carried out with fire balloons as targets. On the following morning the Unit was inspected on parade, when the inspecting officer declared himself " particularly impressed with the general efficiency of the Unit in its technical work."

---

[1] Under the command of Flight Lieutenant J. A. G. Haslam M.C., D.F.C. This Flight was detached from the School of Army Co-operation, Old Sarum, and was stationed at Kenley during the winter months, visiting various aerodromes in the north for co-operation during the summer.
[2] Born 19/7/72, 1st Commission 22/7/92, Colonel 3/6/19, Retired 19/7/29.
[3] This was to be the last occasion on which the plant at Blyth was used. Already, in November 1921, the Superintendent of Blyth cemetery had reported to the authorities that " the uncompleted concrete fort on the links opposite Blyth cemetery was a harbour for undesirable persons," and asked that it should be closed or filled up. In September 1924 the War Office ordered that the Battery was to be given up. The guns and all equipment were returned to Ordnance and the Works were handed over to the Blyth Corporation about April 1925.

P

The Northumberland County Rifle Association meeting, held at Ponteland ranges on the 7th and 8th of September, resulted in the winning by the Tyne Electrical Engineers of the 50th Division Memorial Cup, the Freeman of Newcastle Challenge Shield, the Earl Grey Shield and the Co-operative Union Challenge Cup. To this exceptional success must be added the winning, by individual members of the Unit, of the County Prize and the Chipchase Cup; while before this season closed, the Ladies' Challenge Trophy had been won for the third time in succession.

On the 7th December, Lieut.-Colonel Robinson was appointed Deputy Lieutenant of the County of Northumberland, an honour which was universally welcomed as a well deserved recognition of his valuable military services.

After but two and a half years of existence, the reconstituted Tyne Electrical (Fortress) R.E. (T.A.) was firmly re-established, its ranks full and its technical and military efficiency superior even to that of pre-war days. It is, perhaps, fitting to close the account of this period in the Unit's history in the words of the contemporary press :—

> "The high reputation enjoyed by the Tyne Electrical Engineers R.E. is in a large measure due to his (Colonel Robinson's) skill and energy and, as a hardworking member of the Northumberland Territorial Army Association, he has also done a great deal to advance the strength and usefulness of the Territorial Army units throughout the county."

> (*Newcastle Daily Journal*, 21/12/23)

CHAPTER IX.

1924—1933.

TYNE ELECTRICAL (FORTRESS) R.E. AND TYNE ELECTRICAL ENGINEERS R.E.

Revival of air defence — Camp at Gosport (1924) — Conversion of the
Works Company into the 307th (Tyne) Anti-Aircraft Searchlight Company
— Unity of the Tyne Electrical Engineers — Camp at Gosport (1925) —
" Daily Telegraph " Cup competition — Major Firmin in command —
Presentation to Colonel Robinson — Annual training 1926 — Lewis gun
training commenced — Camps at Tynemouth and Manston (1927) —
Sergeant Knight's gallantry — New Headquarters — Evacuation of
Clifford's Fort — Camps at Kinghorn and Kenley (1928) — Opening of
the new Headquarters — H.M. The King visits Newcastle — Appointment
of full-time adjutant — Camp at Manston (1929) — Major Woodward in
command — Camps at Tynemouth and Manston (1930) — Demonstration
at O.T.C. camp at Strensall — Death of the Duke of Northumberland —
Colonel Robinson appointed Honorary Colonel — Camp at Kinghorn (1931)
— Air manoeuvres — Annual training 1932 — Re-introduction of pipers
— Tynemouth Priory flood-lighted — Re-organisation of coast defences —
No. 1 (E.L. & Works) Company — Camps at Tynemouth and Manston
(1933) — Northern Command Tattoo — The tradition of the Unit.

IN the preceding chapter it has been shewn how, in the two years which followed
the reconstruction of the Unit, most of the features of the annual programme
of pre-war days were revived. In order that the following narrative of the
activities of the Tyne Electrical Engineers during the past decade should not
develop into a mere catalogue of dates and endless recitation of such events as
week-end camps, prize givings and church parades, it is proposed to omit much
of the detail which it has been thought advisable to include in the earlier chapters.

The year 1924 was to be a notable one for the Tyne Electrical Engineers,
as it was during this year that the cherished ambition of returning officially to
anti-aircraft work was achieved. It was, therefore, particularly opportune
that an officer with experience of anti-aircraft searchlights should be appointed
acting-adjutant at this stage. Captain J. M. McSweeney, Coast Battalion, Royal
Engineers, had just completed a period in command of " B " Company of the
recently formed 1st A.A. Battalion, R.E., at Aldershot, when he was posted to
command the 16th (Fortress) Company, R.E., at Tynemouth, and took up the
duties of adjutant to the Tyne Electrical (Fortress) Engineers with effect from
the 22nd December 1923, Captain A. Reed retiring from the service on the 9th
February 1924.

Although it had earlier been asserted that no anti-aircraft units of the Territorial Army were likely to be raised, and coast defence units had been officially

refused permission, as recently as June 1923,[1] to carry out training with anti-aircraft searchlights, a start had, in fact, been made as early as the year 1922 with the reconstruction of the London defences, so hastily swept away at the end of the War. Two air defence brigades, comprising guns, searchlights and signals, had been formed in London, the searchlight units allotted being the 10th and 11th (subsequently the 26th and 27th) A.A. Searchlight Battalions, London Electrical Engineers.[2] Owing to various causes progress with these two units had not been very successful and early in 1924 an inter-departmental committee set to work to draft a detailed plan for again building up an air defence for the Capital. Major-General E. B. Ashmore, C.B., C.M.G., M.V.O., then commanding the 1st (Regular) Air Defence Brigade at Aldershot, was appointed to this committee and given command of the ground troops then in existence and to be raised; he was, in addition, appointed Inspector of Anti-Aircraft.[3]

It was, perhaps, on account of impending developments in the air defences, which were once more to affect the Tyne Electrical Engineers, that on the 12th April 1924 Lieut-.Colonel Robinson's tenure of command was extended for a further year.[4] Much of the preliminary training of the Unit in this year was also devoted to anti-aircraft work, an anti-aircraft searchlight being erected at the Spanish Battery for evening drills, while for week-end camps three further stations were erected in the Tynemouth area.

The annual camp, 1924, for both companies, took place at Gosport. There was a good attendance, every officer being present, and all but five of the other ranks, two of whom were excused on medical grounds and the other three absent at sea or working away from Tyneside. Five A.A. searchlight stations were manned in addition to certain coast defence lights, and the Unit took with it to camp its own mobile searchlight sets. During this camp, an inter-section A.A. searchlight competition was held, in which each section demonstrated its own method for unloading and erecting the plant, no official drill being in existence at this date.

By the middle of this year, the recommendations of the Air Defence Committee had been placed before the War Office and Air Ministry, and the formation of eleven searchlight companies for work with fighter patrols in the defence of London had been authorised. These were in addition to the two London battalions intended primarily for co-operation with A.A. artillery and were to be raised, with one exception, in the country areas round London in which they were to be employed in the event of war—a policy which had proved to be impracticable some twelve years previously in the case of coast defence units. The exception was the company allotted to the Tyne Electrical Engineers, due, no doubt, to their previous experience in this branch of work, and it is important

---

[1] W.O.L. 9/Engrs./8763. (F.W.9.B.) dated 22/6/23.
[2] Two officers of the Tyne Electrical Engineers, Lieutenants C. G. Huntley and H. G. Campbell, transferred to the 10th Battalion, in the rank of captain, on its formation.
[3] This appointment was abolished after only two years' existence.
[4] Lieut.-Colonel Robinson was promoted Brevet-Colonel in the " London Gazette " of 10/10/24 with precedence as from 15/2/21, and his tenure of command was further extended on 6/1/25 to 31/10/25, which date was considered the more satisfactory for a change of command.

to observe that this company was always intended to supplement the London defences in the event of war and was not raised for the local defence of the Tyne.

Accordingly, in September 1924, the Works Company of the Tyne Electrical (Fortress) R.E. was transformed into an anti-aircraft searchlight company, under the following authority :—

> " Approval is given for the conversion, forthwith, of the Works Company of the Tyne Electrical (Fortress) Engineers into an Anti-Aircraft Searchlight Company, Royal Engineers (Tyne Electrical Engineers)."[1]

For some months, however, no new establishment was approved and the change was one simply of title, although courses at the School of Anti-Aircraft Defence at Biggin Hill became available to members of the company, thirty N.C.Os. and men attending during October 1924. At the same time, courses in P.E. lorry driving were commenced at the School of Electric Lighting, four N.C.Os. being sent there for a six weeks' course.

The Director-General of the Territorial Army, Lieutenant-General Sir Hugh S. Jewdwine, K.C.B., K.B.E., who had intended to be present at the annual prize giving on the 5th February 1925 but, at the last moment, was prevented through illness from attending, sent to the Commanding Officer the following telegram which is quoted here as emphasising the national importance of the work which the Unit was now undertaking :—

> " Deeply regret inability to be with you to-night. Had greatly looked forward meeting you and officers and men of Tyne Electrical Engineers and to telling you how important a place you occupy in the forces of the Crown. The additional responsibility in connection with anti-aircraft defence which you are about to assume renders this importance greater still. The fine record of your Unit inspires confidence that any duties which may fall to you in the future will be performed with the zeal and efficiency which you have shewn in the past. The Territorial Army will form the backbone and framework of any National Army which future emergency may compel us to create and I congratulate all ranks on the privilege of belonging to it and of standing now in the forefront of the defence of the Empire."

Army Order No. 7 of 1925 promulgated the designations of the new A.A. companies R.E. (T.A.), the Tyne company receiving the title " 307th (Tyne) Anti-Aircraft Searchlight Company, Royal Engineers (Tyne Electrical Engineers)."[2]

---

[1] Authority : W.O.L. 20/Engrs./5764 (T.A.8.) dated 17/9/24.

[2] The eleven companies were :—

307th (Tyne)	A.A. Searchlight Company R.E.			...	...	North Shields.
309th (Essex)	,,	,,	,,	,,	... ...	Harlow.
310th (Essex)	,,	,,	,,	,,	... ...	Epping.
311th (Essex)	,,	,,	,,	,,	... ...	Brentwood.
312th (Essex)	,,	,,	,,	,,	... ...	Upminster.
313th (Kent)	,,	,,	,,	,,	... ...	Chatham.
314th (Kent)	,,	,,	,,	,,	... ...	Tonbridge.
315th (Surrey)	,,	,,	,,	,,	... ...	Croydon.
316th (Surrey)	,,	,,	,,	,,	... ...	Kingston.
317th (Middlesex)	,,	,,	,,	,,	... ...	The Hyde, N.W.8.
318th (Surrey)	,,	,,	,,	,,	... ...	Guildford.

It should be observed that no company numbered 308 was formed, a fact which has always been held to indicate that it was proposed to allot this Company to the Tyne Electrical Engineers when an opportunity arose.

This change of title introduced a peculiar complication which continued for some years, for the E.L. Company remained "No. 2 Electric Lights Company, Tyne Electrical (Fortress) R.E.," while the 307th Company bore, once more, the pre-War title "Tyne Electrical Engineers." The two companies now fell under different commands for training. The E.L. Company remained directly under the Chief Engineer, Northern Command and the A.A. Company came under the Commander, A.A. Ground Troops, whose headquarters were at Uxbridge. Any suggestion to dissociate the two companies was, however, resisted, both on the grounds that the principle of training coast defence E.L. personnel in A.A. work had, by now, been accepted,[1] and that such a separation would inevitably result in the extinction of the smaller company for lack of recruits. In this view the Chief Engineer, Northern Command, in whose hands the administration of the whole Unit was left, concurred, with the result that a recommendation was forwarded by him to the War Office, stating that :—

"from a practical point of view it is essential to maintain the principle of interchange of all ranks between the various companies of the Tyne Electrical Engineers, and nothing should be done which would tend to split the corps up into distinct and separate units."

If the importance of retaining the interchangeability of personnel so far as the rank and file were concerned was thus accepted, it was very much more important from the officers' point of view that they should be treated as one unit and kept on one promotion list. To this the War Office agreed in the following terms :—

"I am directed to inform you that your recommendation that the officers of the 307th Anti-Aircraft Searchlight Company and those of the Defence Electric Lights Company be kept on one list under the heading Tyne Electrical Engineers, as officers of one unit, is approved."[2]

On the 30th April 1925 the new establishment for the A.A. Company was issued and expansion to the new total of 10 officers and 283 other ranks formally approved.[3] The new establishment contained considerably more N.C.Os than that of the old Works Company, and its appearance was immediately followed by a number of promotions to fill the vacancies, while sixty intending recruits, who had been on the Unit's waiting-list, were forthwith absorbed and the recruiting of 80 more set in hand. At the same time, it became necessary for all ranks in this company (as in all A.A. defence units) to sign army form E.622 by which they undertook to come up for service for purposes of defence at any place in the United Kingdom when called upon to do so under the authority of the Secretary of State, even though no Order calling out the Territorial Army

[1] W.O.L. 9/Engrs./8852 (M.T.) dated 16/4/25.
[2] W.O.L. 26/A.L./50 (M.S.2.T.) dated 11/1/26.
[3] Authority : W.O.L. 20/Engrs./5764 (A.G.7.) dated 30/4/25.

for actual military service was in force at the time. Although no such undertaking was, at that time, required of men serving in the E.L. Company, it became habitual for every member of the Tyne Electrical Engineers to sign this form, thus maintaining fully the interchangeability of personnel between the two companies.

On Saturday, May 2nd 1925, a practice with fire-balloons, which attracted many spectators, was carried out on the Town Moor, Newcastle-upon-Tyne, three 90 cms. searchlight stations being erected at Blue House, Claremont Place and at the Old Aerodrome.

This year the Unit went again to Gosport for annual training, the E.L. Company manning the coast defence stations at Haslar and Monckton and the 307th Company eight A.A. stations, which they erected in the vicinity. Night-flying was provided by the R.A.F. stationed at Fort Grange and control work was practised, while the School of Electric Lighting arranged short courses for the duration of the camp at which members of the Unit attended for instruction in P.E. lorry driving and engine driving. The technical work of the Unit was inspected by the Chief Engineer, A.A. Ground Troops, Colonel J. P. Moir, D.S.O.,[1] and by the Inspector of Anti-Aircraft, and the Director-General of the Territorial Army (General Jewdwine) inspected the Companies on parade on the 30th June. Although Colonel Robinson was still officially O.C. of the 307th A.A. Company, the executive command of this Company was exercised by Major Firmin, the E.L. Company being under Captain Fairbairn-Crawford. Despite the increase of establishment in April, the 307th Company went to camp only six men short of its full complement. The inspecting officer

> " expressed great satisfaction with the general smartness and steadiness on parade of all ranks, and was much impressed by the very large percentage of strength present in camp and keenness shewn, and the *esprit-de-corps* in the Unit."

When the War Office announced the result of the *Daily Telegraph* Cup Competition for the best attendance at annual training of all units in the Territorial Army, it transpired that this Company, with a figure of 98.3%, were runners up to the Tynemouth Heavy Brigade, R.A.[2]

---

[1] Born 28/4/72, 2nd Lt. 12/2/92, Col. 1/2/21, Retired 3/7/29.

[2] The percentage was, at that time, calculated by adding together the number of all officers and other ranks who attended for fifteen days and half the number of officers and other ranks who attended for eight days only. The E.L. Company was ineligible to compete, being below the establishment fixed as a minimum for this competition. The cup was offered for the first time in 1925 and various amendments to the conditions have since been made. The following are the percentage of attendances each company since 1925, calculated to the nearest digit :—

No. 1 E.L. Company.				307th A.A. Company.			
1925	...	...	89%	1925	...	...	98%
1926	...	...	No camp.	1926	...	...	No camp.
1927	...	...	70%	1927	...	...	90%
1928	...	...	89%	1928	...	...	93%
1929	...	...	87%	1929	...	...	93%
1930	...	...	95%	1930	...	...	89%
1931	...	...	100%	1931	...	...	92%
1932	...	...	No camp.	1932	...	...	No camp.
1933	...	...	98%	1933	...	...	98%

The following letter, dated the 2nd November, was received from the G.O.C. in C., Northern Command, Lieutenant-General Sir Charles H. Harington, G.B.E., K.C.B., D.S.O. :—

> "May I be permitted to send my warmest congratulations to you and all ranks of the unit under your command on the place reached in the percentages of attendance in camp this year.
>
> It is indeed a matter of satisfaction to me that, of 21 units which averaged 90%, 14 should be in the Northern Command.
>
> I am grateful to you for your hard work and I wish the unit continued success."

On the 31st October 1925 Colonel Robinson handed over the command to Major Firmin and was transferred to the Territorial Army Reserve of Officers.[1]

The great value, and the extent, of the services rendered by Colonel Robinson will have been apparent from preceding chapters. He brought to the Unit exceptional ability as an organiser and administrator. During the Great War, no duty which fell to his lot and no task which he was asked to undertake, however onerous, proved too great for him to accomplish and it was largely due to his foresight and initiative that the Unit expanded to so remarkable a degree and took such a prominent part in the searchlight work of the army, both at home and abroad.

On his relinquishing the command of the Unit, Colonel Robinson was presented by the officers, past and present, who had served under him, with his portrait in oils by Mr. Cowan Dobson, R.B.A. The presentation was made by the Honorary Colonel, the Duke of Northumberland, on the occasion of the annual reunion dinner on the 13th November 1926.

The winter session of the rifle club 1925/26 was notable as the first occasion on which the Unit was successful in one of the miniature rifle competitions organised by the Northumberland County Rifle Association, winning the Miniature Chipchase Cup with a score of 711 points, the runners-up being the City of Newcastle Rifle Club with 681.

During February 1926 Captain B. T. C. Freeland, R.E.,[2] assistant instructor at the School of A.A. Defence at Biggin Hill, came north and gave a series of lectures to officers and N.C.Os on anti-aircraft training which were well attended and proved of particular interest and value.

In His Majesty's Birthday Honours this year, the M.B.E. (Military Division) was awarded to Coy.-Sjt.-Major J. Chambers, who was invested by the King at Buckingham Palace on July 12th.

---

[1] Colonel Robinson was promoted Commander of the Order of the British Empire (Military Division) in the Birthday Honours, 1925, the announcement being made in the "London Gazette," dated the 3rd June, 1925.

[2] Born 8/6/93, Capt. 5/2/21, Major 5/5/32.

LIEUTENANT-COLONEL N. H. FIRMIN, O.B.E., T.D.,
Commanding Officer, 1925-1929.

To face page 232]

As a result of the general strike, which took place in the spring of this year, all camps of Territorial Army Units involving long rail journeys and due to assemble up to July 15th, which included the annual camps of the Tyne Electrical Engineers, were cancelled. To compensate in some measure for the resulting loss of training, rather larger grants than usual were made to cover additional week-end camps and other forms of local training. In this manner, it was possible to hold two week-end camps at Clifford's Fort and a week's camp at Ponteland. At the Ponteland camp, A.A. searchlight work was carried out, aircraft co-operation being provided by two R.A.F. machines from Folkstone, piloted by Flight-Lieutenant D. E. Don[1] and Flight-Serjeant Pennicott. Three A.A. lights were manned and Captain Freeland was attached for the duration of the training, the camp being visited by Colonel Biddulph, Chief Engineer, Northern Command, and by Captain K. M. Loch, M.C., R.A., Staff Officer to the Commander, A.A. Ground Troops.

In March 1927 the establishment for permanent staff instructors was increased from two to three, of whom two were attached to the A.A. Company and one to the E.L. Company.

This year was the first in which Lewis gun training was given, the A.A. Company being equipped with one servicable gun per section and one D.P. for instructional purposes. The annual course, which included practice on the open range, was restricted to sixteen men.[2]

Annual training took place this year from the 18th June to the 2nd July, No. 1 E.L. Company remaining at Tynemouth under Captain Fairbairn-Crawford, while the 307th A.A. Company camped for the first time at Manston aerodrome, near Margate, which later became the established camping ground of the A.A. searchlight units of the Territorial Army Air Defence Formations.

The year 1927 was a notable one for the marksmen of the unit which, representing but one-tenth of the total strength of Territorial troops in the County of Northumberland, succeeded in winning three trophies, twenty-six medals and about 30% of the prize money offered at the annual meeting of the Northumberland Rifle Association held at Ponteland during the Whitsun holidays. Later in the year, the Unit won the Ladies' Trophy by a margin of twelve points which constituted their ninth successive victory in this competition. At Bisley, in July, Major Firmin was selected to captain the Territorial Army Eight in the United Services competition and he and L/Serjeant T. Wilkinson qualified to shoot among the fifty territorials to compete for the King's Medal.

At a week-end camp in September 1927 a flight of four aeroplanes from No. 17 (Fighter) Squadron, Upavon, co-operated with the A.A. Company. On this occasion the Company manned for the first time five permanent searchlight stations which had recently been established for training purposes at Wallsend, Balkwell, Tynemouth, Whitley Bay and Shiremoor.

---

[1] Pilot to H.R.H. The Prince of Wales.
[2] The full A.A. course was not fired before 1930, as the correct appliances were not available at the range before that year. The inter-section Lewis gun competition was first introduced in the same year, when an individual event with this weapon was also added to the programme of the regimental rifle meeting.

233

On the 10th April 1928 Serjeant F. W. J. Knight, permanent staff instructor to the E.L. Company, rescued a small boy who was in imminent danger of drowning in the Tyne. Serjeant Knight, who was only a moderate swimmer, dived into the river from the Fish Quay, North Shields, having only time to remove his uniform jacket before doing so. He afterwards applied artificial respiration and, in about twelve minutes, restored the child to consciousness. The G.O.C.-in-C., Northern Command, directed that this incident be recorded in the documents of Serjeant Knight as an act of gallantry and the Royal Humane Society awarded him their life-saving certificate on vellum.

It will have been gathered from earlier chapters that Clifford's Fort had long been unsatisfactory as a headquarters owing to its position and to the very limited accommodation available, although the congestion had been somewhat alleviated since 1920 by the use of the small drill hall at Whitley Bay. The acquisition of new headquarters of adequate proportions had been brought within the bounds of possibility by the desire of the Tynemouth Corporation to obtain possession of the site at Clifford's Fort for developments in connection with the North Shields fish quay. Consideration was given to various sites at Tynemouth and Newcastle and, after protracted negotiations, a settlement was reached by which the Corporation transferred to the War Office a site of approximately 8,530 sq. yds. adjoining Tynemouth railway station, together with the sum of £12,000, in exchange for Clifford's Fort. It was intended that the 16th (Fortress) Company, R.E., should also be housed on the new site and the land was divided into two parts according to the contemplated requirements of the Regular and Territorial personnel. Shortly afterwards, however, accommodation became available for the 16th Company in Tynemouth Castle, and the only Regular Army buildings erected on the new site were the R.E., workshops.

Plans for the Tyne Electrical Engineers' new headquarters were put in hand by Major D. Hill, M.C., who held the appointment of architect to the Northumberland County Territorial Army Association, and the first stone was laid in June 1927. The main building includes a drill hall measuring 80 feet by 40 feet, offices, stores, instruction rooms, engine bays, miniature rifle range, officers' and N.C.Os' messes, sappers' recreation room and a quarter for a caretaker. A quarter for a permanent staff instructor and a garage for two lorries adjoin the main building. The buildings are designed in early Italian Rennaissance style of brick, with dressings of pink stone led from Byers Hill quarry, Northumberland, and the roof is covered with Kentmere sea-grey slates. The exterior walls are roughcast with waterproof white cement. The carved panel over the main entrance, incorporating the crest of the Tyne Electrical Engineers, flanked by winged figures of Fame and Glory, is from the chisel of the Scottish sculptor, Mr. Alexander Proudfoot, A.R.S.A. The carved stone tablet over the main gateway into Clifford's Fort was, by kind permission of the Corporation, removed and erected inside the drill hall at Tynemouth. The *Architect and Building News* of October 19th 1928, commenting on the new building, accorded special merit

NEW HEADQUARTERS, TYNEMOUTH.
1928.

to Major Hill's design in that it represented a departure from the erroneous late 19th century convention which was being applied to military architecture and symbolised a return to the older and much better tradition. " It is difficult to see," writes this critic, " how they could have made more skilful use of the architectural opportunity which here presented itself."

The official " marching-out " from Clifford's Fort and entry into possession of the new headquarters took place, during a week-end camp, on Saturday 12th May 1928. At 3.0 p.m. the last parade fell in at Clifford's Fort. The Commanding Officer, in addressing the men and a large crowd of ex-members of the Unit who were gathered to witness the event, referred to the sentimental association which the Unit would always retain for the Fort and reminded his listeners that it had been within those walls that the Tyne Electrical Engineers were cradled and brought through childhood to sturdy maturity, and that the *esprit-de-corps* which was so marked a feature of the Unit had been acquired. The " Last Post " was then sounded, the flag slowly lowered and the Unit marched out. With this simple ceremony Clifford's Fort was evacuated after 256 years as a military station. On arrival at the new headquarters " Reveille " was sounded and the flag hoisted.

Annual training in 1928, from the 16th June to the 1st July, found the E.L. Company, under Lieutenant Leeson, manning coast defence lights at Kinghorn, on the north side of the Firth of Forth. The A.A. Company went to Kenley aerodrome in Surrey, which was under the command of Wing Commander R. P. Willock. Co-operation was afforded by the Fighter squadrons stationed at the aerodrome and also by machines of No. 7 (Bomber) Squadron which flew over for each night practice from Worthydown. The Unit owes a debt of gratitude to Wing Commander C. F. A. Portal, D.S.O., M.C., who commanded the Bomber Squadron, for the special arrangements he made to enable officers to fly at night for observation purposes.

The Company was inspected in camp by the newly appointed Commander of the Air Defence Formations, Major-General C. A. Ker, C.B., C.M.G., C.B.E., D.S.O., and later, by the Director-General of the Territorial Army, Lieutenant-General Sir Reginald B. Stephens, K.C.B., C.M.G., who " expressed his pleasure at the soldierly bearing of the personnel of the Unit and congratulated all ranks on their excellent turn-out on parade and on the high standard attained during training."

The new headquarters were formally opened by Major-General C. A. Ker on Saturday July 21st in the presence of a large and representative gathering. After inspecting the Guard of Honour which received him, the General was welcomed by Colonel Sir Francis D. Blake, Bt., C.B., D.L., chairman of the Northumberland Territorial Association, and was presented by Major Hill, the architect, with a silver paper-weight fashioned to represent the crest of the Tyne Electrical Engineers. The General was then handed a pair of scissors (for which, in obedience to superstition, he paid one penny !) and cutting a ribbon of Royal Engineers' colours stretched across the centre engine-bay he declared the premises open.

When on October 10th H.M. the King visited Newcastle-upon-Tyne in order to open the New Bridge, the streets were lined by detachments of R.N.V.R. and various units of the Territorial Army in the County of Northumberland. The Tyne Electrical Engineers' detachment, numbering approximately 100, was stationed in Northumberland Street. The following letter, from Northern Command, was subsequently received by the Units who had provided detachments :—

> " The General Officer Commanding-in-Chief has been commanded to convey to all concerned His Majesty the King's appreciation of the military arrangements made for his visit to Newcastle upon Tyne and Durham on the 10th October.
>
> His Majesty was very favourably impressed by the smart appearance and soldierly bearing both of the troops lining the streets and those forming the Guards of Honour.
>
> The General Officer Commanding-in-Chief congratulates the troops who took part in this parade and the officers who commanded them on receiving this praise from His Majesty. Great credit is due to all concerned."

It was in this year that the bandsmen of the Unit were provided with full dress uniform for the first time since pre-war days.

On the 19th March 1929 Captain Sergeant, having attained the age limit, relinquished his commission and, the appointment of quartermaster to the Tyne Electrical Engineers having been officially discontinued, he was not replaced.

The Unit was singularly fortunate in being individually represented on the occasion of the dedication of the Memorial Church at Ypres on Sunday 24th March, Captain Woodward having been selected by ballot as the Territorial officer included among the five representatives of the Royal Engineers who assisted Major-General R. N. Harvey, C.B., C.M.G., D.S.O. (deputising for General Sir Bindon Blood, G.C.B., G.C.V.O.), at the presentation of Communion Plate on behalf of all branches of the Corps.

On the 1st June the adjutancy of the Unit was taken over from Major McSweeney by Captain L. E. C. M. Perowne, R.E.,[1] who was posted from " B " Company of the 1st A.A. Battalion, R.E., at Blackdown. Captain Perowne was gazetted adjutant, and not acting-adjutant, this being the first time that the Unit, since its formation, obtained the undivided services of a regular officer.

In pursuance of the principle of training coast defence E.L. personnel in A.A. searchlight work, and of an understanding that the two companies of the Tyne Electrical Engineers might be permitted once in every three years to carry

---

[1] Born 11/6/02, 2nd Lt. 31/1/23, Lt. 31/1/25, Temp. Capt. in T.A. 1/6/29.

MANSTON CAMP, 1929.
OFFICERS.

Back Row, Left to Right—2nd Lieut. E. M. Robinson, Lieut. E. J. Parnall, Lieut. D. A. Williamson, Lieut. A. S. Richards, Lieut. H. H. Mullens.

Centre Row, Left to Right—Lieut. J. L. Batey, Lieut. H. Ash, R.A.O.C. (O.M.E.), Lieut. R. P. Winter, M.C., Captain and Qr.-Mr. J. Harlin (L.E.E.), Captain C. K. Davies, R.E. (S.A.A.D.), Lieut. B. H. Leeson, Captain H. Sherlock, T.D.

Seated, Left to Right—Captain R. Sharp, O.B.E., Captain L. E. C. M. Perowne, R.E. (Adjt.), Bt.-Lieut.-Col. N. H. Firmin, O.B.E., T.D. (Cmdg.), Maj.-Gen. C. A. Ker, C.B., C.M.G., C.B.E., D.S.O. (Cmdr. A.D.F.), Colonel T. T. Grove, C.M.G., D.S.O. (Chief Engr., A.D.F.), Lieut.-Col. T. C. Newton, D.S.O., O.B.E. (Staff Offr. A.D.F.), Major E. H. E. Woodward, M.C.

In front—2nd Lieut. J. Boyle, " Searchlight Champus," 2nd Lieut. E. H. Sadler.

ANTI-AIRCRAFT SEARCHLIGHT DETACHMENT.

To face page 236]

out their annual training in camp together, the whole Unit entrained on Saturday, 22nd June 1929 for Manston. No one who attended this camp will easily forget the unexpectedly long and exhausting march which awaited the Unit on its arrival at Margate at 7.0 a.m. the following day. Owing to an unfortunately misplaced confidence in the topographical knowledge of the bandmaster, the column arrived almost on the outskirts of Broadstairs before it was discovered that a wrong turning had been taken. Eventually, a cross-country route brought the hungry and exhausted companies into the camp, at the opposite end to that at which the new Chief Engineer, Colonel T. T. Grove, C.M.G., D.S.O.,[1] and the quarter-guard waited anxiously to welcome them!

At this camp, sixteen searchlights were manned, one of which was operated by the E.L. Company, and air-craft co-operation was furnished by No. 9 (Bomber) Squadron, commanded by Wing-Commander W. V. Strugnell, M.C.

On the 1st November Lt.-Col. Firmin, who had just completed four years in command, left the district to take up an appointment in London and transferred to the Reserve. He was succeeded by Major Woodward as Commanding Officer.

During the period of Lt.-Col. Firmin's command noticeable progress was made in the technical training of the Unit, particularly on the anti-aircraft side. When this officer took over command, the plant available for practice purposes was very limited and methods of operation and tactical handling were far from standardised. Gradually, however, more plant became available and definite instructions as to tactics and operation were received and, before he vacated the command, the technical training had been brought into line with post war conditions. The Unit also owes much to Lt.-Col. Firmin for his great interest in rifle shooting. A prominent markman himself, he communicated his enthusiasm to others and the successes of the Unit in this direction, both during and since his command, have been very largely due to him.

In 1930 an innovation was made inasmuch as the periods of the camps of the two companies were not simultaneous, but overlapped by one week, a system which afforded the Commanding Officer and the adjutant a better opportunity of being present at both camps. The E.L. Company remained at Tynemouth under the command of Major Sherlock, and camped on the W.D. land adjoining the headquarters. They were inspected by Colonel P. E. Hodgson, D.S.O.,[2] Chief Engineer, Northern Command. The A.A. Company again camped at Manston where they manned 14 lights and operated in conjunction with No. 9 (Bomber) Squadron, R.A.F. During this camp a lorry convoy of 4 officers and 72 other ranks visited Chatham, where they were shown over the Electrical School at the S.M.E. and the R.E. Museum.

Shortly after the completion of annual training, at the request of Northern Command, a detachment of the Unit under the command of Lieut. D. A.

---

[1] Born 23/12/79, 2nd Lt. 23/6/98, Lt. 1/4/01, Capt. 23/6/07, Major 30/1/15, Bt. Lt. Col. 1/1/18, Lt. Col. 12/8/20, Col. 1/1/22.

[2] Born 26/9/74, first Commission 25/7/93, Col. 3/6/22, retired 1/5/31.

Williamson gave a demonstration in anti-aircraft searchlight work to the contingents of the Junior Division of the Officers Training Corps who were in camp at Strensall. In a letter subsequently received by the Commanding Officer, the G.O.C.-in.C, Northern Command, expressed his appreciation of the work done by all ranks of the detachment and stated that their " presence in camp was of the highest value and interest to the cadets, giving them an insight into what is becoming an increasingly important part of warfare."[1]

On Saturday, August 23rd, the Duke of Northumberland, who had been ill for some time, died at his home in London in his fifty-first year. The great esteem in which the Duke was held may be judged by the following extract from a speech made by Colonel Robinson when presiding at a meeting of the Territorial Army Association of the County of Northumberland on the 5th September 1930 :—

> " The Duke took a great interest in the Territorial Units of the County, especially those of which he was Honorary Colonel. My experience, while in command of the Tyne Electrical Engineers, was that he was at all times ready and willing to give all possible assistance and to further the interests of the unit in every way in his power, often at personal inconvenience. He was a great nobleman, an able president, a gallant soldier and a wise and fearless counsellor."

The acceptance by Colonel Robinson of the post of Honorary Colonel, in succession to the late Duke, met with the enthusiastic approval of all ranks, serving and retired, and was duly promulgated in the *London Gazette* of the 24th October 1930.

In 1931, No. 1 E.L. Company, under Major Sherlock, camped at Kinghorn with the Tynemouth Heavy Brigade, R.A. from the 21st June to the 5th July. Here, at full strength, they manned the searchlights covering the north channel of the Firth of Forth between the island of Inchkeith and the mainland and, in addition, trained to some extent in elementary fieldworks. This Company was inspected by the Chief Engineer, Northern Command, Colonel H. N. North, D.S.O.,[2] who expressed himself well satisfied with its efficiency both at works and on parade.

In order to participate in the air defence manoeuvres, the 307th Company did not go to camp until the 11th July when it moved to Biggin Hill, near Westerham, Kent. The strength in camp was 10 officers and 198 other ranks and fourteen lights, one section control and a sector control were manned, the controls being furnished with double reliefs. The Company was inspected on works by Major-General H. F. Salt, C.M.G., D.S.O., (who had succeeded Major-General Ker as Commander, Air Defence Formations, T.A.), and Colonel Grove. The following

---

[1] C.R.N.C. 51887/G. dated 5/8/30.
[2] Born 11/6/83, 2nd Lt. 21/12/00, Lt. 21/12/03, Capt. 21/12/10, Major 2/11/16, Bt. Lt. Col. 3/6/18, Lt. Col. 10/7/26, Col. 3/6/22, Brig. 9/10/33.

COAST DEFENCE SEARCHLIGHTS ON THE FIRTH OF FORTH.
KINGHORN CAMP, 1931.

To face page 238]

letter,[1] with regard to the camp, was subsequently received by the Commanding Officer :—

> " The Commander, Air Defence Formations, T.A., has been pleased to see that a schedule of claims preferred in connection with sites occupied at this year's annual training of the A.A. Searchlight Units, T.A., contains no claim in connection with those occupied by your 307th Company.
>
> He considers this highly creditable to the Unit and appreciates it as a testimony of its good discipline and the good conduct of your detachments.
>
> <div align="right">(<i>Signed</i>) T. T. GROVE,
Colonel,
Chief Engineer,
Air Defence Formations, T.A."</div>

The year 1932 will long be remembered as the year of national economy, during which, following the financial crisis of the preceding August, the hastily formed National Government strove, by every means in its power, to curtail expenditure and to balance the budget. As one such measure it had been decided to cancel, for this year only, the annual camps of the Territorial Army and this decision had been made public under the title of the *Territorial Army (Annual Training) Order*, 1932 in the *London Gazette* of December 18th 1931.

In place of the usual fortnight's camp, a local camp of one week's duration was held at Tynemouth during Race week (June 20th—26th) which was attended by 10 officers and 108 other ranks for the whole period, and by a further 1 officer and 58 other ranks for the week-end only. Both coast defence and A.A. search-light practice were obtained, the latter in co-operation with aircraft of No. 36 (A.C.) Squadron, R.A.F., who established a temporary night landing-ground at Usworth, Co. Durham. The fine weather which prevailed throughout the week enabled a programme of training to be carried out which, although it did not entirely compensate for the loss of the normal camp, sufficed to keep the Unit together during this difficult year and to provide a climax to the preliminary training period.

If, however, the Territorial Army was to go short of training, it was not allowed to suffer from lack of interest displayed by higher authority. This year, the Unit was visited on five occasions, four of them falling during the short week's camp at headquarters. Thus, on the 21st June arrived the Chief Engineer, Air Defence Formations, (Colonel Grove) ; on the 23rd, the Chief Engineer, Northern Command, (Colonel North) inspected the E.L. Company, and on the same date the Unit was visited at work by the General Officer-Commanding-in-Chief, Northern Command, Lieutenant-General The Hon. Sir J. Francis Gathorne-Hardy, K.C.B., C.M.G., D.S.O. On Friday, 24th June, the Commander Air Defence Formations (Major-General Salt) paid a visit of inspection to the 307th A.A. Company. Finally, on the 13th August, the Inspector of Royal Engineers, Major-General H. L. Pritchard, C.B., C.M.G., D.S.O., inspected a detachment

---

[1] A.D.F. No. 43/883 (C.E.) dated 30/10/31.

of 6 officers and 87 other ranks from both companies on the parade ground at the headquarters, Tynemouth, and subsequently witnessed representative parties at work with A.A. and C.D. searchlights. In his report, the Inspector observed that, " whenever the War Office desired more Fortress or Anti-Aircraft Engineers, Newcastle, Tynemouth, Manchester and Liverpool are places where they can be obtained."

General Pritchard's visit was notable as the first occasion since volunteer days on which the Unit paraded with pipers, the enlistment of two sappers within the existing establishment of the E.L. Company being authorised by the War Office for this purpose, with permission for the wearing of the appropriate dress, provided that no expense fell on the public.[1] The pattern of uniform previously used by the pipers of the Tyne Division, Royal Engineers, Volunteers, was copied as closely as possible, and the approval of General Sir Charles Fergusson of Kilkerran, Bt., G.C.M.G., K.C.B., M.V.O., was sought and obtained for the wearing once more of the tartan of the clan of which he was the Head. The only appreciable changes in the uniform from the original pattern lay in the details of the various badges, the designs of which were now modified to conform to the modern title of the Unit and to incorporate its present crest.

In September of this year (1932), *H.M.S. Malaya* visited the Tyne and the assistance of the Tyne Electrical Engineers was sought by the Tynemouth Corporation to flood-light Tynemouth Priory in celebration of the event. This eminently peaceful project was carried out by a detachment of the Unit, employing two searchlights and a number of small floodlight-projectors, and operated by them nightly from the 16th to the 21st September inclusive.[2]

Decisions with regard to the coast defence of the County had recently been reached which resulted in a reduction of the Regular Fortress Companies, R.E., and the handing over of the full responsibility for coast defence to the Territorial Army. This change took effect on the 1st October when the 16th (Fortress) Company, R.E., at Tynemouth, was disbanded and its personnel absorbed into the 4th (Fortress) Company, R.E., at Gosport. The organisation and establishment of No. 1 E.L. Company now underwent certain alterations. To enable it fully to carry out the functions required of a combined fortress works and electric light company, a works section was added to it, bringing the war establishment up to 4 officers and 72 other ranks as against the existing figures of 4 officers and 40 other ranks. A restricted recruiting establishment was, however, introduced, limiting the strength to 4 officers and 50 other ranks, including a C.Q.M.S.[3] The establishment of permanent staff for this company was at the same time increased to two.

---

[1] Authority : W.O.L. Engrs./2961 (M.G.O.7.B.) dated 21st June 1932.

[2] The local Press, commenting upon this event, observed that this was not the first occasion on which the Priory had been flood-lighted as in the year 1662 a coal fire was burned in the Priory garden, which illuminated the structure and served as a landmark to mariners. This coal fire was superceded by an oil light on March 11th 1802 and was discontinued on August 31st 1898 when the lighthouse at St. Mary's Island was established.

[3] Excluding 2 plant attendants, enlisted into the Unit, but retained " supernumerary " and not undergoing any training.

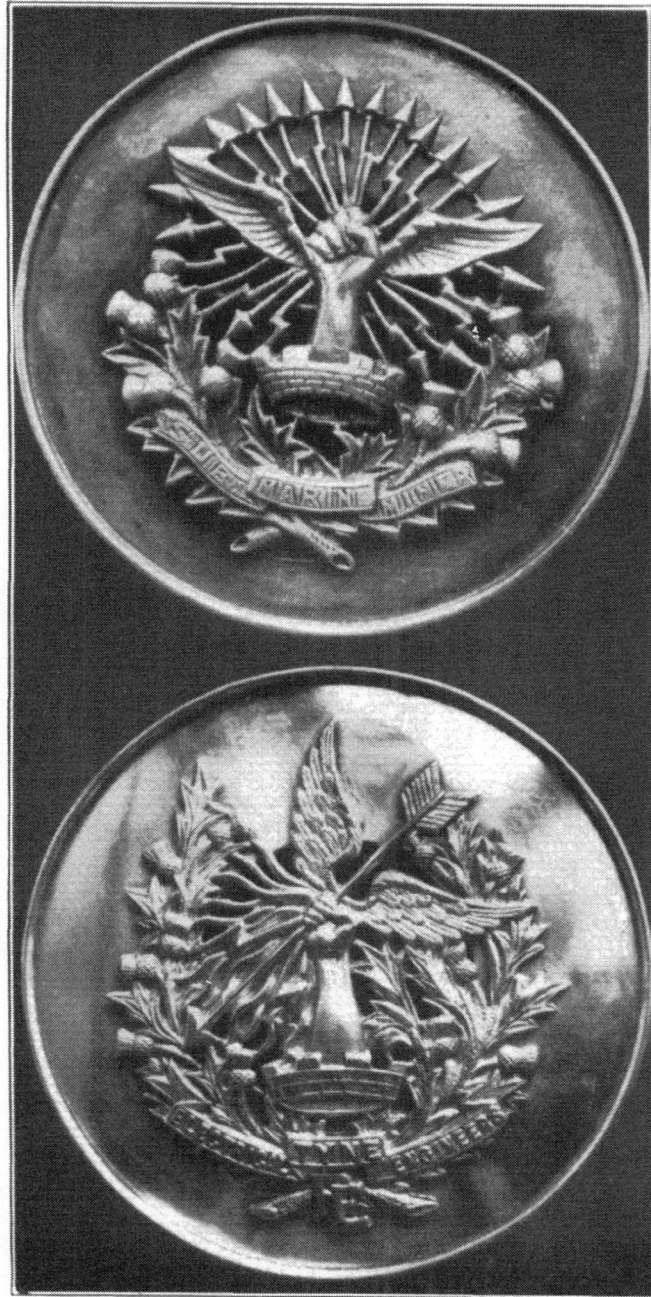

PIPERS' SHOULDER-BROOCHES.

Above—That of 1900, incorporating the crest of the Royal Engineers,
Submarine Miners.
Below—That of 1932, incorporating the new crest of the Tyne Electrical
Engineers.

To face page 240]

Consequent upon the re-organisation of the coast defences, the title of No. 1 E.L. Company, Tyne Electrical (Fortress) R.E., was altered to " No. 1 (E.L. and Works) Company, Tyne Electrical Engineers, R.E."[1] with effect from October 1st 1932. Thus the official title of each company once more included the pre-war name of the Unit, " Tyne Electrical Engineers," and the opportunity was taken to ask for the grouping of the two companies under one heading in the army list. The granting of this request removed the confusion as to the identity of the Unit which had arisen out of the previously existing difference in designation, and resulted in its re-appearance in the Army List for July, 1933, under the old heading " Electrical Engineers " as follows :—[2]

ROYAL ENGINEERS.
II(b) ELECTRICAL ENGINEERS.
Tyne Electrical Engineers. R.E.
No. 1 (E.L. & Works) Company.
307th (Tyne) A.A.S.L. Company.

Annual training in camp this year followed very normal lines, No. 1 (E.L. & Works) Company camping in the grounds of the headquarters at Tynemouth from the 10th to the 24th June. During this camp inspections were made by the G.O.C.-in-C., Northern Command, (Lieut.-General the Hon. Sir Francis Gathorne-Hardy) and by the C.E., Northern Command, (Col. North). The Company was also visited on several occasions by Lt.-Col. J. N. Cash, M.C.[3] C.R.E., Northumbrian area, who had recently been made responsible to the C.E., Northern Command, for the training of the fortress companies in the Command.

The 307th A.A. Company, leaving Tynemouth on Saturday 17th June, went again to Manston, where, with a strength of 10 officers and 186 other ranks, they manned 12 searchlights, two visual plotting stations and the operations room and co-operated with a flight of No. 10 (Bomber) Squadron, R.A.F., under Flight Lieutenant D. Dickson, by night, and No. 2 (A.C.) Squadron R.A.F., under Sqdr.-Ldr. P. F. Fullard, D.S.O., M.C., A.F.C., by day. The work in camp was watched throughout by Colonel A. B. Ogle, O.B.E.,[4] who had succeeded Colonel Grove as Chief Engineer, Air Defence Formations, and the Company was inspected on parade by Major-General Salt and returned to the north on Saturday July 1st.

Immediately after the conclusion of the Manston camp, a detachment of 32 other ranks, selected from volunteers from both companies, was dispatched to York under Lieut. Boyle, to operate the searchlights for the Northern Command Tattoo. Although on former occasions some of the Unit's technical equipment had been used and, in 1932, it had provided two drivers for its P.E. lorries for the Tattoo held at York and Leeds, the duty of operating the lights had hitherto been assigned to the 16th (Fortress) Company, R.E. This year, however, consequent upon the disbandment of the latter company, the Tyne Electrical

---

[1] A.O. 176/1932.
[2] Authority : W.O.L. 26/A.L./376 (C.3.A.L.) dated 24/6/33.
[3] Born 28/8/87, 2nd Lt. 18/12/07, Lt. 2/5/10, Capt. 30/10/14, Bt. Major 3/6/18, Major 22/10/24, Lt. Col. 16/5/32.
[4] Born 12/4/83, 2nd Lt. 21/12/01, Lt. 3/8/04, Capt. 21/12/12, Major 21/12/16, Lt. Col. 1/5/28, Col. 1/5/32.

Q

Engineers were called upon to undertake the work. The detachment operated 11 lorries and engines and 15 searchlights under the supervision of Captain F. J. R. Heath, R.E.,[1] the officer in charge of R.E. works and lighting. At the conclusion of the Tattoo, Lieut. Boyle received the following letter from the Chief Engineer (Colonel North) :—

> " I wish to thank you and those under your command for the invaluable work done in connection with the Tattoo.
>
> The hard work involved in the assembly and erection of the plant, the technical difficulties caused by engine failures and the long hours of continuous running were met cheerfully and efficiently. The lighting itself, on which the Tattoo depends, has never been better carried out.
>
> In all respects your detachments maintained to the fullest degree the very high reputation of the Tyne Electrical Engineers, as one of the most efficient units of His Majesty's Corps of Royal Engineers."

With this gratifying reference to the latest addition to the Unit's varied activities, this record of the Tyne Electrical Engineers must be brought to a close. In the foregoing pages we have traced the growth of the Unit from small beginnings to the enormous dimensions it had assumed by the end of the Great War and thence pursued its history, through the reductions of the age of disarmament, to the present day. We have trained with it in drill hall and camp, played with it on football ground and cricket pitch and fought beside it in the stern reality of war ; we have laughed with it in its humours and mourned with it in its losses ; worked with it on sea and land and in the air and we have witnessed its participation in the charitable pageantry of peace. Throughout the whole length of the variegated history of the Unit we have observed the building up and maintenance of an exceptional reputation for readiness, the highest technical standard and a military proficiency beyond the average. The readiness, born in the days of its first Commanding Officer, to undertake any and every duty required of it, even though not directly connected with its work ; the exceptionally high standards, since unswervingly maintained, which were so carefully fostered during the decade which preceded the Great War, and the remarkable successes with the rifle, attained particularly in the years that followed it ; these are probably the most outstanding characteristics of a Unit of which Sir Francis Blake was led to say, at the opening of the new Headquarters on the 21st July, 1928, " Whether in the depths of the sea, or in the height of the clouds, it is equally great and equally efficient."

Such, then, is the tradition and this the record with which the Unit embarks upon the fiftieth year of its existence.

---

[1] Born 9/3/00, 2nd Lt. 18/12/19, Lt. 18/12/21, Capt. 18/12/30.

LIEUTENANT-COLONEL E. H. E. WOODWARD, M.C., T.D., M.I.E.E.
Commanding Officer, 1929.

To face page 242]

APPENDIX I.

———

CHRONOLOGICAL TABLE OF ANNUAL CAMPS.

CHRONOLOGICAL TABLE OF ANNUAL CAMP.

Year.	Camps.	In Command.
1888	Willington Quay Shipyard	Major Wm. Johnson
1889	Tynemouth, Spanish Battery	Major Wm. Johnson
1890	North Shields, Clifford's Fort...	Major Wm. Johnson
1891	North Shields, Preston Avenue	Major Wm. Johnson
1892	Whitley Bay, Rockcliffe Cricket Ground	Major Wm. Johnson
1893 to 1899	Clifford's Fort, Preston Avenue, or Rockcliffe ...	Major Wm. Johnson
1900	North Shields, Clifford's Fort	Hon. Lt. Col. Wm. Johnson, V.D.
	Sheerness (Thames Defences)	Lt. G. A. Bruce
1901	North Shields, Clifford's Fort (1 week)	Hon. Lt. Col. Wm. Johnson, V.D.
	Hull (Humber Defences) Weymouth (Portland Defences) } (1 week) Portsmouth	
1902	Portsmouth	Hon. Lt. Col. Wm. Johnson, V.D.
1903	Stokes Bay, Portsmouth	Lt. Col. Wm. Johnson, V.D.
	Weymouth (Portland Defences)	Captain A. Blackburn
1904	Weymouth (Portland Defences) ...	Captain C. Johnson
	Sheerness (Thames Defences)	Captain G. A. Bruce
	Portsmouth and Isle of Wight (Totland } (1 week) Bay)	Major F. G. Scott
	North Shields, Preston Avenue ...	Lt. Col. Wm. Johnson, V.D.
1905	Gosport (Fort Monckton) and Isle of Wight... ...	Lt. Col. Wm. Johnson, C.B., V.D.
	Sheerness (Thames Defences)	Captain G. A. Bruce
	Weymouth (Portland Defences)	Captain C. Johnson
1906	Gosport (Fort Monckton)	Lt. Col. Wm. Johnson, C.B., V.D.
	Salisbury Plain (Detachment only)	2nd Lt. K. S. M. Scott
1907	Gosport (Fort Monckton) and Isle of Wight (1 week) North Shields, Clifford's Fort (1 week) ... }	Lt. Col. Wm. Johnson, C.B., V.D.
1908	Gosport (Fort Monckton) and Isle of Wight... ...	Lt. Col. Wm. Johnson, C.B., V.D.
1909	Durham Fortress R.E. (T.F.), Tynemouth	Captain C. Johnson
	Tyne Division Electrical Engineers, Southsea Castle...	Lt. Col. Wm. Johnson, C.B., V.D.
1910 1911 }	Northumberland Fortress R.E. (T.F.), Tynemouth...	Captain C. Johnson
	Tyne Electrical Engineers, Gosport (Fort Monckton)	Lt. Col. Wm. Johnson, C.B., V.D.
1912	Gosport (Fort Monckton)	Lt. Col. F. G. Scott, V.D.
1913	No. 1, 2 and 4 Coys. Gosport (Fort Monckton) ...	Lt. Col. F. G. Scott, V.D.
	No. 3 Coy., Tynemouth (Monkhouse Farm) ...	Captain G. A. Bruce

CHRONOLOGICAL TABLE OF ANNUAL CAMP.—*Continued*

Year.	Camps.	In Command.
1914	No. 1 Coy., Tynemouth (Monkhouse Farm)	Captain A. K. Tasker
	Nos. 2, 3 and 4 Coys., Gosport (Fort Monckton) ...	Lt. Col. F. G. Scott, V.D.
1921	Hartley (Roberts Battery)	Lt. Col. E. Robinson, O.B.E., T.D.
1922 ⎫ 1923 ⎭	Monkseaton (Beverley Estate)...	Lt. Col. E. Robinson, O.B.E., T.D., D.L.
1924 ⎫ 1925 ⎭	Gosport (Haslar Barracks)	Bt. Col. E. Robinson, C.B.E., T.D., D.L.
1926	Ponteland Rifle Range	Major N. H. Firmin, O.B.E., T.D.
1927	No. 1 Coy., Kinghorn, Fife, N.B. (Pettycur)... ...	Lt. B. H. Leeson
	307th Coy., Manston Aerodrome, Kent	Major N. H. Firmin, O.B.E., T.D.
1928	No. 1 Coy., North Shields (Clifford's Fort) ...	Major I. F. Fairbairn-Crawford, T.D.
	307th Coy., Kenley Aerodrome, Surrey	Bt. Lt. Col. N. H. Firmin, O.B.E., T.D.
1929	Manston Aerodrome, Kent	Bt. Lt. Col. N. H. Firmin, O.B.E., T.D.
1930	No. 1 Coy., Tynemouth (Headquarters)	Major H. Sherlock, T.D.
	307th Coy., Manston Aerodrome, Kent	Major E. H. E. Woodward, M.C.
1931	No. 1 Coy., Kinghorn, Fife, N.B. (Pettycur) ...	Major H. Sherlock, T.D.
	307th Coy., Biggin Hill (S.A.A.D.), Kent ...	Major E. H. E. Woodward, M.C., T.D.
1932 (Voluntary Camp)	Tynemouth (Headquarters)	Major E. H. E. Woodward, M.C., T.D.
1933	No. 1 Coy., Tynemouth (Headquarters)	Major H. Sherlock, T.D.
	307th Coy., Manston Aerodrome, Kent	Bt. Lt. Col. E. H. E. Woodward M.C., T.D.

Appendix II.

COMPLETE LIST OF OFFICERS HOLDING COMMISSIONS IN, OR ATTACHED TO, THE UNIT.

A. ... ... General List.

B. ... ... Honorary Colonels.

C. ... . Commanding Officers.

D. ... ... Adjutants.

E. ... ... Quartermasters.

F. ... ... Medical Officers.

G. ... ... Chaplains.

INDEX TO ROLL OF OFFICERS.

INDEX TO ROLL OF OFFICERS.—*Continued.*

NOTES AND ABBREVIATIONS.

(*a*) This roll does not include the names of officers of the original submarine mining company of the 1st Newcastle and Durham, Royal Engineers, Volunteers, who did not subsequently transfer to the Tyne Division, Royal Engineers, Volunteers, Submarine Miners.

(*b*) List "A." contains only the names of officers actually having held commissions in the Unit and does not include attached officers. This list is arranged in strict order of seniority of first commission irrespective of whether an officer was first commissioned into this Unit or another.

(*c*) The dates of temporary rank are not given where substantive ranks were subsequently ante-dated prior to the date of temporary rank.

(*d*) In order to avoid complicating the list unnecessarily, transfers between the various classes of the Reserve are not indicated.

(*e*) The following abbreviations are used :—

A/	Acting rank.
a/d.	ante-dated.
A.L.	Army list.
Bt/	Brevet rank.
H/	Honorary rank.
L.E.E.	London Electrical Engineers.
N.D.V.R.E.	1st Newcastle and Durham (Volunteers) R.E.
N.F.	Northumberland Fusiliers.
R.M.S.M.	Royal Marines Submarine Miners.
S/	Surgeon.
T/	Temporary rank.
T.A.R.O.	Territorial Army, Reserve of Officers.
T.E.E.	Tyne Electrical Engineers.
T.F.R.O.	Territorial Force, Reserve of Officers.
‡	Service in any theatre of war overseas.

All other abbreviations are conventional.

A.—General List of Officers holding Commissions in, or Attached to, the Unit.

A.	Names.		Second Lieutenant.	Lieutenant.	Captain.	Major.	Lt.-Colonel.	Colonel.	Remarks.
1 (C1)	Johnson, W.	C.B., V.D.	28/10/74 1st	28/1/77 N.D.V.R.E.	14/3/77	3/3/88	H/5/12/94 1/4/03	—	In Command 3/3/88 to 31/10/11. Retired 31/10/11. Died 28/5/20.
2	Allan, T. J.	—	(1st N.D.V.R.E.	10/11/83	3/3/88	—	—	—	Retired 13/7/89.
3	Baguley, C.	—	(1st N.D.V.R.E.	13/3/86	3/3/88	—	—	—	Resigned commission 26/7/90.
4	Hartness, J. G. R.	—	(1st N.D.V.R.E.	9/1/86 3/3/88	—	—	—	—	Retired 9/1/92.
5	Scott, J. R.	—	3/3/88	—	20/7/89	—	—	—	Resigned commission 18/3/96. Died 11/4/21.
6 (C2)	Scott, F. G.	V.D.	3/3/88	—	20/7/89	H/18/9/01 1/4/03	H/3/3/08 14/11/11	T/5/2/15 R.M.S.M.	In command 1/11/11 to 4/2/15. Transferred to R.M.S.M. 4/2/15. Died 15/8/22.
7	Cay, A.	—	17/3/88	—	—	—	—	—	Resigned commission 21/7/88.
8	Wesencraft, R. H.	—	17/3/88	17/8/89	4/10/90	—	—	—	Resigned commission 22/7/96. Died 2/4/29.
9	Heaviside, A. W.	—	22/12/88	20/12/90	—	—	—	—	Resigned commission 4/7/96.
10	Nicholson, Sir John R.	Kt., C.M.G. M.I.C.E.	17/8/89	—	—	—	—	—	Resigned commission 25/3/93.
11	Winter, R. P.	—	5/4/90	27/2/92	—	—	—	—	Resigned commission 11/12/95. Died 14/9/26.
12 (C3)	Toomer, C. R.	V.D.	20/12/90	17/9/92	8/7/96	4/6/04	5/2/15	—	In command 5/2/15 to 30/8/18. Lt.-Col. Reserve 30/8/18. Retired (age limit) 30/7/21. Died 29/12/25.
13	Macfadyen, F. E.	—	15/8/91	—	—	—	—	—	Resigned commission 21/1/93.
14	Stephenson, R.	—	12/3/92	18/3/96	26/8/96	—	—	—	Resigned commission 8/5/01. Died –8/13.
15	Towers, E. (Jun.)	—	4/2/93	26/8/96	11/8/00	H/29/1/08	—	—	Resigned commission 30/1/08. Died 14/3/14.

A.—GENERAL LIST.—*Continued.*

A.	Names.		Second Lieutenant.	Lieutenant.	Captain.	Major.	Lt.-Colonel.	Colonel.	Remarks.
16	Green, W.	—	12/4/93	—	—	—	—	—	Resigned commsision 7/4/94.
17	Swan, G.	—	8/8/94	—	—	—	—	—	Resigned commission 28/11/94.
18	Hopkinson, J. (Jun.)	—	4/3/96	—	—	28/4/97 The Electrical Engineers	—	—	Transferred to Electrical Engineers 28/4/97. Resigned commission 8/98.
19	Hopkinson, B.	C.M.G., F.R.S., M.A., M.I.E.E.	4/3/96	— —	25/8/97	5/9/03 The 1/4/07 London 17/7/08 Unattached List T.F. Cambridge University O.T.C. (R.E. Fortress)	Electrical Engineers on Electrical Engineers	—	Staff Capt., D. of Mil. Aer. 25/9/15 Employed Air Board (Ass. Dir. Aircraft Equipment) 27/3/16. Died 23/8/18.
20	Blackburn, A.	—	4/7/96	2/3/98	11/8/00	—	—	—	Resigned commission 28/9/10.
21	Johnson, C.	—	4/7/96	2/3/98	13/7/01 9/11/15	—	—	—	Capt. Reserve 18/2/12. Retired (medical board) 23/4/18.
22	Bruce, G. A.	T.D., M.I.E.E.	12/5/97	11/8/00 (R.M.S.M.)	13/6/03 T/1/4/15	5/2/15 T/1/6/15	—	—	Major Reserve 29/6/20. Retired (age limit) 1/9/20.
23	White, J.	—	1/12/97	—	—	—	—	—	Resigned commission 29/3/99.
24	Winlaw, W. W.	—	1/12/97	—	—	—	—	—	Resigned commission 13/6/00.
25	Johnson, G.	—	21/9/98	11/8/00	—	—	—	—	Resigned commission 26/3/02.
26	Warburton, F. J.	—	4/7/00	23/11/01	—	—	—	—	Resigned commission 21/5/04.
27 (B3) (C4)	†Robinson, E.	C.B.E., T.D., D.L.	11/8/00	23/11/01	13/6/03	14/11/11	T/6/2/15 1/6/16	Bt./12/4/24	In Command 31/8/18 to 31/10/25. Bt. Col. Reserve 31/10/25. Resigned commission 23/3/27. Hon. Col. 25/10/30.
28	Towers, G.	—	11/8/00	23/11/01	4/8/04	—	—	—	Resigned commission 22/5/07. Died 16/6/30.
29	Readhead, S.	—	11/8/00	21/5/02	4/6/04	—	—	—	Resigned commission 23/1/07. Died 20/11/26.

A.—GENERAL LIST.—Continued.

A.	Names.		Second Lieutenant.	Lieutenant.	Captain.	Major.	Lt.-Colonel.	Colonel.	Remarks.
30	Fernandes, C. B. L.	—	11/8/00 25/7/16	13/6/03 T/29/10/16 26/1/18	—	—	—	—	Resigned commission 23/1/07. Retired (age limit) 22/7/21. Died.
31	Arnison, C.	—	13/10/00	13/6/03	1/9/11	T/6/2/15 1/6/16	—	—	Retired (age limit) 25/9/18.
32	Bertram, H.	Medjidie 4th Class	16/2/01	13/6/03	—	—	—	—	Died in Brazil of yellow fever 7/12/07.
33	Dunford, E. S.	—	4/1/02	—	—	—	—	—	Resigned commission 7/2/03.
34	Tasker, A. K.	T.D., D.L.	7/5/02	13/2/04	14/11/11	T/5/2/15 1/6/16	31/8/18	—	Lt. Col. Reserve 26/7/21. Retired (age limit) 26/3/31.
35	Short, O. M.	O.B.E., T.D.	14/3/03	4/6/04	18/2/12	T/7/2/15 1/6/16	—	—	Major Reserve 26/7/21. Retired (age limit) 2/11/27.
36	Swift, E.	—	27/5/03	4/6/04	2/12/13	T/8/2/15 1/6/16	—	—	Major Reserve 23/7/18. Retired (age limit) 27/8/21. Died 20/11/21.
37	Peverley, A. W.	—	21/5/03	4/6/04	—	—	—	—	Resigned commission 13/5/09.
38	Scott, K. S. M.	M.B.E., B.Sc. (Eng.)	27/5/03 7/11/14	T/8/2/15	T/25/4/16 1/6/16	—	—	—	Resigned commission 13/5/09. Capt. Reserve 21/3/25. Bt. Major Reserve 9/1/26.
39	Schaeffer, P. P.	—	27/5/03	—	—	—	—	—	Resigned commission 23/1/07.
40	Ching, J. T. L.	—	27/5/03	—	T/19/2/15 R.M.S.M.	—	—	—	Resigned commission 23/1/07.
41	Bolster, S. F.	—	13/6/03	—	—	—	—	—	Resigned commission 23/1/07.
42	‡Webb-Bowen, H. E.	C.M.G., D.S.O., T.D., A.M.I.Mch.E. A.M.I.E.E.	13/6/03 (/12/07)	1/4/08 L.E.E.	3/3/09	T/R.E. 4/8/15	—	T/R.E. 8/9/17	Col. Reserve 3/1/22.

A.—General List.—Continued.

A.	Names.		Second Lieutenant.	Lieutenant.	Captain.	Major.	Lt.-Colonel.	Colonel.	Remarks.
43	Elder, W. H.	—	6/2/04	—	—	—	—	—	Resigned commission 23/1/07.
44 (C6)	‡Woodward, E. H. E.	M.C., T.D., B.Sc.(Eng.), M.I.E.E.	(14/3/06	6/10/06 1st Vol. Bn. & 4th Bn. (T.F.) (T/1/10/14) 4th City of	T/4/12/14 T/25/11/16 23/1/17	1/5/29	Glos. Regt.) Bris. Bn. Glos. Regt.)	—	Resigned commission 0/3/10. In Command 1/11/29.
45 (C5)	‡Firmin, N. H.	O.B.E., T.D.	25/3/07	1/9/11	T/6/10/14	T/23/10/15 1/6/16	Bt./1/1/28	—	Employed Admiralty Ass. Dir. Experiments and Research 0/3/18—0/3/19. In Command 1/11/25—31/10/29. Bt. Lt. Col. Reserve 1/11/29.
46	Buck, M. W.	T.D.	16/4/07	3/5/12	T/6/10/14 1/6/16	23/7/18	—	—	Served with 40th Coy. R.E. at Hong Kong 1914-1919. Major Reserve 29/10/20. Died 9/4/24.
47	Forster, C. M.	O.B.E., T.D., A.M.I.E.E.	30/8/11	17/5/13	T/6/10/14 1/6/16	Bt./3/6/18 31/8/18	—	—	Major Reserve 26/2/27. Retired (age limit) 25/5/29.
48	‡Fairbairn-Crawford, I. F.	T.D., A.M.I.Mch.E.	30/8/11	2/12/13	T/6/10/14 1/6/16	11/1/28	—	—	Employed 1915-1918 on R.F.C. and Admiralty (Airship Construction). Major Reserve 1/5/29. Retired (age limit) 4/1/33.
49	Campbell, C. M.	A.M.I.E.E.	30/8/11	2/12/13	T/6/10/14 1/6/16	T/11/7/16	—	—	Major Reserve 26/7/21. Retired (age limit) 16/4/31.
50	Ward, W. G.	A.M.I.E.E.	30/8/11	2/9/14	T/6/10/14 1/6/16	A/24/9/17	—	—	Capt. T.F.R.O. 2/9/20.
51	‡Russell, G. L. L.	A.M.I.E.E.	4/3/12	6/10/14	T/5/2/15 1/6/16	—	—	—	Died of influenza 1/3/19.
52	Pyne, F. D.	O.B.E., A.M.I.E.E., A.M.I.Mch.E.	1/3/13	6/10/14	T/6/2/15 1/6/16	A/4/3/18	—	—	Died 2/12/19 from injuries received in a motor car accident.

A.	Names.		Second Lieutenant.	Lieutenant.	Captain.	Major.	Lt.-Colonel.	Colonel.	Remarks.
53	Mountain, K. A.	A.M.I.E.E.	1/3/13	6/10/14	T/7/2/15 1/6/16	—	—	—	Relinquished commission (ill-health) 30/9/21.
54	†Bowers, F. H.	—	17/5/13	6/10/14	T/25/5/16	—	—	—	Killed in a motor-cycle accident 12/6/17.
55	†Newnam, P. E.	—	(5/7/13)	22/12/14 26/9/17	1/6/16 (0/5/17 N. Cyclists Bn.) — 4th Bn. E. Yorks. Rgt.)	—	—	—	Lieut. Reserve 13/10/20. Relinquished commission 3/2/22.
56	†Monkhouse, S. E.	M.I.E.E.	2/12/13	6/10/14	T/8/2/15 1/6/16	Bt./3/6/18	—	—	Employed R.E. (S.E.L.) -/11/16— -/4/19. Lt. Col. Reserve 3/1/20.
57 (D6)	Sharp, R.	O.B.E., T.D.	11/3/14	6/10/14	T/24/10/15 1/6/16	—	(A/R.E. 7/9/17)	—	Adjt. 1/1/17—6/6/19. Capt. Reserve 1/11/29.
58	Young, J.	—	11/3/14	6/10/14	T/23/10/15 1/6/16	—	—	—	Employed Ministry of Munitions and Air Ministry 22/7/17—13/10/20. Capt. Reserve 13/10/20. Transferred to General List, R.E.T.F. Reserve 1/0/21.
59	†Wood, R. L.	M.C., D.L.	1/10/14 (50th (Northumbrian) Div. R.E.	5/2/15	T/R.E. 31/1/16 1/6/16 21/5/20	T/R.E.15/2/18	23/5/20	—	Employed R.E. 22/10/15—24/4/19. C.R.E. 50th (Northumbrian) Div. R.E. 20/4/23—19/4/27. Lt. Col. Reserve 20/4/27. Bt. Col. Reserve 19/3/25.
60	†Hall, W.	D.Sc., F.R.A.S.	6/10/14	5/2/15 (T/R. Marines -/11/18)	T/22/4/16 1/6/16	—	—	—	Employed R. Marines -/11/18. Retired (age limit) 7/7/21.
61	†Hunter, J. F. S.	M.B.E.	6/10/14	T/2/9/15	T/26/4/16 1/6/16	—	—	—	Capt. Reserve 27/8/21.
62	†Williamson, A. B.	—	6/10/14	T/3/9/15	T/27/4/16 1/6/16	—	—	—	Capt. Reserve 13/10/20.

A.—GENERAL LIST.—*Continued.*

A.	Names.		Second Lieutenant.	Lieutenant.	Captain.	Major.	Lt.-Colonel.	Colonel.	Remarks.
63	†James, M. C.	—	6/10/14	T/7/2/15	T/24/4/16 1/6/16	—	—	—	Employed R.E. -/11/16) and Ministry of Shipping until 30/5/19. Capt. Reserve 27/8/21. Killed in motor car accident 1/6/30.
64	Abbott, J. R.	A.M.I.E.E.	6/10/14	T/4/10/15	1/6/16	—	—	—	Capt. Reserve 13/10/20.
65	†Pinkney, R. B. T.	—	(5/11/14)	T/8/1/15	1/6/16 22/6/16 20th (Ser.) Bn. N.F., 1st Tyneside Scot.	—	—	—	Relinquished commission 11/10/21.
66	†Anderson, R. W.	M.C., M.I.Min.E.	6/11/14	T/6/2/15	1/6/16	A/15/6/18 (1/10/20 50th (North'b). Div. Sigs.)	—	—	Employed Signal Service R.E. 19/10/15—28/2/19. Retired -/-/22.
67	Rogerson, H. O.	—	6/11/14	T/5/9/15	1/6/16	—	—	—	Resigned commission 19/2/21.
68	†Henderson, T.	M.C., A.F.C., El Nadha, F.R.G.S., F.R.S.A., F.G.S., A.M.I.A.E.	6/11/14	T/6/9/15 1/4/16	1/6/16	—	—	—	Seconded R.F.C. 23/7/15. Relinquished Territorial commission 31/7/23. R.A.F. permanent commission 31/7/23. Resigned commission 1/12/27.
69	Sherlock, H.	T.D.	6/11/14	T/6/9/15	1/6/16	1/11/29	—	—	Employed R.E. (S.E.L.) 15/1/17—4/4/19.
70	Scott, C. F.	—	7/11/14	T/7/9/15	1/6/16	—	—	—	Seconded R.E. and served at Malta 1918–1919. Capt. Reserve 12/7/21. Perished in the S.S. "Hartley" disaster 27/11/24.
71	†Sharp, W.	O.B.E.	7/11/14	T/7/9/15	1/6/16	A/R.E. 26/2/17 Bt./3/6/18 29/3/20	—	—	Employed R.E. 1/6/15—30/11/21. Major Reserve 18/3/22. Retired (age limit) 17/5/33.
72	†Gibbon, E. H.	—	7/11/14	T/8/9/15	1/6/16	—	—	—	Seconded R.F.C. 15/2/16—22/8/16. Capt. Reserve 13/10/20.
73	†Deane, G. I. N.	—	7/11/14	T/8/9/15	1/6/16	—	—	—	Seconded R.F.C. 27/9/15—0/3/19. Capt. Reserve 16/12/20.

A.—GENERAL LIST.—*Continued.*

A.	Names.		Second Lieutenant.	Lieutenant.	Captain.	Major.	Lt.-Colonel.	Colonel.	Remarks.
74 (D7)	†Williamson, D. A.	—	15/11/14	T/5/2/15, 1/11/25, a/d 1/3/23	T/3/12/15 } 1/6/16	—	—	—	Seconded R.F.C. 27/11/16—9/17. Adjt. 7/8/19—20/11/19. Capt.Reserve 27/8/21—31/10/25. Transferred to Active List 1/11/25.
75	†Myles, D.	Croix-de-Guerre (France) —	(21/10/14, 25/11/14	5th Bn. N.F.) T/9/9/15	1/6/16	—	—	—	Capt. Reserve 13/10/20.
76	†Phillips, R.	—	(14/12/14	7/1/15 23rd (Service) Bn. N.F., 4th Tyneside Scottish) 1/5/17	T/21/6/17 a/d 1/6/16 }	—	—	—	Employed R.E. 5/7/19—30/4/20. Lieut. Reserve 14/10/20.
77	†Danby, C. D.	M.C.	26/1/15	—	T/27/5/16, 1/6/16	—	—	—	Seconded R.F.C. 3/8/15. Killed in action 18/7/18.
78	Suthery, F. B. C.	A.M.I.E.E.	26/1/15	T/10/9/15 (T/Royal Marines –/10/18)	1/6/16	A/R.E. 12/3/19	—	—	Employed R. Marines 10/18—31/1/19. EmployedR.E.1/2/19—15/11/19. Major Reserve 13/10/20. Retired (age limit) 5/11/30.
79	Tucker, T. T.	—	26/1/15	T/18/10/15, 1/6/16	T/17/7/16, 23/7/18	—	—	—	Capt. Reserve 13/10/20. Resigned commission 3/2/22.
80	†Winkworth, L. S.	A.M.I.E.E. A.M.I.Mch.E.	(7/4/15	T/21/9/15, 1/6/16, 2/2/17	2/9th Bn. Manchester Regt.) 31/8/18	—	—	—	Capt. Reserve 27/8/21.
81	James, W. H.	—	25/4/15	T/22/4/16, 1/6/16	T/18/7/16	—	—	—	Capt. Reserve 13/10/20. Resigned commission 13/2/22.
82	Horton, L.	—	(3/5/15, 4/10/16	13th (Reserve) Bn. 5/4/18	R. Warwickshire Regt.)	—	—	—	Resigned commission 3/12/20.
83	†Ogilvy, J. A.	—	(28/5/15, 29/12/16	6/9/15, 30/6/18	5th (City of Glasgow) Bn. H.L.I.)	—	—	—	Relinquished commission on account of ill-health 9/3/19.
84	Lawther, J.	—	10/6/15	T/23/4/16, 1/6/16	T/19/7/16	—	—	—	Capt. Reserve 20/11/20. Resigned commission 3/2/22.

B

A.—GENERAL LIST.—*Continued.*

A.	Names.		Second Lieutenant.	Lieutenant.	Captain.	Major.	Lt.-Colonel.	Colonel.	Remarks.
85	Hollobone, N. G.	—	21/6/15	—	—	—	—	—	Resigned commission (medical grounds) 28/1/16. Served subsequently with the R.N.V.R. and on the Corn Commission (Min. of Food) till 1919.
86	‡Aitkens, C. A. C.	—	23/6/15	1/6/16	—	—	—	—	Transferred to Special Brigade R.E. 15/6/16. Died of wounds 10/7/16.
87	‡Rooksby, R. H.	—	23/6/15 (1/11/23)	T/25/4/16 1/6/16 1/11/25	T/5/8/16 5th Bn.N.F.)	—	—	—	Capt. Reserve 13/10/20—31/10/23. Capt. Reserve (N.F.) 18/1/28. Died 11/6/32.
88	Ripley, H. S.	M.B.E.	15/7/15	T/26/4/16 1/6/16	T/23/10/16	—	—	—	Employed W.O. (A.A.D.) 23/10/16 —10/1/18. Employed W.O. (F.W.8) 10/1/18 —11/19. Capt. Reserve 27/8/21.
89	‡Buxton, H. G.	—	(21/7/15 (3/9/15 22/9/16	Alexandra Princess of Wales' Own Yorkshire Regt.) London Div. R.E.) 1/11/16	T/R.E. 5/9/19	—	—	—	Lieut. Reserve 16/10/20.
90	‡Baird, J. O.	M.C.	20/8/15	T/27/4/16 1/6/16	T/R.E. 6/8/16	A/R.E. 14/8/18	—	—	Major Reserve 13/10/20.
91 (E3)	‡Souter, W. A.	Hon. Lieut. and Qr. Mr. 10/6/15	1/9/15	T/1/10/15 1/6/16	T/29/10/16	—	—	—	Capt. Reserve 13/10/20. Retired (age limit) 18/5/29.
92	‡Rendell, E. F.	M.C., Mil. Order of Aviz (Portugal) M.C.	4/9/15	T/28/4/16 1/6/16	A/R.E. 2/2/18	A/R.E. 30/6/18	—	—	Major Reserve 13/10/20.
93	‡Hunter, R. Ll.	—	5/9/15	T/29/4/16 1/6/16	—	—	—	—	Capt. Reserve. Resigned commission 19/10/21.
94	‡Winter, R. P.	M.C.	6/9/15	T/30/4/16 1/6/16	T/R.E. 12/4/18 1/11/29	—	—	—	—

A.—General List.—Continued.

A.	Names.		Second Lieutenant.	Lieutenant.	Captain.	Major.	Lt.-Colonel.	Colonel.	Remarks.
95	†Findlay, R. P.	M.C.	7/9/15 (1st North'b'n R.F.A.	T/1/5/16 1/6/16 (T) 8/9/20	A/24/3/18 A/R.E. 25/2/19	—	—	—	Employed R.E. 4/17—1919. Lieut. Reserve (R.A.).
96	†White, H. G.	—	8/9/15	1/6/16	T/11/3/16	—	—	—	Employed R.E. (S.E.L.) 15/1/17—1919. Capt. Reserve 15/12/20. Retired (age limit) 2/11/27.
97	†Edwards, W. G.	—	9/9/15	T/2/5/16 1/6/16	A/R.E. 30/8/18	—	—	—	Capt. Reserve 20/11/20.
98	Crawford, T. W.	—	10/9/15	T/25/5/16 1/6/16	—	—	—	—	Resigned commission 4/12/20.
99	Russell, C.	—	10/9/15	T/3/5/16 1/6/16	—	—	—	—	R.F.C. 29/10/17. Relinquished commission (ill-health) 23/11/18.
100	†Ellis, R. G.	A.M.I.E.E.	15/9/15	T/26/5/16 1/6/16	—	—	—	—	Lieut. Reserve 27/8/21.
101	†Erskine, C. W.	—	16/9/15	T/27/5/16 1/6/16	—	—	—	—	Resigned commission 1/2/22.
102	†Dickson, A. G.	—	22/9/15	T/28/5/16 1/6/16	—	—	—	—	Employed R.E. 10/19. Lieut. Reserve 20/11/20.
103	Rintoul, W.	—	(24/9/15 Northern Signal Coys. R.E.(T.F.) 24/7/17	25/1/19	—	—	—	—	Resigned commission 6/10/20.
104	†Huntley, C. G.	A.M.I.E.E.	28/9/15 (10th A.A. Bn. R.E.(T.)	T/29/5/16 1/6/16	21/11/22	Subsequently 26th A.A. Bn.)	—	—	Lieut. Reserve 24/8/21—21/11/22. Capt. Reserve (L.E.E.).
105	†Power, J. A.	—	6/10/15	T/30/5/16 1/6/16	—	—	—	—	Resigned commission 1/2/22.
106	†Watson, W.	—	21/10/15	T/31/5/16 1/6/16	—	—	—	—	Lieut. Reserve 13/10/20.
107	Plunket, A. J. L.	—	22/10/15	1/6/16	—	—	—	—	Resigned commission 1/2/22.
108	‡Williamson, C. B.	—	15/11/15	1/6/16	—	—	—	—	Lieut. Reserve 24/2/26. Resigned commission 20/2/32.

A.—GENERAL LIST.—*Continued.*

A.	Names.		Second Lieutenant.	Lieutenant.	Captain.	Major.	Lt.-Colonel.	Colonel.	Remarks.
109	Raven, C. H. R. E.	—	1/12/15	1/6/16	—	—	—	—	Employed at Min. of Munitions 28/11/16. Seconded R.A.F. 1/4/18.
110	†Robinson, J.	—	2/12/15	1/6/16	—	—	—	—	Lieut. Reserve 13/10/20.
111	†White, E. H.	A.I.E.E.	3/12/15	1/6/16	—	—	—	—	Employed R.E. 31/10/19—11/23. Resigned commission 1/2/22.
112	†Campbell, H. G.	—	4/12/15 (10th A.A. Bn. R.E. (T.)	1/6/16	22/11/22	Subsequently 26th A.A. Bn.)	—	—	Major Reserve (L.E.E.) 2/1/25.
113	†Baldwin, E. V.	A.M.I.C.E.	24/12/15	1/6/16	A/R.E. 22/8/18	—	—	—	Capt. Reserve 25/8/21.
114	Harrison, E.	—	25/12/15	1/6/16	—	—	—	—	Seconded R.A.F. 24/12/17. Resigned Territorial commission 1/4/18. Lieut. R.A.F. 1/4/18. Transferred to Unemployed List R.A.F. 17/6/19.
115	†Batey, J. L.	—	(1/1/16 22/2/16	15thBn.N.F.) 1/6/16	—	—	—	—	Lieut. Reserve 21/5/30.
116	Hedley, O. W. E.	O.B.E., D.L.	6/1/16	1/6/16	—	—	—	—	Resigned commission 3/12/20.
117	Wilson, W. W.	—	6/1/16	1/6/16	—	—	—	—	Lieut. Reserve 13/10/20. Resigned commission 3/2/22.
118	Micklam, C. B. S.	—	11/1/16	1/6/16	—	—	—	—	Lieut. Reserve 13/10/20. Resigned commission 1/9/21.
119	†Fox, W.	O.B.E., Croix-de-Guerre (France)	25/1/16	1/6/16	A/R.E. 17/7/18	—	—	—	Capt. Reserve 20/11/20. Resigned commission 3/2/22.
120	Murray, J. B.	—	28/1/16	1/6/16	—	—	—	—	Capt. Reserve 13/10/20.

A.	Names.		Second Lieutenant.	Lieutenant.	Captain.	Major.	Lt.-Colonel.	Colonel.	Remarks.
121	Graham, C.	—	5/2/16	1/6/16	—	—	—	—	Resigned commission 2/12/20.
122	‡Ross, D.E.	—	18/4/16	T/21/7/16 19/10/17	—	—	—	—	Lieut. Reserve 14/9/21.
123	‡Burdis, A. S.	A.M.I.Mch.E.	29/5/16	T/22/7/16 30/11/17	—	—	—	—	Lieut. Reserve 20/11/20.
124	‡Emerson, J. R. T.	—	2/6/16	T/23/7/16 3/12/17	—	—	—	—	Lieut. Reserve 20/11/20.
125	‡Hamilton, F. T.	M.B.E.	14/6/16	T/24/7/16 15/12/17	—	—	—	—	Capt. Reserve 13/10/20. Employed R.E. (W.O.F.W.8.) Retired (age limit) 13/11/25.
126	Bird, L.	—	30/6/16	T/23/10/16 31/12/17	—	—	—	—	Resigned commission 4/12/20.
127	‡Hutchinson, H.	—	30/6/16	T/24/10/16 31/12/17	—	—	—	—	Resigned commission 2/9/21.
128	‡Dixon, W.	—	5/7/16	T/25/10/16 6/1/18	—	—	—	—	Employed R.E. (S.E.L.) 26/8/18 —7/2/19. Resigned commission 2/9/21.
129	Anderson, D. S.	—	9/7/16	T/26/10/16 10/1/18	—	—	—	—	Lieut. Reserve 20/11/20.
130	‡Adendorff, F. L.	M.C.	9/7/16	T/27/10/16 10/1/18	—	—	—	—	Employed R.E. 16/5/16. Lieut. Reserve 13/10/20. Resigned Commission 3/2/22. Died 17/12/31 in Johannesburg.
131	Rea, T. S.	A.M.Inst. C.E.	11/7/16	T/28/10/16 12/1/18	—	—	—	—	Employed R.E. (S.E.L.) 5/2/17 —1919. Capt. Reserve 8/9/21. Retired (age limit) 18/5/29.
132	Stephens, C. A.	A.M.I.E.E.	17/8/16	T/30/10/16 18/2/18	—	—	—	—	Resigned commission 1/9/21.
133	Owen, H. V.	—	25/8/16	T/31/10/16 26/2/18	—	—	—	—	Lieut. Reserve 24/8/21. Retired (age limit) 23/7/32.

A.—GENERAL LIST.—*Continued.*

A.	Names.		Second Lieutenant.	Lieutenant.	Captain.	Major.	Lt.-Colonel.	Colonel.	Remarks.
134	Howell, A.	A.M.I.E.E.	25/8/16	T/1/11/16 26/2/18	—	—	—	—	Lieut. Reserve 20/11/20.
135	‡Jardine, R. E.	—	25/8/16	T/2/11/16 26/2/18	—	—	—	—	Lieut. Reserve 26/11/20.
136	‡Corby, F. S., The Rev.	—	25/8/16	T/3/11/16 26/2/18	—	—	—	—	Lieut. Reserve 27/8/21. Retired (age limit) 28/3/33.
137	‡Ingman, H. J.	—	31/8/16	T/4/11/16 1/3/18	—	—	—	—	Lieut. Reserve 13/10/20. Resigned commission 3/2/22.
138	Joseph, H.	—	17/9/16	T/5/11/16 18/3/18	—	—	—	—	Lieut. Reserve 20/11/20. Retired (age limit) 3/2/22.
139	‡Porter, R. O.	—	27/9/16	T/7/11/16 28/3/18	T/R.E. 1/8/18	—	—	—	Capt. Reserve 13/10/20.
140	‡Newman, S. W.	—	8/10/16	9/4/18	—	—	—	—	Lieut. Reserve 27/11/20. Resigned commission 3/2/22.
141	Elliot, C. B.	—	14/10/16	15/4/18	—	—	—	—	Resigned commission 1/2/22.
142	‡Phillips, A. D.	—	24/10/16	25/4/18	—	—	—	—	Employed R.E. 1919—14/11/19. Lieut. Reserve 27/11/20.
143	Watson, H. S.	—	25/10/16	—	—	—	—	—	Resigned Territorial commission 23/1/18. Lieut. (temp. Capt.) R.A.F. 1/6/18. Lieut. (Act. Major) R.A.F. 27/5/19. Transferred Unemployed List R.A.F. 1/2/20.
144	Brigham, E. R.	—	25/10/16	—	—	—	—	—	Retired (ill-health) 14/3/18.
145	‡Gladwin, W. A.	—	25/10/16	26/4/18	T/R.E. 21/8/19	—	—	—	Employed R.E. (S.E.L.) 1917—7/4/19. Employed R.E. 7/4/19—1921. Capt. Reserve 14/9/21.

A.—GENERAL LIST.—*Continued.*

A.	Names.		Second Lieutenant.	Lieutenant.	Captain.	Major.	Lt.-Colonel.	Colonel.	Remarks.
146	Marshall, T. S.	—	26/10/16	—	—	—	—	—	Killed in a motor cycle accident at Colchester 1/8/17.
147	Winstanley, W. M.	—	27/10/16	28/4/18	—	—	—	—	Lieut. Reserve 20/11/20. Resigned commission 3/2/22.
148	†MacDonald, H. E.	—	2/11/16	3/5/18 (23/6/21 50th (North'mbrian) Div. R.E.T.A.)	—	—	—	—	Employed R.E. 1/18—5/19. Lieut. Reserve 13/10/20—23/6/21. Lieut. Reserve (50th Div. R.E.) 20/9/32.
149	Woods, A.	—	2/11/16	3/5/18	1/2/22	—	—	—	Employed R.E. (S.E.L.) -/-/—1919. Resigned commission 1/2/22.
150	Elliot, W. H.	—	17/11/16	18/5/18	—	—	—	—	Lieut. Reserve 13/10/20. Resigned commission 3/2/22.
151	Scullard, G. T. B.	—	6/12/16	7/6/18	—	—	—	—	Resigned commission 1/2/22.
152	Thompson, F.	—	6/12/16	7/6/18	—	—	—	—	Lieut. Reserve 20/11/20.
153	Hampton, E. L.	—	8/12/16	9/6/18	—	—	—	—	Lieut. Reserve 20/11/20.
154	Algar, H.	—	12/12/16	—	—	—	—	—	Died at Haslar 12/4/17.
155	†Robertson, J. N.	—	24/12/16	25/6/18	—	—	—	—	Resigned commission 1/2/22.
156	†Leeson, B. H.	M.I.E.E.	4/1/17	5/7/18	—	—	—	—	—
157	†Smith, E. F.	A.M.I.E.E.	5/1/17	6/7/18	—	—	—	—	Lieut. Reserve 29/8/21.
158	Rendell, E.	—	12/1/17	13/7/18	—	—	—	—	Lieut. Reserve 26/11/20. Resigned commission 3/2/22.
159	Welsh, J. E. P.	—	27/1/17	28/7/18	—	—	—	—	Resigned commission 1/2/22.
160	Davis, A. A.	—	31/1/17	1/8/18	—	—	—	—	Lieut. Reserve 10/9/21. Resigned commission 5/8/20. Retired (age limit) 25/6/32.

A.—GENERAL LIST.—*Continued.*

A.	Names.		Second Lieutenant.	Lieutenant.	Captain.	Major.	Lt.-Colonel.	Colonel.	Remarks.
161	Swan, W.	—	4/2/17	5/8/18	—	—	—	—	Lieut. Reserve 20/11/20. Resigned commission 3/2/22.
162	Bond, F. W.	A.M.I.Mch.E.	13/2/17	14/8/18	—	—	—	—	Employed R.E. (S.E.L.) -/-/-1919. Lieut. Reserve 27/8/21.
163	‡Preston, E. B.	—	15/2/17	16/8/18	—	—	—	—	Resigned commission 3/2/22.
164	‡Horler, P. V.	—	7/3/17	8/9/18	—	—	—	—	Lieut. Reserve 13/10/20.
165	Thompson, N. K.	—	7/3/17	8/9/18	—	—	—	—	Resigned commission 6/10/20.
166	Wallis, R. P.	—	6/4/17	7/10/18	—	—	—	—	Resigned commission 1/2/22.
167	Birrell, E.	M.B.E., A.M.I.Mch.E.	17/4/17	18/10/18	—	—	—	—	Resigned commission 1/2/22.
168	‡Braithwaite, F.	—	17/4/17	18/10/18	—	—	—	—	Lieut. Reserve 13/10/20.
169	‡Parnall, E. J.	B.Sc.(Eng.), A.M.I.E.E.	5/5/17	6/11/18 28/2/23	—	—	—	—	Lieut. Reserve 12/8/21. To Active List 28/2/23.
170	‡Armstrong, C.	—	31/5/17	1/12/18	—	—	—	—	Resigned commission 3/12/20.
171	McConnell, G.	—	12/6/17	13/12/18	—	—	—	—	Resigned commission 1/2/22. Died 8/2/22.
172	Barrett, E. B.	—	26/6/17	—	—	—	—	—	Died at Gosport 10/7/17.
173	Watts, W. S.	—	5/7/17	5/1/19	—	—	—	—	Kidnapped and believed killed near Cork 15/11/20 while serving with 33rd Fortress Coy. R.E.
174	Rood, C. W.	—	6/7/17	7/1/19	—	—	—	—	Lieut. Reserve 25/8/21.
175	Banner, S. V.	—	22/7/17	23/1/19	—	—	—	—	Resigned commission 1/2/22.

A.—GENERAL LIST.—*Continued.*

A.	Names.		Second Lieutenant.	Lieutenant.	Captain.	Major.	Lt.-Colonel.	Colonel.	Remarks.
176	Robertson, R. D.	—	17/8/17	18/2/19	—	—	—	—	Lieut. Reserve 20/11/20. Resigned commission 3/2/22.
177	Wait, C. S.	—	18/8/17	19/2/19	—	—	—	—	Lieut. Reserve 13/10/20. Resigned commission 3/2/22.
178	Lang-Hyde, J. R.	—	6/11/17	7/5/19	—	—	—	—	Lieut. Reserve 26/6/20.
179	Fawcett, P. F.	—	1/12/17	2/6/19	—	—	—	—	Employed R.E. 4/20—12/20. Lieut. Reserve 28/2/23.
180	Adams, W. H.	—	1/12/17	2/6/19	—	—	—	—	Lieut. Reserve 20/11/20. Resigned commission 3/2/22.
181	Armitage, C.	—	5/12/17	6/6/19	—	—	—	—	Lieut. Reserve 20/11/20.
182	Poland, H.	—	12/12/17	13/6/19	—	—	—	—	Lieut. Reserve 13/10/20.
183	Hunter, G.	—	20/12/17	21/6/19	—	—	—	—	Lieut. Reserve 13/10/20. Resigned commission 3/2/22.
184	Metcalf, J. A.	—	1/2/18	2/8/19	—	—	—	—	Lieut. Reserve 13/10/20.
185	†Lidiard, H. D. St. J.	—	12/3/18	13/9/19	—	—	—	—	Lieut. Reserve 27/11/20.
186	Batey, F. S.	—	(13/7/18 R.E.) 12/6/23 30/3/25 (1/11/26	14/10/24 —	—	—	—	—	Lieut. Reserve 31/3/26.
187	Sherlock, O.	—	30/3/25 (1/11/26	30/3/27	18/12/31	26th A.A. Bn, L.E.E.)	—	—	—
188	Mullens, H. H.	—	(17/2/18 25/3/26	R.N.A.S.) 25/3/29	—	—	—	—	—
189	Richards, A. S.	—	26/3/26	26/3/29	—	—	—	—	Lieut. Reserve 30/5/31.
190	Robinson, E. M.	—	1/12/26	11/3/31	—	—	—	—	—
191	Boyle, J.	—	17/6/27	10/11/31	—	—	—	—	—
192	Sadler, E. H.	—	11/4/29	11/4/32	—	—	—	—	—

A.—General List.—*Continued.*

A.	Names.	Second Lieutenant.	Lieutenant.	Captain.	Major.	Lt.-Colonel.	Colonel.	Remarks.
193	Fraser, M. A.	8/2/30	8/2/33	—	—	—	—	
194	Morpeth, G.	3/5/30	3/5/33	—	—	—	—	
195	Booth, N.	14/6/30	14/6/33	—	—	—	—	
196	Burnett, J. E.	21/5/32	—	—	—	—	—	
197								
198								

B.—Honorary Colonels.

B.	Names.	Second Lieutenant.	Lieutenant.	Captain.	Major.	Lt.-Colonel.	Colonel.	Remarks.
1	Palmer, Sir Chas. M., Bart. V.D., D.L.	—	(1st N. & D.V.R.E. —		20/10/75	28/3/(78)	25/2/88	Hon. Commandant 25/2/88. Died 4/6/07.
2	‡Northumberland, The Duke of	—	Gren. Gds. 13/6/03	S.R. Gren. Gds. 14/2/12	—	T/29/12/16 Bt./14/2/18	—	Hon. Colonel 20/3/12—23/8/30. Died 23/8/30.
3 (A27) (C4)	‡Robinson, E. C.B.E., T.D., D.L.	11/8/00	23/11/01	13/6/03	14/11/11	T/6/2/15 1/6/16	Bt./12/4/24	In Command 31/8/18—31/10/25. Bt./Col. Reserve 31/10/25. Resigned commission 23/3/27. Hon. Colonel 25/10/30—
4								

C.—COMMANDING OFFICERS.

C.	Names.		Second Lieutenant.	Lieutenant.	Captain.	Major.	Lt.-Colonel.	Colonel.	Remarks.
1 (A1)	Johnson, W.	C.B., V.D.	(28/10/74 —	28/1/77 —	14/3/77 1st N.&D.V.R.E.)	3/3/88	H/5/12/94 1/4/03	—	In Command 3/3/88—31/10/11. Retired 31/10/11. Died 28/5/20.
2 (A6)	Scott, F. G.	V.D.	3/3/88	—	20/7/89	H/18/9/01 1/4/03	H/3/3/08 14/11/11 (R.M.S.M.	T/5/2/15)	In Command 1/11/11—4/2/15. Transferred to R.M.S.M. 4/2/15. Died 15/8/22.
3 (A12)	Toomer, C. R.	V.D.	20/12/90	17/9/92	8/7/96	4/6/04	5/2/15	—	In Command 5/2/15—30/8/18. Lt. Col. Reserve 30/8/18. Retired (age limit) 30/7/21. Died 29/12/25.
4 (A17) (B3)	‡Robinson, E.	C.B.E., T.D., D.L.	11/8/00	23/11/01	13/6/03	14/11/11	T/6/2/15 1/6/16	Bt./12/4/24	In Command 31/8/18—31/10/25. Bt./Col. Reserve 31/10/25. Resigned commission 23/3/27. Hon. Colonel 25/10/30—
5 (A45)	‡Firmin, N. H.	O.B.E., T.D.	25/3/07	1/9/11	T/6/10/14	T/23/10/15 1/6/16	Bt./1/1/28	—	Employed Admiralty 1918—1919. In Command 1/11/25—31/10/29. Bt./Lt. Col. Reserve 1/11/29.
6 (A44)	‡Woodward, E. H. E.	M.C., T.D., B.Sc.(Eng.), M.I.E.E.	(14/3/06	6/10/06 1st Vol. Bn. & (T/1/10/14	T/6/10/14 T/4/12/14 T/25/11/16 23/1/17	4th Bn. (T.F.) Glo'shire (Regt.) 4th (City of Bristol) Bn., Glos. Regt.) 1/5/29	Bt./1/1/33		Resigned commission 3/10. In Command 1/11/29—

D.—ADJUTANTS.

D.	Names.		Second Lieutenant.	Lieutenant.	Captain.	Major.	Lt.-Colonel.	Colonel.	Remarks.
1	Pring, J.	Coast Bn.R.E.	R.E.Q.M.	H/13/6/85 7/3/86	—	—	—	—	A/Adjutant 3/88—1890. Retired 9/9/96.
2	Martin, J.	Coast Bn.R.E.	—	18/1/90	18/1/99 (R.M.S.M.	18/1/07 1/3/15	T/1/3/15	—	A/Adjutant 1890—15/4/07. Retired 15/4/07.

D.—ADJUTANTS.—*Continued.*

D.	Names.		Second Lieutenant.	Lieutenant.	Captain.	Major.	Lt.-Colonel.	Colonel.	Remarks.
3	†Burton, H. E.	O.B.E. Coast Bn. R.E.	—	4/6/02	A/20/6/10 4/6/11	Bt./5/4/15 R.E.5/9/15	—	—	A/Adjutant 16/4/07—2/15. Retired 15/9/19. Adjutant Tyne E. & M. Coys. 4/21—1/8/21. Adjutant 50th (Northumbrian) Div. R.E. 1/1/22—1/11/25.
4	Collins, J.	Coast Bn. R.E.	—	13/1/03	21/12/12	—	—	—	A/Adjutant 2/15—3/17. Retired 25/10/19.
5	Barr, W.	Coast Bn. R.E.	—	18/11/08	T/19/3/15 18/11/17	Bt./1/1/19 18/11/23	—	—	A/Adjutant 3/17—19/11/19. Retired
6 (A57)	Sharp, R.	O.B.E., T.D.	11/3/14	T/6/10/14	T/24/10/15 1/4/16	—	—	—	Adjutant 1/1/17—6/6/19. Capt. Reserve 1/11/29.
7 (A73)	†Williamson, D.A.	—	15/11/14	T/5/2/15 1/11/25 a/d.1/3/23	T/3/12/15 1/6/16	—	—	—	R.F.C. 1916—1917. Adjutant 7/6/19—20/11/19. Capt. Reserve 27/8/21—31/10/25. To Active list 1/11/25.
8 (E4)	Reed, A.	R.E.	—	H/3/9/15	Q.M.3/9/15 3/9/18	—	—	—	Qr. Mr. 3/9/15—5/1/21. A/Adjutant 20/11/19—5/1/21. Adjutant 6/1/21—21/12/23. Retired (age limit) 9/2/24.
9	McSweeney, J. M.	Coast Bn. R.E.	—	25/3/14	T/10/6/18 25/3/22	25/3/29	—	—	A/Adjutant 22/12/23—31/5/29. Retired 23/6/30.
10	Perowne, L. E. C. M.	R.E., O.St.J., A.M.I.E.E.	31/1/23	31/1/25	T.A/1/6/29	—	—	—	Adjutant 1/6/29—31/10/33.
11	Treays, W. H.	R.E.	R.A.17/7/19	R.A.17/7/21 R.E.19/1/24 a/d.17/7/21	—/17/7/30	—	—	—	Adjutant 1/11/33—
12									
13									
14									

E.—QUARTERMASTERS.

E.	Names.	Second Lieutenant.	Lieutenant.	Captain.	Major.	Lt.-Colonel.	Colonel.	Remarks.
1	Rioch, D.	—	H/7/3/91	H/13/4/01	—	—	—	Qr. Mr. 7/3/91—13/8/07. Retired (age limit) 14/8/07. Appointment discontinued. Died 15/10/11.
2	Aitken, J.	M.C. (Durham Fortress R.E. 4/6/15	H/1/12/14 20/2/16)	—	—	—	—	Qr. Mr. 1/12/14—4/6/15. Resigned commission 4/6/15.
3 (A90)	‡Souter, W. A.	—	H/10/6/15	—	—	—	—	Late 1st N/T R.G.A. Vols. Qr. Mr. 10/6/15—2/9/15. Commissioned 2nd Lieut. T.E.E. 3/9/15.
4 (D8)	Reed, A.	R.E.	H/3/9/15	Q.M.3/9/15 3/9/18	—	—	—	Qr. Mr. 3/9/15—5/1/21. A/Adjutant 20/11/19—5/1/21. Adjutant 6/1/21—21/12/23. Retired (age limit) 9/2/24.
5	Sargeant, A. J.	R.E., M.B.E.	H/19/10/15	Q.M. 19/10/18	—	—	—	Qr. Mr. 19/10/15—19/3/29. Retired (age limit) 19/3/29. Appointment discontinued.

F.—MEDICAL OFFICERS.

F.	Names.		Second Lieutenant.	Lieutenant.	Captain.	Major.	Lt.-Colonel.	Colonel.	Remarks.
1	Tait, R. K.	—	—	—	—	—	—	—	Acting Surgeon (4/9/86) 5/5/88. Surgeon 1/2/89. Retired 24/1/91. Died.
2	Brown, W. H.	T.D. R.A.M.C.(T)	—	—	S/21/4/94	S/17/1/03 R.A.M.C. -/-/11	—	—	Surgeon 25/5/89—23/8/12. Died 23/8/12 at North Shields.
3	Gibbon, F. W.	V.D., T.D. R.A.M.C.(T)	—	(Durham V. R.E. A/Sgn. 17/11/88)	S/1/2/89	S/26/9/00, S/13/5/03	H/3/3/06 S/31/3/08 R.A.M.C. 1/4/08	—	M.O. 13/5/03—2/19. Retired 11/21.
4	‡Williamson, J. B.	M.B. R.A.M.C.(Mil.)	—	—	7/9/14	12/2/22	—	—	M.O. 16/2/20—31/12/29.
5	‡Gabriel, W. L. M.	M.B. R.A.M.C.(T)	(15/8/14	15/2/15 S.R. 8/4/18	3rd Bn. W. S.R. 8/4/19 12/6/23	Yorks. Regt.) 5/11/32			M.O. 25/5/32—9/9/32. Transferred to 149th (Northumbrian) Fd. Amb. R.A.M.C. 9/9/32
6									
7									
8									

G.—CHAPLAINS.

G.	Names.		Second Lieutenant.	Lieutenant.	Captain.	Major.	Lt.-Colonel.	Colonel.	Remarks.
1	Bott, Rev. H.	M.A.	—	—	—	—	—	—	A/Chaplain 2/5/88. Retired –/7/09. Died 21/3/13.
2	Pearson, Rev. S.	—	—	—	—	—	—	—	A/Chaplain 25/2/93. Retired –/7/09. Died
3	Lloyd, Rev. H. L.	B.A.	—	—	4th Class T.F. 15/3/12	3rd Class T.F. 10/3/22	—	—	Chaplain 13/3/12.— Retired 28/5/24.
4									
5									

271

DETAILED ESTABLISHMENTS AT DIFFERENT PERIODS.

EXTRACTS FROM PEACE ESTABLISHMENTS.

	Lieut.-Colonel	Majors	Captains	Lieutenants	2nd-Lieuts.	Qr.-Master	M.O.	Chaplains	Total Officers	Qr.-Mr.-Serjt.	Coy.-Sjt.-Major	Serjeants	Total Serjeants	Buglers	Corporals	2nd Corporals	Sappers	Total Rank and File	Total all Ranks
Submarine Mining Coy. 1st N/T & Durham Volunteer Engineers ...1884—1888...	—	—	1	2	—	—	—	—	3	—	1	2	3	2	3	3	49	55	63
Tyne Submarine Mining Coy. ...1887—1888...	—	—	1	2	—	—	—	—	3	—	1	2	3	2	3	3	49	55	63
Tyne Division V.S.M. (3 Coys.) ...1888—1889...	—	1	2	6	—	—	—	—	9	—	3	6	9	6	9	9	147	165	189
,, ,, ,, (4 Coys.)[1] ...1889—1890...	—	1	4	8	1	1[2]	—	—	15	1	4	8	13	8	12	12	196	220	256
,, ,, ,, (3 Coys.) ...1890—1895...	—	1	3	6	1	1	—	—	12	1	3	7	11	6	10	10	147	167	196
,, ,, ,, (3 Coys.) ...1895—1897...	—	1	3	6	1	1	—	—	12	1	3	10	14	6	10	10	144	164	196
,, ,, ,, (3 Coys.) ...1897—1900...	—	1	3	6	1	1	1	—	13	1	3	10	14	6	10	10	143	163	196
,, ,, ,, (5 Coys.) ...1900—1903...	—	1	5	10	1	1	1	—	19	1	5	17	23	10	17	17	238	272	324
,, ,, ,, (7 Coys.) ...1903—1904...	1	1	6	19	1	1	1	—	30	1	7	25	33	14	24	24	332	380	457
,, ,, ,, (7 Coys.) ...1904—1907...	1	2	7	17	1	1	1	—	30	1	7	25	33	14	24	24	332	380	457
,, ,, V.E.E. (5 Coys.) ...1907—1908...	1	1	5	13	1	1	1	—	23	1	5	20	26	10	20	20	268	308	367
E.L. Coy. Durham (F) R.E.T.F.[3] ...1908—1909...	—	—	1	1	1	—	—	—	3	—	1	4	5	2	6	6	69	81	91
Northumberland (F) R.E.[3] (1 Coy.) ...1909—1912...	—	—	1	1	1	—	—	—	3	—	1	4	5	2	6	6	69	81	91
Tyne Division V.E.E.[4] (5 Coys.) ...1908—1912—																			
Headquarters ...	1	—	—	—	—	—	—	—	1	1	—	—	1	—	—	—	—	—	2
For Humber ...	—	—	1	1	—	—	—	—	2	—	1	2	3	1	2	2	27	31	37
For Portsmouth ...	—	1	4	5	5	—	—	—	15	—	4	15	19	8	15	15	207	237	279
TOTAL ...	1	1	5	6	5	—	—	—	18	1	5	17	23	9	17	17	234	268	318
Tyne Elec. Engineers R.E. TF. (4 Coys.) ...1912—1914—																			
Headquarters ...	1	—	—	—	—	—	—	—	1	1	—	—	1	—	—	—	—	—	2
For Portsmouth and I.W. ...	—	1	2	3	2	—	—	—	8	—	2	11	13	4	11	11	111	133	158
For Tyne ...	—	—	1	1	1	—	—	—	3	—	1	6	7	2	8	8	77	93	105
For Special Purposes, etc. ...	—	1	1	—	1	—	—	—	3	—	1	6	7	2	8	8	75	91	103
TOTAL ...	1	2	4	4	4	—	—	—	15	1	4	23	28	8	27	27	263	317	368
Tyne Elec. (F) R.E. TA. (2 Coys.)... ...1920—1924—																			
Works Coy.[7] ...	—	1	3	3	3	1	—	—	11	1	1	7	9	2	14[5]	—	120[6]	134	156
E.L. Coy.[7] ...	—	—	1	1	2	—	—	—	4	—	—	2	2	2	6[8]	—	29[9]	35	43
TOTAL ...	—	1	4	4	5	1	—	—	15	1	1	9	11	4	20	—	149	169	199
Tyne Elec. (F) R.E. TA. (2 Coys.)... ...1924—1927—																			
No. 1 (E.L.) Coy.[7] ...	—	—	1[14]	3	—	1	—	—	5	—	—	2	2	2	6[8]	2	27	35	44
307th (Tyne) A.A.S.L. Coy.[4]...	—	1	1	8	—	—	—	—	10	—	1	18[15]	19	—	20[13]	2	244	264	293
TOTAL ...	—	1	2	11	—	1	—	—	15	—	1	20	21	2	26	2	271	299	337
Tyne Elec. (F) R.E. TA. (2 Coys.)... ...1927—1932—																			
No. 1 (E.L.) Coy.[7] ...	—	—	1[14]	3	—	1	—	—	5	—	—	2	2	2	6[8]	2	27	35	44
307th (Tyne) A.A.S.L. Coy.[11]...	—	1	1	8	—	—	—	—	10	—	1	18[15]	19	—	20[13]	16	157	193	222
TOTAL ...	—	1	2	11	—	1	1	—	15	1	1	20	21	2	26	18	184	228	266
Tyne Elec. Engineers R.E. TA. (2 Coys.) ...1932—																			
No. 1 (E.L. & Works) Coy.[17]...	—	—	1[14]	3	—	—	—	—	4	—	—	4[18]	4	—	6[19]	4	36	46	54[21]
307th (Tyne) A.A.S.L. Coy.[11]...	—	1	1	8	—	1	—	—	11	—	1	17[20]	18	—	21[13]	16	157	194	223[21]
TOTAL ...	—	1	2	11	—	1	1	—	15	—	1	21	22	—	27	20	193	240	277

1 The 4th Company does not appear to have been formed as it disappears from the establishment in the following year.
2 A second surgeon could be appointed if deemed necessary.
3 Grouped with Durham (F) R.E. for A/Adjutant and A/R.S.M.I. The Coy. had, in addition, 2 P.S.I. Serjts.
4 "To be retained pending raising of local Units."
5 Includes 7 L/Sjts.
6 Includes 10 L/Cpls.
7 1 P.S.I. (Serjt.).
8 Includes 3 L/Sjts.
9 Includes 2 L/Cpls.
10 P.S.I. 1. (Serjt.).
11 P.S.I. 2. (1 C.S.M., 1 Sjt.).
12 Includes 3 L/Sjts.
13 Includes 12 L/Sjts.
14 The senior in the group Tyne, Tees, and Humber may be a Major.
15 Includes 2 C.Q.M.S. and 4 Mechanists.
16 An Adjutant was allowed from June 1929 onwards.
17 P.S.I. 2. (Serjts.).
18 Includes 1 C.Q.M.S.
19 Includes 2 L/Sjts.
20 Includes 1 C.Q.M.S. and 4 Mechanists.
21 These are restricted recruiting establishments. The full Peace Establishments to which the Unit would expand on Mobilisation are :—

Tyne Elec. Engrs. R.E., T.A. (2 Coys.) 1932																			
No. 1 (E.L. & Works) Coy.[17] ...	—	—	1[14]	3	—	—	—	—	4	—	—	4[18]	4	—	6[19]	5	49	64	68
307th (Tyne) A.A.S.L. Coy.[11] ...	—	1	1	8	—	1	—	—	11	—	1	17[20]	18	—	21[13]	16	228	283	294
TOTAL ...	—	1	2	11	—	1	—	—	15	—	1	21	22	—	27	21	277	347	362

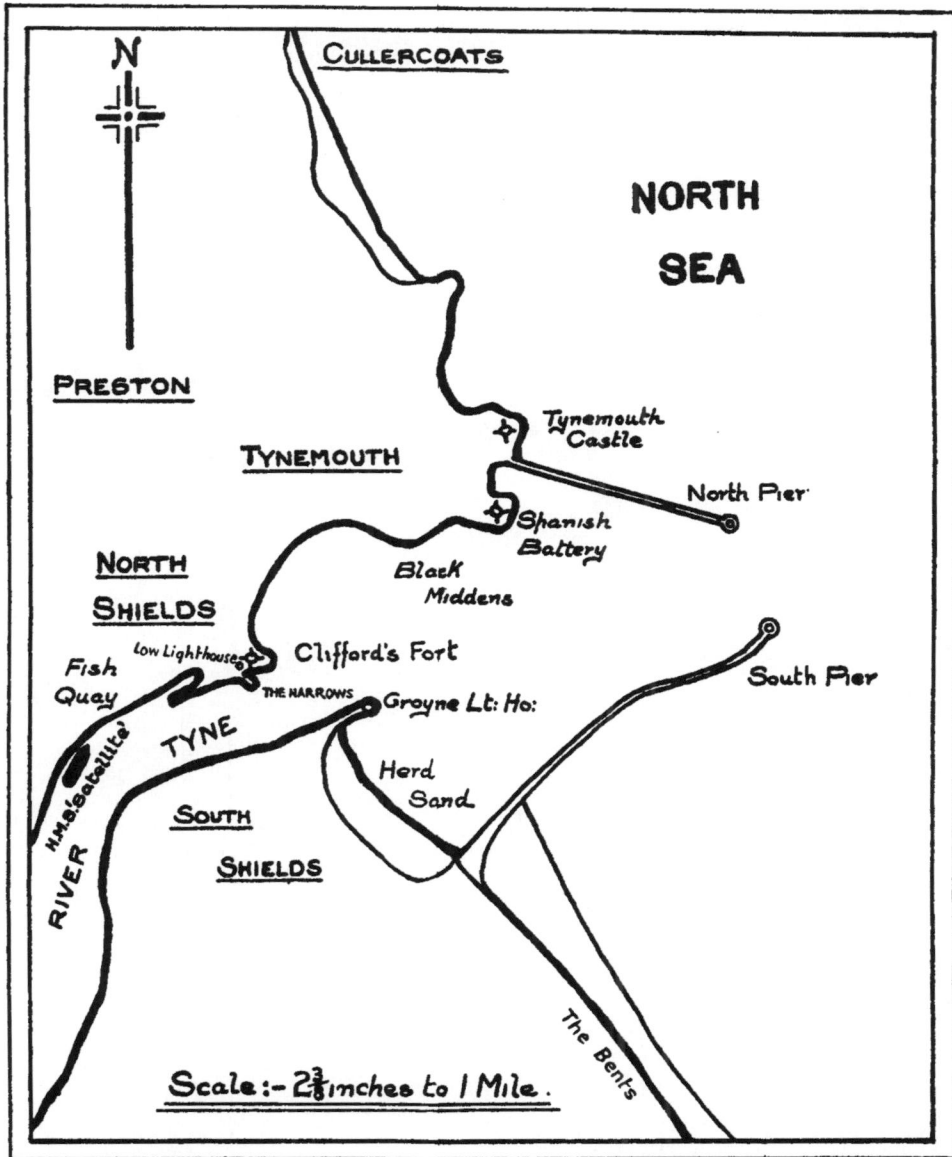

N

CULLERCOATS

NORTH
SEA

PRESTON

TYNEMOUTH

Tynemouth
Castle

North Pier

Spanish
Battery

NORTH
SHIELDS

Black
Middens

Low Lighthouse
Clifford's Fort

Fish
Quay

THE NARROWS

Groyne Lt: Ho:

South Pier

TYNE

H.M.S.Satellite

Herd
Sand

SOUTH

RIVER

SHIELDS

The Bents

Scale:- 2¾ inches to 1 Mile.

CLIFFORD'S FORT, NORTH SHIELDS.

From a plan, made about 1900;
formerly in the possession of the
O.C. 1st Volunteer Company.
— Scale 60 feet to 1 Inch —

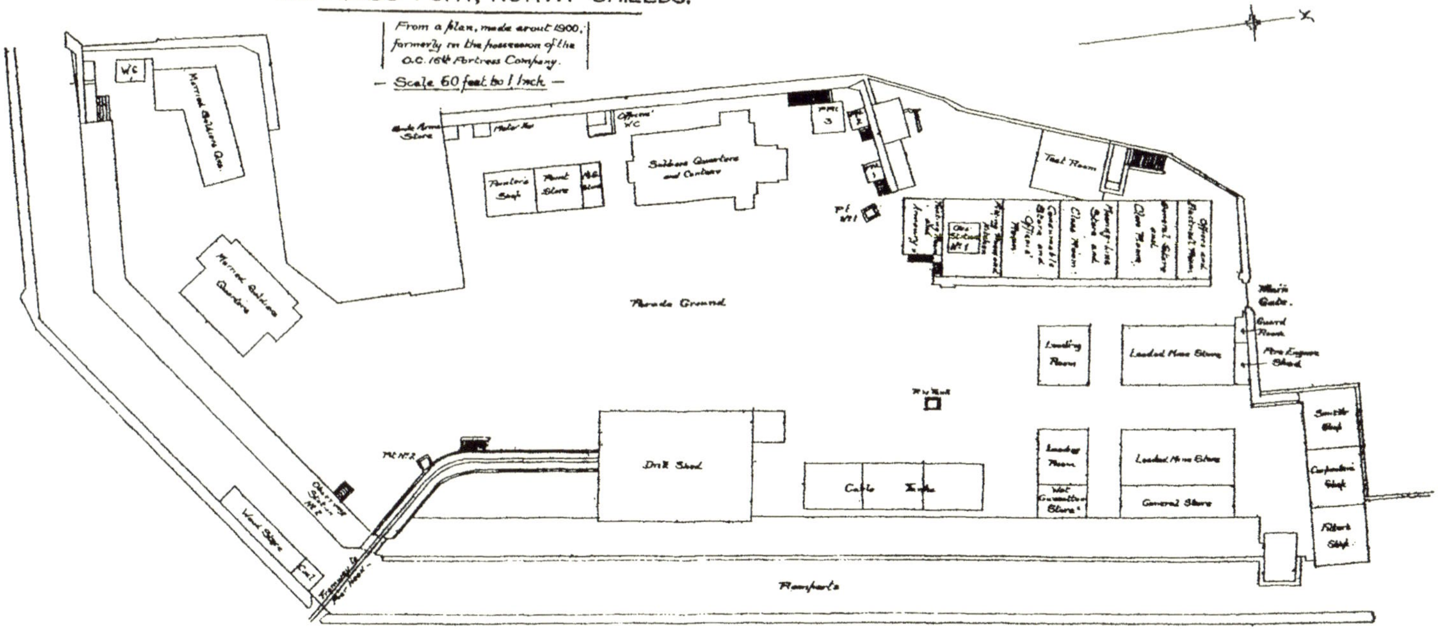

Offrs. and
Electrical Room.

General Store
and
Class Room.

Hearing-Line
Store and
Class Room.

Consumable
Store and
Officers'
Room.

Firing Room and Kitchen

Testing Room

Testing Room and
Armoury

Soldiers Quarters
and Cookhouse

Parade Ground

Drill Shed

Ramparts

Loaded Mine Store

Leather & Iron Store

General Store

Testing
Room

Testing
Room

Detonator
Store

Loading
Room

Cells

Guard
Room

Carpenters
Shop

Artificers
Shop

Smiths
Shop

Magazine

Married Soldiers Qrs.

Married Soldiers
Quarters

Coal Store

Vent Store

W.C.

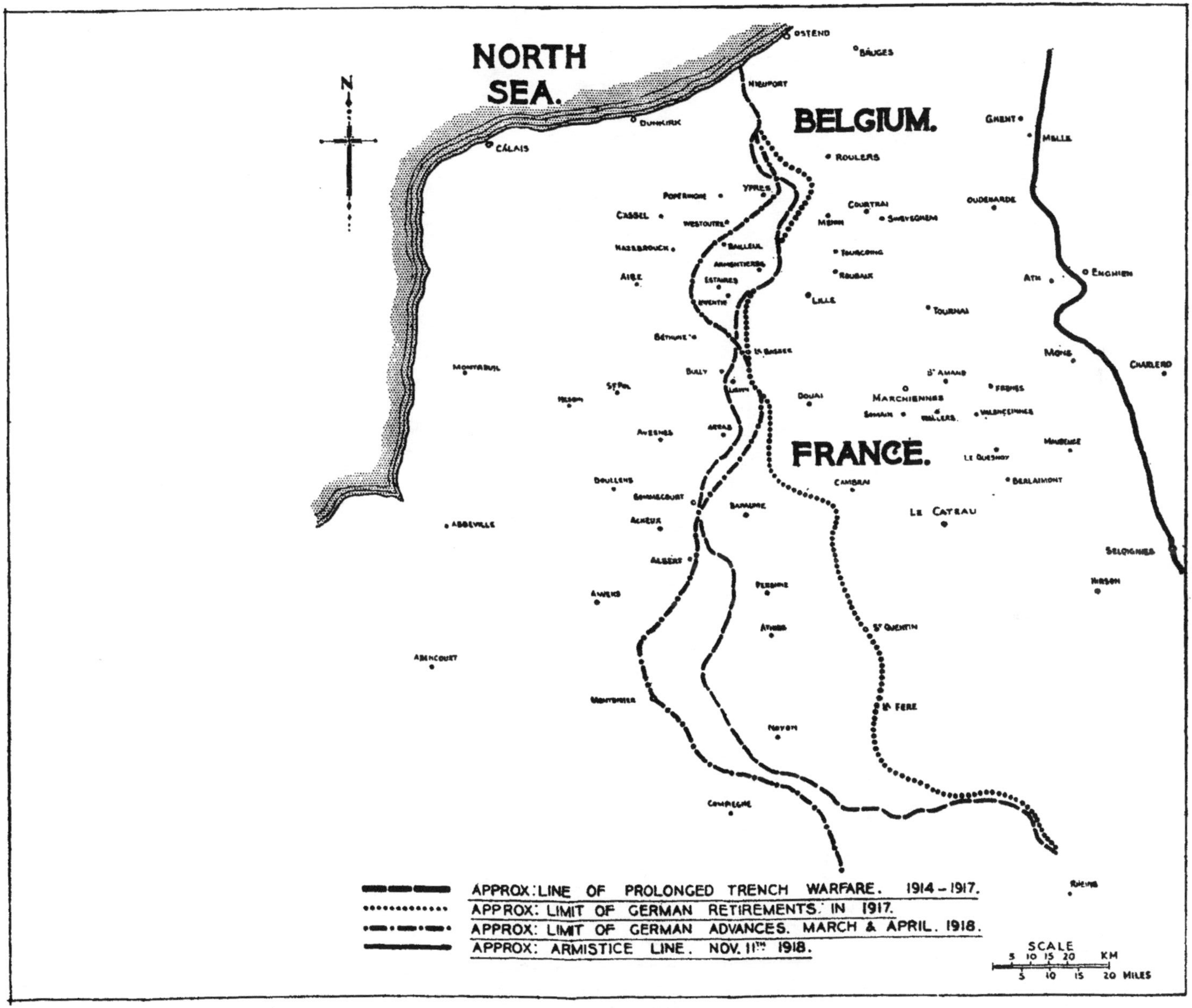

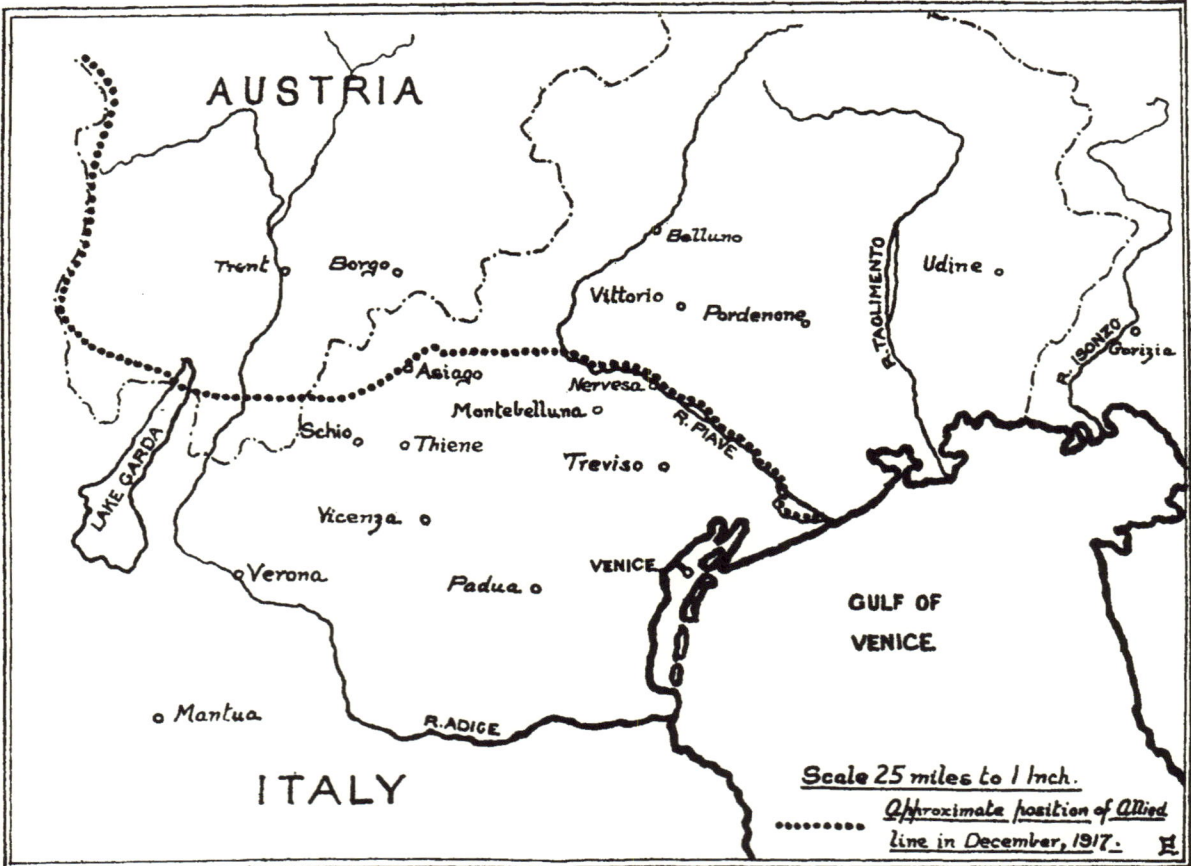

AUSTRIA

Trent Borgo Belluno Udine

Vittorio Pordenone

R. TAGLIAMENTO

R. ISONZO Gorizia

Asiago Nervesa

LAKE GARDA

Schio Montebelluna R. PIAVE

Thiene

Treviso

Vicenza

VENICE

Verona Padua GULF OF VENICE

Mantua R. ADIGE

ITALY

Scale 25 miles to 1 Inch.

•••••••• Approximate position of Allied
line in December, 1917.

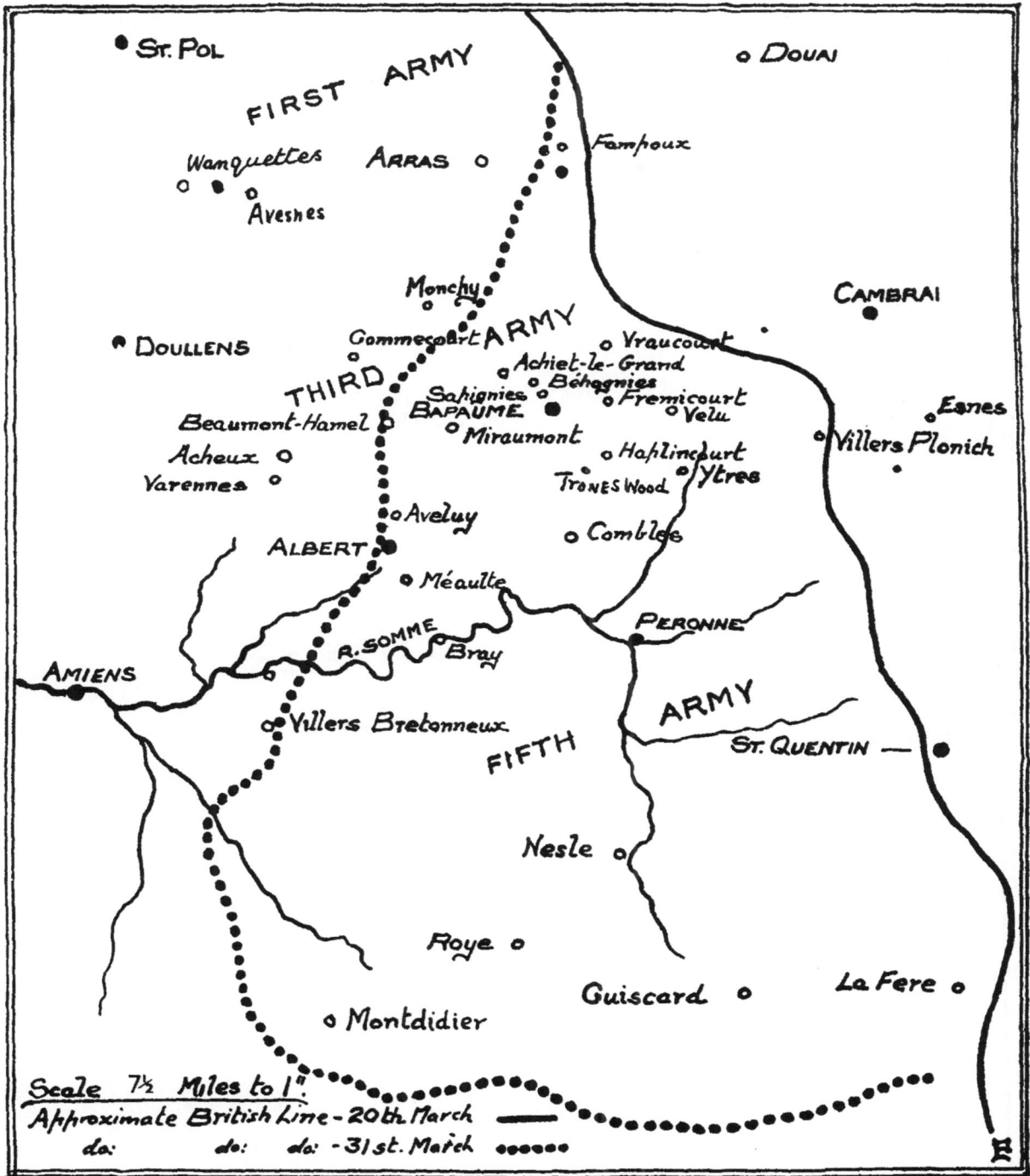

St. Pol

Douai

FIRST ARMY

Wanquettes   Arras   Fampoux

Aveshes

Monchy   Cambrai

Doullens   Gommecourt   THIRD ARMY   Vraucourt

Achiet-le-Grand   Esnes

Sahignies   Béhagnies   Fremicourt   Velu

Beaumont-Hamel   BAPAUME   Villers Plonich

Miraumont

Acheux   Haplincourt

Varennes   Tronse Wood   Ytres

Aveluy

ALBERT   Comblee

Méaulte

R. SOMME   Bray   PERONNE

AMIENS   ARMY

Villers Bretonneux   St. Quentin

FIFTH

Nesle

Roye   La Fere

Guiscard

Montdidier

Scale 7½ Miles to 1".

Approximate British Line - 20th March

do: do: do: - 31st March

SKETCH MAP OF N.E. FRANCE & PART OF FLANDERS.

SCALE := APPROX 1" = 12 MILES.

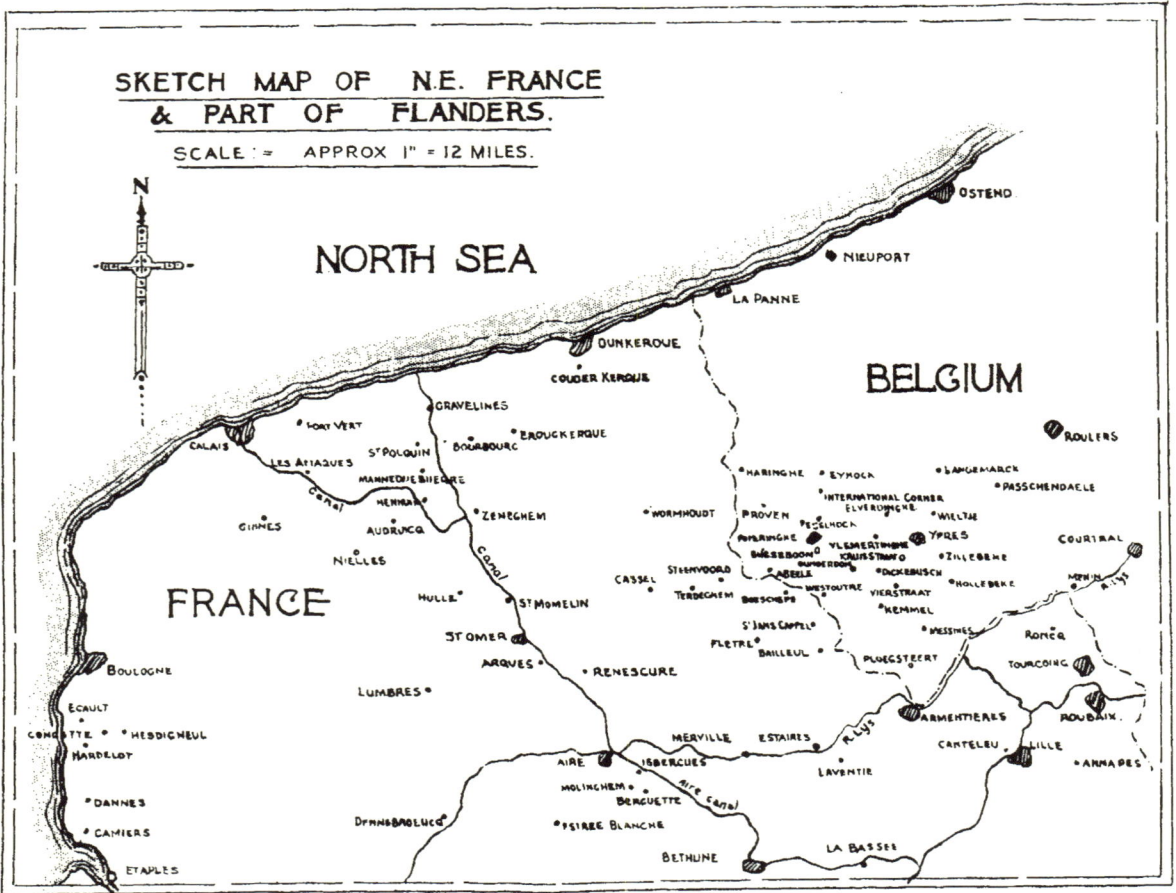

N

NORTH SEA

OSTEND

NIEUPORT

LA PANNE

DUNKERQUE

COUDER KERQUE

BELGIUM

ROULERS

SORY VERT

GRAVELINES

EROUCKERQUE

CALAIS

LES ATTAQUES

St POLQUIN

BOURBOURG

HARINGHE

EYKOCA

LANGEMARCK

PASSCHENDAELE

MANNEQUEBILEGRE

HENNAN

ZENEGHEM

PROVEN

ELVERDINGHE

INTERNATIONAL CORNER

WIELTJE

GINNES

AUDRUCQ

WORMHOUDT

PELDHOCA

POPERINGHE

YPRES

COURTRAL

NIELLES

BRISSEBOOM

VLEMERTINGHE

KRUISSTRANTO

ZILLEBEKE

OUDERDOM

DICKEBUSCH

HOLLEBEKE

MENIN

FRANCE

CASSEL

STEENVOORD

TERDEGHEM

ABEELE

WESTOUTRE

VIERSTRAAT

RLYS

HULLE

St MOMELIN

BRESCHEPS

BELS

VELS

KEMMEL

RONCA

ST OMER

ARQUES

RENESCURE

St JANS CAPPEL

FLETRE

BAILLEUL

MESSINES

TOURCOING

LUMBRES

PLOEGSTEERT

BOULOGNE

ARMENTIERES

ROUBAIX

ECAULT

MERVILLE

ESTAIRES

RLYS

CONDETTE

HESDIGNEUL

AIRE

ISBERGUES

LAVENTIE

CANTELEU

LILLE

HARDELOT

MOLINGHEM

BERGUETTE

ANNAPES

DANNES

PSIREE BLANCHE

CAMIERS

DANNERROEUCO

BETHUNE

LA BASSEE

ETAPLES

Printed in Great Britain
by Amazon.co.uk, Ltd.,
Marston Gate.

THE HISTORY

OF THE

TYNE ELECTRICAL ENGINEERS,
ROYAL ENGINEERS.

1884-1933.